MONITORING, CONTROL AND EFFECTS OF AIR POLLUTION

Edited by **Andrzej G. Chmielewski**

Monitoring, Control and Effects of Air Pollution
Edited by Andrzej G. Chmielewski

Published by InTech
Janeza Trdine 9, 51000 Rijeka, Croatia

Publishing Process Manager Ivana Lorkovic

Technical Editor Teodora Smiljanic

Cover Designer InTech Design Team

Additional hard copies can be obtained from orders@intechopen.com

Monitoring, Control and Effects of Air Pollution, Edited by Andrzej G. Chmielewski
p. cm.
ISBN 978-953-307-526-6

For EU product safety concerns:
IN TECH d.o.o., Prolaz Marije Krucifikse Kozulić 3, 51000 Rijeka, Croatia,
info@intechopen.com or visit our website at intechopen.com.

Meet the editor

Prof. Andrzej G. Chmielewski, Ph.D., D.Sc., is a director of the Institute of Nuclear Chemistry and Technology, Warsaw, Poland and a professor at the Department of Chemical and Process Engineering, Warsaw University of Technology. In the years of 1976/77 he was employed by the University of Tennessee, Knoxville, USA and in 2003/05 by IAEA, Vienna, Austria, then lectured at the universities in Sao Paulo, Brazil; Hefei, China and Pavia, Italy. As an UN expert undertook missions to over thirty countries all over the world. He is a member of Academy of Engineering and in the year 1999 was nominated a "Gold Engineer of the Year" by the Polish Federation of Engineering Associations. He is an author of more than 150 publications and books, and over 60 international patents, some of this inventions were the bases for construction of the big installations for air and water pollution control built in industry. Eleven Ph.D. and over fifty M. Sc. theses were carried out under his supervision.

Contents

Preface XI

Part 1 Emission and Transformation of Air Pollutants 1

Chapter 1 **Generation and Dispersion of Total Suspended Particulate Matter Due to Mining Activities in an Indian Opencast Coal Project 3**
Ratnesh Trivedi, M. K. Chakraborty and B. K. Tewary

Chapter 2 **Secondary Acidification 15**
Mizuo Kajino and Hiromasa Ueda

Part 2 Air Pollution Monitoring and Modelling 39

Chapter 3 **Gas Sensors for Monitoring Air Pollution 41**
Kwang Soo Yoo

Chapter 4 **Development of Low-Cost Network of Sensors for Extensive In-Situ and Continuous Atmospheric CO2 Monitoring 59**
Kuo-Ying Wang, Hui-Chen Chien and Jia-Lin Wang

Chapter 5 **Aerosol Optical Thickness and PM10 Study by Using a Handheld Spectroradiometer Over Penang 73**
Tan Fuyi, Lim Hwee San, Mohd. Zubir Mat Jafri and K. Abdullah

Chapter 6 **Remote Sensing of PM2.5 Over Penang Island from Satellite Measurements 81**
Sim Chong Keat, Lim Hwee San, Khiruddin Abdullah and Mohd. Zubir Mat Jafri

Chapter 7 **Photoacoustic Measurements of Black Carbon Light Absorption/Scattering Coefficients and Visibility Degradation in Jordan During 2007/2008 93**
Khadeejeh M. Hamasha

Chapter 8 **$PM_{2.5}$ Source Apportionment Applying Material Balance and Receptor Models in the MAMC 103**
V. Mugica, R. Vallesa, J. Aguilar, J. Figueroa and F. Mugica

Part 3 Air Pollution in Office and Public Transport Vehicles 119

Chapter 9 **Emission and Formation of Fine Particles from Hardcopy Devices: the Cause of Indoor Air Pollution 121**
David D. Massey and Ajay Taneja

Chapter 10 **In-Office Dispersion and Exposure to Contaminants Originating from an Unfolded Letter 135**
Alfred D. Eisner, Russell W. Wiener and Jacky Rosati

Part 4 Effect of Air Pollutants on Historical Objects 151

Chapter 11 **Air Pollution and Cultural Heritage: Searching for "The Relation Between Cause and Effect" 153**
Eleni Metaxa

Chapter 12 **Effect of Air Pollution on Archaeological Buildings in Cairo 179**
Mohamed Kamal Khallaf

Chapter 13 **Influence of Air Pollution on Degradation of Historic Buildings at the Urban Tropical Atmosphere of San Francisco de Campeche City, México 201**
Javier Reyes, Francisco Corvo, Yolanda Espinosa-Morales, Brisvey Dzul, Tezozomoc Perez, Cecilia Valdes, Daniel Aguilar and Patricia Quintana

Part 5 Plasma Technologies for Air Pollution Control 227

Chapter 14 **Plasma-Based Depollution of Exhausts: Principles, State of the Art and Future Prospects 229**
Ronny Brandenburg, Hana Barankova, Ladislav Bardos, Andrzej G. Chmielewski, Miroslaw Dors, Helge Grosch, Marcin Hołub, Indrek Jõgi, Matti Laan, Jerzy Mizeraczyk, Andrzej Pawelec and Eugen Stamate

Preface

The municipal and industrial activities of the man affects human health, leads to degradation of the environment and even degradation of the objects built by himself or his predecessors. The pollutants are emitted to the atmosphere with off-gases from industry, power stations, residential heating systems and vehicles, some additional emissions come from the natural sources as volcanoes as well. Fossil fuels, which include coal, natural gas, petroleum, shale oil and bitumen are the main source of heat, electrical energy and are the basic duels for the engines of the transport vehicles. All these fuels contain beside major constituents (carbon, hydrogen, oxygen) other materials as metal, sulfur and nitrogen compounds. During the combustion process different pollutants as fly ash, sulfur oxides ($SO_x = SO_2$ and SO_3), nitrogen oxides ($NO_x = NO_2 + NO$) and volatile organic compounds are emitted. Fly ash contains different trace elements (heavy metals). Gross emission of pollutants is tremendous all over the world. These pollutants are present in the atmosphere in such conditions that they can affect man and his environment. The books addresses the subjects related to the selected aspects of pollutants emission, monitoring and their effects. The most of recent publications concentrated on the review of the pollutants emissions from industry, especially power sector. In this one emissions from opencast mining and transport are addressed as well. Beside of SO_x and NO_x emissions, small particles and other pollutants (e.g. VOC, ammonia) have adverse effect on environment and human being. The natural emissions (e.g. from volcanoes) has contribution to the pollutants concentration and atmospheric chemistry governs speciation of pollutants, as in the case of secondary acidification. The methods of ambient air pollution monitoring based on modern instrumentation allow the verification of dispersion models and balancing of mass emissions. The comfort of everyday human's activity is influenced by indoor and public transport vehicles interior air contamination, which is effected even by the professional appliances operation. The outdoor pollution leads to cultural heritage objects deterioration, the mechanism are studied and the methods of rehabilitation developed. However to prevent emissions the new technologies are being developed, the new class of these technologies are plasma processes, which are briefly reviewed at the final part of the book.

Prof. **Andrzej G. Chmielewski**
Institute of Nuclear Chemistry and Technology
Warsaw, Poland

Part 1

Emission and Transformation of Air Pollutants

1

Generation and Dispersion of Total Suspended Particulate Matter Due to Mining Activities in an Indian Opencast Coal Project

Ratnesh Trivedi, M. K. Chakraborty and B. K. Tewary
Scientists, Central Institute of Mining and Fuel Research, India

1. Introduction

The knowledge of ambient air quality plays an important role in assessing the environmental scenario of the region. The ambient air quality status in the vicinity of the mining activities forms an indispensable part of the Environmental Impact Assessment Studies. The quality of ambient air depends upon the concentrations of specific contaminants, the emission sources and meteorological conditions. The mining activities contribute to the problem of air pollution directly or indirectly (Trichy ,1996, Corti and Senatore, 2000, Baldauf *et al.*, 2001 and Collins *et al.*, 2001). Coal dust is the major pollutant in the air of open cast coal mining areas. (Kumar *et al.*, 1994, Vallack and Shillito, 1998. and CIMFR, 1998) The primary source of fugitive dust at fully operational surface mine may include overburden (OB) removal, blasting, mineral haulage, mechanical handling operations, minerals stockpiles and site restoration (Appleton *et. al.* 2006). Major air pollutants due to opencast mining are total suspended particulate matter and respirable particulate matter whereas concentration of SO_2 and NO_X is negligible (Sinha and Banerjee, 1997, CIMFR, 1998, Banerjee, 2006, and Trivedi *et. al.*, 2009).

Transportation of materials is the major source of TSPM generation in the mining areas. The vehicle and haul road intersection has been identified as the most critical source producing as much as 70% of total dust emitted from surface coal mines (Muleski and Cowherd, 1987, Sinha and Banerjee, 1997, Ghose and Majee, 2002), while it was accounted to be 80-90% of the PM_{10} emission (Cole and Zapert, 1995). Maximal concentrations of particulate matter are generally occurred during winter and minimal in the rainy season.(Ghose and Majee, 2000, Tayanc, 2000, Reddy and Ruj, 2003). However, in certain urban areas maximal concentrations of particulate matters are also observed in summer season (Crabbe *et al.*, 2000, Almbauer *et al.*, 2001, Triantafyllou *et al.*, 2002 and Triantafyllou, 2003). The dispersion of particulate matter follows the annual predominant wind direction of an area (Corti and Senatore, 2000, Baldauf *et al.*, 2001 and Pandey *et. al.*, 2008).

Such a large amount of dust generated cause safety and health hazards such as poor visibility, failure of mining equipment, increased maintenance cost etc which ultimately lowers the productivity. A prolonged exposure to air borne dust may cause to damage of lung tissues of the miners which may further lead to pneumoconiosis or black lung disease. The maximum tissue damage is caused by the dust of 5 microns lesser sizes since such particles reach the alveoli of the lung (Peavey et. al, 1985). These air pollutants reduce air quality and this ultimately affects people, flora and fauna in and around mining areas (Crabbe *et al.*, 2000, Wheeler *et al.*, 2000). Implementation of effective air quality control measures by the mining company are needed and green belts development can be devised wherever necessary (Kapoor and Gupta, 1984, Sharma and Roy, 1997, Shannigrahi and Sharma, 2000, Chaulya, 2004).

In the present study, an attempt has been made to generate ambient air quality data, micro-meteorological data, source-wise emission inventory data for an Indian coal mine namely Padampur Opencast Coal Project (O.C.P.) of Western Coalfields Ltd. (W.C.L.), India. The status of TSPM and PM_{10} concentration in ambient air has been monitored through a well defined at monitoring network. In the light of micro-meteorological data such as wind speed, stability class etc, dispersion coefficients of the dust for vertical as well as horizontal direction have been estimated. A correlation between TSPM and PM_{10} concentration has been sought out. Emission inventory data for all point, area and line sources of TSPM at Padampur OCP have been generated for the determination of emission rates. Air Pollution modeling has also been attempted using Fugitive Dust Model (FDM) developed by United States Environment protection Agency (USEPA). FDM has been used for the validation of the study by comparing predicted and observed values. FDM is a computerized Gaussian Plume Air Quality Model, specifically designed for the estimation of the concentration and deposition impacts from fugitive dust sources. FDM employs an advance transfer particle deposition algorithm.(USEPA, 1995). The model gives hourly average, long term concentration and deposition of particulate matters at all user selected receptor locations. FDM represents the behavior of particles in the atmosphere most accurately. Since terrain features are not included in FDM, it can be used only for local scale.

2. Materials and method

2.1 Field settings at OCP

As mentioned earlier that Padampur OCP is selected for the study purpose. Padampur OCP is located at Chandrapur district in Maharastra State of India. The Project is located between latitudes 20^0 2′ N to 20^0 3′ N and Longitudes 79^0 17′ E to 79^0 19′ E and is covered by Survey of India Toposheet No 55 P/8. Geologically the area forms the central part of eastern limb of Regional anticline structure of Wardha Valley Coalfield of Western Coalfields Limited (WCL). The area is undulating with few isolated ridges of Kamthi Sandstone. The area covers two separate and adjoining geological blocks namely Padampur and Motaghat blocks. The net geological reserve of Padampur OCP is about 43.5 Million Tones. The annual production is 1 Million Tones with an average stripping ratio of 3.7 m^3/tones. The coal produced from the mine is of non-coaking type with 'D' and 'E' grade. The shovel dumper combination is being used to excavate the overburden as well as coal. The shovels of 4.6 m^3 bucket capacity and dumpers are of 35 Ton capacities have been deployed in the field. Backfilling is also practiced simultaneously with the production of coal.

Micrometeorological data collection is an indispensable part of any air pollution study. The data collected during air quality survey are used for proper interpretation of existing ambient air quality status. The ambient air quality monitoring was carried out through reconnaissance followed by air quality surveillance program and micrometeorological study of the area. A weather monitoring station and SODAR have been installed at study site. The weather monitoring station measures ambient air temperature in degree centigrade, wind speed in km. per hour and wind direction in degrees from north. It also measures relative humidity, barometric pressure, and total rainfall. Site specific relevant parameters like mixing height and stability class have been accurately measured by SODAR. The amount of turbulence in the ambient air has a major effect upon the rise and dispersion of air pollutant plumes. The amount of turbulence is categorized into "stability classes". The most commonly used categories are the Pasquill stability classes A, B, C, D, E, and F. Class A denotes the most unstable or most turbulent conditions and Class F denotes the most stable or least turbulent conditions. The most common procedure for estimating the dispersion coefficients was introduced by Pasquill (1961), modified by Gifford (1961) and adopted by the U.S. Public Health Service (Turner, 1970).

Meteorological data has been collected from the nearest Indian Meteorological Department (IMD) station at Nagpur, India. The climate of the area is tropical. Summer is well defined from April to June, followed by rainy season from July to September and winter from December to February. May is the hottest month with temperature rising to a maximum of around 48^{0}C. December is the coldest month when the temperature falls down to about 10^{0}C. The mean annual rainfall is around 1250 mm. Wind direction is generally from North and Northwest, with velocities up to 6-7 Km./hour during monsoon and about 3-4 Km./hour in winter. Relative humidity varies from 74-83% during August and September and is about 15-20% during summer. Wind rose diagram during the study period is illustrated in Fig.1.

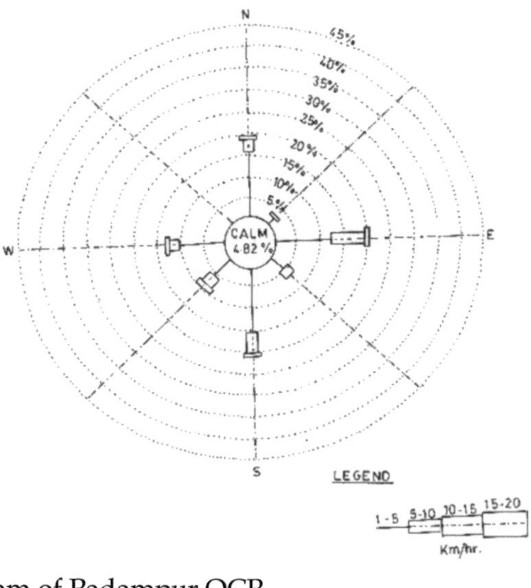

Fig. 1. Wind Rose Diagram of Padampur OCP

2.2 Ambient air quality monitoring

The ambient air quality status in the impact zone was assessed through a network of ambient air quality monitoring locations. The studies on air environment include identification of specific air pollutants for assessing the impacts of proposed mining projects including other activities. Air quality monitoring was carried out in winter season. Among the ambient air quality parameters, Total Suspended Particulate Matter (T.S.P.M.) and Respirable Particulate Matter (PM_{10}.) have been measured at 8 hours interval for 24 hours using the High Volume Sampler with Respiratory Particulate Matter measurement arrangement with the standard methods as shown in Table 1. Other air qualities parameters are not considered because of their concentrations are found much below the threshold value in the study area.

The existing status of air environment was assessed through a systematic air quality surveillance program in which five ambient air quality stations have been selected to know the air quality of the area. The measured data of TSPM and PM_{10} are shown in Table 2 along with the arithmetic mean and standard deviation (S.D.) of the measured data.

Parameter	Time weighted Avg.	Concentration in Ambient Air	Method	Instruments
TSPM	Annual 24 hours	430 µg/ m³ 600 µg/ m³	IS-5182 Part XIV	High Volume Sampler with PM_{10} Measurement arrangement (Av. Flow rate not < 1.1 m³/min)
PM_{10}	Annual 24 hours	215 µg/ m³ 300 µg/ m³	IS-5182 Part XIV	High Volume Sampler with PM_{10} Measurement arrangement (Av. Flow rate not < 1.1 m³/min)

(Source: Central Pollution Control Board Notification, 1994)

Table 1. Air Pollutant Analysis Methods: Coal Mine Standards

Sl. No.	Sample Site	TSPM (µg/ m³)		PM_{10} (µg/ m³)		PM_{10} (Percent of TSPM
		mean	S.D.	mean	S.D.	
1	Filter Plant	294.30	21.67	120.34	18.90	40.89
2	SAM Office	631.72	58.44	120.03	21.55	19.00
3	Kitadi Village	390.30	32.20	103.03	25.90	26.39
4	Padampur Village	654.38	54.55	130.88	12.15	20.00
5	Manager Office Sec-IV	1078.10	75.56	226.40	35.60	20.99

Table 2. Table Showing Ambient Air Quality of Padmapur OCP

2.3 Source-wise emission inventory details

Emission inventory details have been collected by installing two High Volume Samplers at down wind sides and, to know the back ground concentration, one High Volume Sampler at up wind side of the TSPM source. High Volume Samplers are placed at a distance nearly 100m from the source Emission data have been generated for various mining activities such as overburden loading, coal loading, haul road transportation, unloading of overburden, unloading of coal, stock yard, exposed overburden dumps, coal handling plant, exposed pit face and workshop. Blasting being an instantaneous source was monitored separately which is not included in the present study.

The modified Pasquill and Gifford formula for ground level emission has been used to calculate the emission rate.

$$C(x,0) = \frac{Q}{\Pi u \, \sigma y \, \sigma z}$$

Where, C(x,0) =DN max - UP

C (x, 0), Difference in pollutant concentration , µg/.m; DN max, maximum concentration in down wind direction; UP, back ground concentration in up wind direction; Q, Pollutant emission rate, µg/s; Π, 3.14159; u, Mean wind speed, m/s; σy, Standard deviation of horizontal plume concentration, evaluated in terms of downwind distance x, m (as shown in Fig. 2); σz, Standard deviation of vertical plumes concentration, evaluated in terms of downwind distance x, m (as shown in Fig. 3).

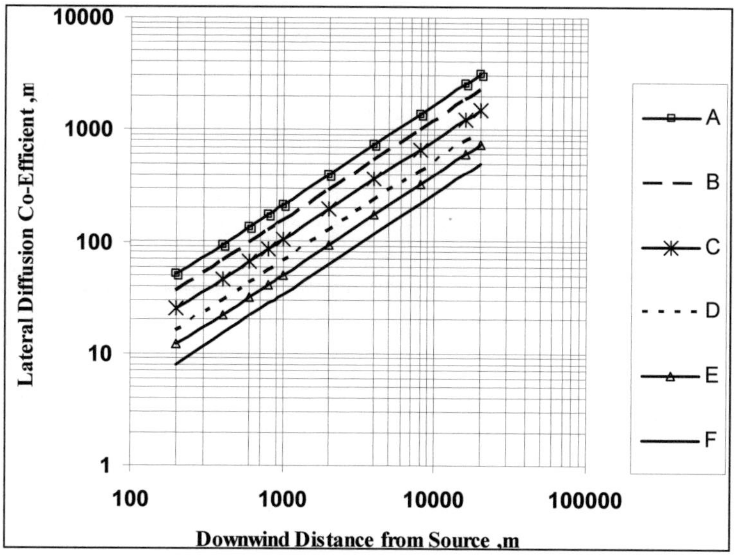

Fig. 2. Lateral Diffusion Co-Efficient Vs Downwind Distance from Source (Source : Turner, 1970).

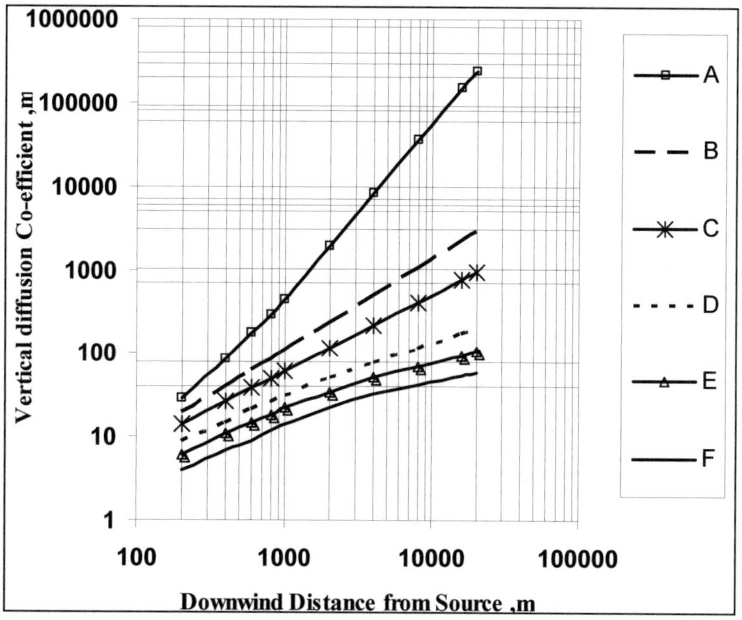

Fig. 3. Vertical diffusion Co-efficient Vs Downwind Distance from Source (Source : Turner, 1970).

Source-wise emission inventory has been shown in Table 3. Source-wise emission properties such as moisture content silt content etc. measured from the samples collected during field study have been placed in Table 4. Secondary data affecting TSPM emission such as frequency of drilling, vehicle movement on the haul road and transport road etc have also been collected as shown in Table 4.

TSPM source	TSPM Concentration (μg/m³)				Wind velocity, m/s	Diffusion coefficient		Emission Rate	
	DN Min	DN Max	UP	DN_{max} -UP		σy, m	σz, m	Unit	Value
Drilling	1340	1758	1233	525	2.1	14	8	g/s	0.3877
Overburden Loading	1234	1660	1108	552	2.4	14	8	g/s	0.4659
Coal Loading	1648	2092	1377	715	2.1	14	8	g/s	0.5218
Haul Road	1963	2498	1336	1162	3.1	14	8	g/ms	0.0127
Transport Road	2015	2605	1387	1218	3.1	14	8	g/ms	0.0132
Overburden Unloading	1195	1605	942	663	1.9	18	12	g/s	0.8544
Coal Unloading	1438	1897	1135	762	2.0	14	8	g/s	0.5360
Exposed Overburden dump	1030	1387	1002	385	2.5	24	16	g/m²s	0.0000363
Stockyard	1482	1872	1027	845	1.8	14	8	g/m²s	0.0001981
Workshop	1062	1478	1040	438	1.7	25	15	g/m²s	0.0000878
Exposed pit surface	1015	1357	985	372	1.0	15	32	g/m²s	0.0000160
Overall mine	469	713	365	348	2.4	95	60	g/m²s	0.0000108

Table 3. Source-Wise TSPM Emission Inventory at Padmapur OCP

TSPM sources	Source type	Moisture content, %	Silt content, %	Emission rate		Remarks
				Unit	Value	
Drilling	Point	7.4	38.0	g/s	0.443	Hole dia 160 mm;12 hole/day
Overburden Loading	Point	7.6	13.6	g/s	0.4867	drop height 1.4 m; frequency 23 no/hr
Coal Loading	Point	8.1	10.9	g/s	0.5783	drop height 0.9 m; frequency 23 no/hr
Haul Road	Line	12.4	34.5	g/ms	0.0144	Frequency 18 no/hr; average speed 2.6m/sec;
Transport Road	Line	9.8%	30.0	g/ms	0.0146	Frequency 27 no/hr; Average speed 10 m/sec.
Unloading of Overburden	Point	7.2	14.2	g/s	1.2740	Frequency 10 no/hr ;drop height 14.3m.
Unloading of coal	Point	8.0	11.2	g/s	0.7333	Frequency 7 no/hr; drop height 2.5 m
Exposed Overburden dump	Area	7.4	8.2	g/m²s	0.00004	dump area 0.029 sq. km
Stock yard	Area	6.0	12.5	g/m²s	0.00024	Unloading freq. 3 No/hr; loading freq. 12.0 No/hr
Work shop	Area	12.4	31.8	g/m²s	0.0001	Area 5000 sq. m
Exposed pit	Area	8.1	7.8	g/m²s	0.00002	Exposed area 0.03 sq.km.

Table 4. Source Wise TSPM Emission Properties

3. Results and discussion

Air pollution modeling has been exercised with the help of Fugitive Dust Model (FDM). The input parameters include source types, dust concentration near sources, hourly meteorological data such as wind speed and direction, temperature atmospheric stability and receptor locations. All data depicted the average value for the study period. Emission source has been demarcated in three categories of sources like point, line and area sources using mine plan. All these sources have been numbered for preparation of data sheet. Emission rate has been assigned to each activity as per the field measurement data in the mine. From the modeling exercise, TSPM concentrations at certain receptor locations have been predicted. The receptor locations have been selected such that these are exactly same of one where ambient air quality measurement was carried out. The predicted values at receptor locations have been added to regional background levels to get the total predicted TSPM concentration. Regional background data are the average of the monitored data in no activity zone. The predicted and observed TSPM concentrations at receptor locations for different mines are listed in Table 5 and Table 6.

Field observations of ambient air quality of Padampur OCP have been placed in the Table 2. The Ambient Air quality at the five sites of Padmapur is well within the limit except Manager Office sec IV The higher value of TSPM and PM_{10} at Manager Office sec IV may be contributed by the presence of main transport road and other industries nearby. The 24-hr average of TSPM concentrations ranged from 294.3 to 1078.1 $\mu g\ m^{-3}$ in industrial area i.e. mining area and from 390.3 to 654.38 $\mu g\ m^{-3}$ in residential area respectively. The 24-hr average of PM_{10} concentrations ranged from 120.34 to 226.4 $\mu g\ m^{-3}$ in industrial area and from 103.03 to 130.88 $\mu g\ m^{-3}$ in residential area respectively as shown in Table 2. On average the PM_{10} in the ambient air constituted 19.00 % to 40.89 % of the TSPM in mining area and 20.00 % to 26.39 % of the TSPM in residential area. The concentration of particulate matters vary with the meteorological parameters and a relation also exist between TSPM and PM_{10} (Tayanc, 2000, Jones *et al.*, 2002, Triantafyllou *et al.*, 2002, Triantafyllou, 2003, Chaulya, 2004). The case under study also reveals that there exists a relationship between TSPM and PM_{10} concentrations. Linear regression correlation coefficient (R^2) between TSPM with PM_{10} has been found to be 0.8116 as shown in Fig. 4. With the help of FDM, TSPM concentration at five monitoring stations has been predicted. The variation between measured and predicted values, as shown in Table 5 and fig. 6 , may be due to non-accountability of emission from various other sources like non-mining area activities, domestic use of fuels, transportation network nearby thermal power plant , cement plant etc. The value of coefficient of correlation between observed values of TSPM Concentration and predicted values by FDM have been calculated to be 0.969

Source-wise emission inventory data placed in Table 3. Stability classes have been found to be B, C & D. It is clearly evident from the Table 3 that among the point sources namely drilling, overburden loading, overburden unloading and coal unloading highest value of emission rates (g/s) has been found in case of unloading of overburden. Among the area sources namely exposed OB dump, stockyard, workshop, exposed pit surface, highest values of emission rate ($gm/m^2/s$) has been found in case of exposed OB dump. Among the line sources, emission rates have been in case of haul found and transport road to be 0.0127 gm per meter per second and 0.0132 gm per meter per second respectively. In terms of overall TSPM pollution line sources contribute more than other sources because of their lengths and nature of mining operations. This very fact again confirms that the vehicle and haul road intersection is the major source of dust in opencast mines (Muleski and Cowherd, 1987, Sinha and Banerjee, 1997, Ghose and Majee, 2002) Emission rate for whole mine is found 0.0000108 gm per sq. meter per second.

With the help of FDM, TSPM concentration has been predicted at various distances in down wind direction as shown in Table 6. As far as rate of fall in concentration of TSPM with the distance from the source is concerned, an exponential fall in the TSPM concentration with the distance from the source has been observed which can be clearly seen in the Fig. 5. Maximal concentrations of TSPM and PM_{10} have been found to occur within the mine. Again the dust generated due to mining activities does not contribute to ambient air quality in surrounding areas beyond 500 meters in normal meteorological condition as shown in the Table 6. Thus the result matches with the findings of the Several researchers (Hanna *et al.*, 1982, Chaulya *et al.*, 2001 and Jones *et al.*, 2002, Chaulya, 2004) that maximal concentrations of TSPM and PM_{10} are found in a mining area and the concentrations are gradually diminished with increase in distance due to transportation, deposition and dispersion of particles. The value of coefficient of correlation between observed values of TSPM Concentration and predicted values by FDM have been calculated to be 0.9957.

From the emission study, it is quite clear that haul road and transport road were the major contributor to the pollution load of ambient air quality. Therefore proper dust suppression arrangement is to be made. The prevailing practice of water sprinkling does not seem to be adequate. Therefore, the installation of continuous atomized spraying system for haul roads should be used. Exposed overburden dump is another major contributor of pollution load. These dumps not only contribute to air pollution by way of wind erosion but also spread the dump it self. Therefore, judicious, plantation on these dumps is highly recommended. These plantations will not only stabilize the dump but also attenuate the dust emission. Biological reclamation of overburden dumps and wastelands is also essential. Effective control measures at the coal handling plant, excavation area and overburden dumps should also be implemented to mitigate the TSPM emissions at source.

Green belt development around the mining area is highly recommended. The capacity of plants to reduce air pollution is well known (Kapoor and Gupta, 1984, Sharma and Roy, 1997, Shannigrahi and Sharma, 2000, Chaulya, 2004). A Few plant species can be grown around highly polluted areas where dust (TSPM) is the main pollutant as given Table 7. These species not only reduces air pollutants but also retards water and soil pollution.

S. No.	**Sample site**	**Observed value (μg/m³)**	**Predicted value (μg/m³)**
1	Filter plant	294.30	250
2	SAM Office	631.72	540
3	Kitadi Village	390.30	325
4	Padmapur Village	654.38	564
5	Manager Office Sec-IV	1078.10	910

Table 5. Comparison between observed and predicted values of TSPM Concentration

TSPM Source	**Predicted values (μg/m³)**					
	At source	**100 m**	**200 m**	**300 m**	**400 m**	**500 m**
Padampur mine as a whole	698	505	375	260	165	125

Table 6. Predicted values of TSPM Concentration along down Wind Direction

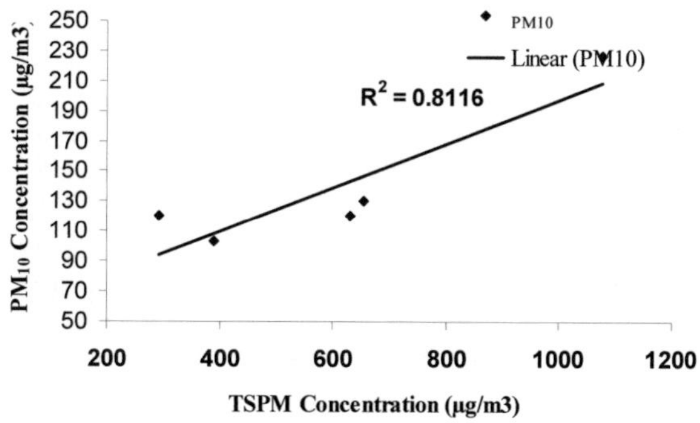

Fig. 4. Correlation between TSPM and PM_{10} Concentration.

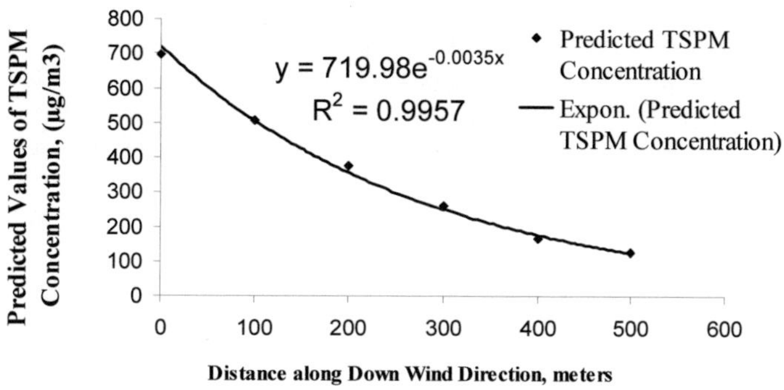

Fig. 5. Relation of TSPM Concetration with Distance from OCP.

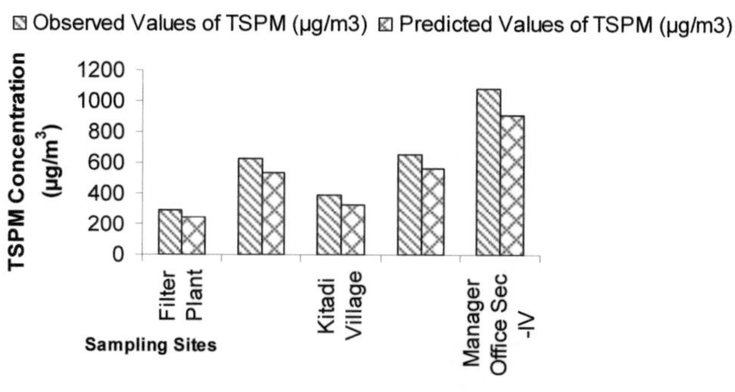

Fig. 6. Comparision between Observedvalues and Predicted Values of TSPM.

Species Name	Family	Local Name of Plants	Evergreen (E) or deciduous
Butea monsperma	Moraceae	Palas	Deciduous
Spathodea companulata	Bignoniaceae	Sapeta	Evergreen
Fiscus infectoria	Moraceae	Pakur	Evergreen
Cassia fistula	Caesalpiniaceae	Amaltas	Deciduous
Anthocephalus cadamba	Rubiaceae	Kadam	Deciduous
Cassia siamea	Caesalpiniaceae	Minjari	Deciduous

Table 7. Recommended pollution retarding plant species for green belt development

4. Conclusions

TSPM and PM_{10} are the major sources of emission from various opencast coal mining activities. The predicted values of TSPM using FDM are 70 percent to 94 percent of observed values. The difference between observed values and predicted values of TSPM indicates that there are non-mining sources of emission viz. domestic transportation network near by mine sites and other industries etc. Fugitive Dust Model (FDM) has been found to be most suitable for modeling of dispersion pattern of fugitive dust at Padampur Opencast Coalmine Project of W.C.L. PM_{10} is the main focus of concern for human health. Correlation between PM_{10} and TSPM would help in predicting the PM_{10} concentration by knowing the concentration of TSPM for a similar mining site. Maximal concentration of TSPM is found in a mining area and the concentrations falls exponentially with increase in distance due to transportation, deposition and dispersion of particles.

Of the various sources of TSPM pollution, line sources contribute more than other sources because of their lengths and nature of mining operations. Among the line sources, emission rates have been in case of haul found and transport road to be 0.0127 gm per meter per second and 0.0132 gm per meter per second respectively. Emission rate for whole mine is found 0.0000108 gm per sq. meter per second. Various management strategies are evaluated for reduction of dust emission at the source and design of green belt with few recommended species is also very effective tool to mitigate air pollution. Proper dust suppression arrangement is to be made including installation of continuous atomized spraying system for haul roads and transport roads. As exposed overburden dump is another major contributor of pollution load, judicious, plantation on these dumps is highly recommended. However, for achieving the effective result to bring down the air pollution level in the mining area a constructive measure at political level is also highly essential. This would lead to an eco-friendly mining and better habitat for all those living in the area.

5. Acknowledgements

Authors are grateful to the Director, Central Institute of Mining and Fuel Research (CIMFR), Dhanbad, India for giving permission to publish this article. Authors are also thankful to M/s Western Coalfields Limited, Nagpur for sponsoring this study and providing necessary facilities.

6. References

Almbauer, R.A., Piringer, M., Baumann, K., Oettle D., & Sturm P.J. (2001). Analysis of the daily variations of winter time air pollution concentrations in the city of Graz, Austria., *Environmental Monitoring and Assessment,* Vol. 65, pp. 79–87.

Appleton, T.J., Kingman, S.W., Lowndes I.S., & Silvester, S.A. (2006). The development of a modeling strategy for the simulation of fugitive dust emissions from in-pit quarrying activites: a UK case study, International *Journal of Mining, Reclamation and Environment,* Vol. 20, P. 57-82.

Baldauf, R.W., Lane D.D., & Marote, G.A. (2001). Ambient air quality monitoring network design for assessing human health impacts from exposures to air-borne contaminants, *Environmental Monitoring and Assessment,* Vol. 66, pp. 63–76.

Banerjee, S.P (2006). TSP emission factors for different mining activities for air quality impact prediction as collated from different sources, Minetech, Vol. 27, pp. 3-18.

CPCB, *Central Pollution Control Board Notification, India,* (1994).

Chaulya, S.K., Chakraborty, M.K., & Singh R.S.(2001). Air pollution modelling for a proposed limestone quarry. *Water, Air, and Soil Pollution* Vol. 126, pp. 171–191.

Chaulya, S. K. (2004). Assessment and management of air quality for an opencast coal mining area, *Journal of Environmental Management*, Vol. 70, No. 1, pp. 1-14

CIMFR. Central Institute of Mining and Fuel Research (erstwhile Central Mining Research Institute) Report (1998). Determination of Emission Factor for Various Opencast Mining Activities, *GAP/9/EMG/MOEF/97,* Dhanbad, India

Collins, M.J., Williams P.L, & MacIntosh, D.L (2001). Ambient air quality at the site of a former manufactured gas plant. *Environmental Monitoring and Assessment* Vol. 68, pp. 137–152.

Corti, A. & Senatore, A. (2000). Project of an air quality monitoring network for industrial site in Italy. *Environmental Monitoring and Assessment,* Vol. 65, pp. 109–117

Crabbe, H., Beaumont R. & Norton, D. (2000). Assessment of air quality, emissions and management in a local urban environment. *Environmental Monitoring and Assessment,* Vol. 65 , pp. 435–442.

Cole, C.F. & Zapert, J.C. (1995).Air Quality Dispersion Model Validation at Three Stone Quarries, Washington DC, National Stone Association.

Ghose, M.K. & Majee, J. (2000). Assessment of dust generation due to opencast coal mining—an Indian case study. *Environmental Monitoring and Assessment,* Vol. 61, pp. 255–263.

Ghose, M.K. & Majee, S.R. (2000). Assessment of the impact on the air environment due to opencast coal mining -an Indian case study, *Atmospheric Environment* ,Vol. 34, pp. 2791-2796

Gilford, F.A. (1961). Uses of Routine Meteorological Observations for estimating atmospheric Dispersion, *Nuclear Safe*, Vol. 2

Grundnig, P.W., Höflinger, W., Mauschitz, G., Liu, Z., Zhang G., & Wang, Z. (2006). Influence of air humidity on the suppression of fugitive dust by using a water-spraying system, *China Particuology,* Vol. 4, No. 5, pp. 229-233.

Hanna, S.R., Briggs, G.A. &. Hosker, R.P. (1982). Handbook on Atmospheric Diffusion, DOE/TIC-11223, *US Department of Energy, Technical Information Center*

Jones, T. Blackmore, P. Leach, M. Matt, B.K. Sexton K. &. Richards, R. (2002). Characterisation of airborne particles collected within and proximal to an opencast coalmine: South Wales. UK *Environmental Monitoring and Assessment,* Vol. 75, pp. 293–312.

Kapoor, R.K. &. Gupta, V.K., (1984). A pollution attenuation coefficient concept for optimization of green belt. *Atmospheric Environment,* Vol. 18, pp. 1107–1117.

Karaca, M., Tayanc M. &. Toros, H. (1995). The effects of urbanization on climate of Istanbul and Ankara: a first study. *Atmospheric Environment,* Vol. , pp. 3411–3429.

Kumar, C.S.S., Kumar, P., Deshpande, V.P., &. Badrinath S.D. (1994). Fugitive dust emission estimation and validation of air quality model in bauxite mines, *Proceedings of International Conference on Environmental Issues in Minerals and Energy Industry, IME Publications, New Delhi, India,* pp. 77–81.

Muleski G.E. &. Cowherd, C. (1987). Evaluation of the effectiveness of Chemical dust Suppressants on Unpaved Roads, EPA/600/2-87.102. *U.S. Environmental Protection Agency, Research Triangle Park N.C.,* pp-81.

Pandey, S.K., Tripathi, B.D. &. Mishra, V.K. (2008). Dust deposition in a sub-tropical opencast coalmine area, India, *Journal of Environmental Management,* Vol. 86, No. 1, pp. 132-138.

Pasquill, F. (1962). *Atmospheric Diffusion, Van Nostrand Co. Ltd. Londan*

Peavy, H.S., Rowe, D.R. &. Obanoglous, G. Tech (1985). *Environmental Engineering, Megraw Hill, New York*, pp. 668-670.

Reddy, G.S. &. Ruj, B. (2003). Ambient air quality status in Raniganj-Asansol area, India. *Environmental Monitoring and Assessment,* Vol. 189, pp. 153-163.

Roney J. A. &. White, B. R. (2006). Estimating fugitive dust emission rates using an environmental boundary layer wind tunnel, *Atmospheric Environment*, Vol. 40, pp. 7668-7685

Shannigrahi, A.S. &. Sharma, R.C. (2000). Environmental factors in green belt development-an overview. *Indian Journal of Environmental Protection* , Vol. 20, pp. 602-607.

Sharma, S.C. &. Roy, R.K. (1997). Green belt—an effective means of mitigating industrial pollution. *Indian Journal of Environmental Protection* Vol. 17, pp. 724-727.

Sinha, S. &. Banerjee, S.P. (1997). Characterisation of haul road in Indian open cast iron ore mine. *Atmospheric Environment,* Vol. 31, pp. 2809-2814.

Tayanc, M. (2000) An assessment of spatial and temporal variation of sulphur dioxide levels over Istanbul, Turkey. *Environmental Pollution,* Vol. 107, pp. 61-69.

Tichy, J. (1996). Impact of atmospheric deposition on the status of planted Norway space stands: a comparative study between sites in Southern Sweden and the North Eastern Czech Republic. *Environmental Pollution,* Vol. 93, pp. 303-312

Triantafyllou, A.G. (2003). Levels and trends of suspended partcles around large lignite power station. *Environmental Monitoring and Assessment,* Vol. 89, pp. 15-34.

Triantafyllou, A.G., Kyros E.S. &. Evagelopoulos, V.G. (2002). Respirable particulate matter at an urban and nearby industrial location: concentrations and variability, synoptic weather conditions during high pollution episodes. *Journal of Air and Waste Management Association,* Vol. 52, pp. 287-296.

Trivedi, R. Chakraborty M. K., &. Tiwary, B.K. (2009). Dust Dispersion Modeling Using Fugitive Dust Model at an Opencast Coal Project of Western Coalfields Limited, India, *Journal of Scientific and Industrial Research,* Vol. 68, pp71-78

Turner, D.B. (1970).*Workbook of atmospheric Dispersion Estimates,* U.S.E.P.A.,Washington, DC.

USEPA, *United States Environmental Protection Agency* (1995). User's guide for the fugitive dust model (FDM), vol. 1, User Instructions, Region 10, 1200 sixth Avenue, Seattle, Washington, USA.

Vallack, H.W. &. Shillito, D.E. (1998). Suggested guidelines for deposited ambient dust, *Atmospheric Environment,* Vol. 32, No. 16, pp. 2737-2744.

Wheeler, A.J., Williams, I. Beaumont, R.A., &. Manilton, R.S. (2000). Characterisation of particulate matter sampled during a study of children's personal exposure to air borne particulate matter in a UK urban environment. *Environmental Monitoring and Assessment,* Vol. 65, pp. 69-77.

2

Secondary Acidification

Mizuo Kajino[1] and Hiromasa Ueda[2]

[1]*Meteorological Research Institute, Japan Meteorological Agency,*
[2]*Toyohashi Institute of Technology,*
Japan

1. Introduction

Secondary acidification (Kajino et al., 2008), also referred to as indirect acidification (Kajino et al., 2005; Kajino & Ueda, 2007), is a process that involves accelerated acid deposition associated with changes in gas–aerosol partitioning of semivolatile aerosol components, such as nitric acid (HNO_3), hydrochloric acid (HCl), and ammonia (NH_3), even though emissions of these substances and their precursors (e.g., NO_x) remain unchanged. HNO_3, HCl, and NH_3 are thermodynamically partitioned into gas and aerosol (particulate) phases in the atmosphere. This partitioning depends on temperature, humidity, and the presence of other components such as sulfuric acid (H_2SO_4) and crustal cations (Na^+, Mg^{2+}, Ca^{2+}, and K^+). Among acidic components in the air, H_2SO_4 has an equilibrium vapor pressure very much lower than that of other acids. When H_2SO_4 concentrations increase, NO_3^- and Cl^- in the aerosol phase shift to the gas phase, which causes the concentrations (fractions) of gaseous HNO_3 and HCl to increase, although total nitrate (t-NO_3 = HNO_3 + NO_3^-) and total chloride (t-Cl = HCl + Cl^-) remain unchanged. The deposition velocities of the highly reactive gaseous phases of HNO_3 and HCl are larger than those of their aerosol phases. For example, measured dry deposition velocities of HNO_3 gas are 20 times those of NO_3^- aerosols (Brook et al., 1997). Moreover, HNO_3 and HCl gases both readily dissolve into cloud and rain droplets. For solution equilibrium, their Henry's law constants are 2.1×10^5 and 727 mol L^{-1} atm^{-1}, respectively, which are extremely large values compared with those of SO_2 and NO_2 (1.23 and 0.01 mol L^{-1} atm^{-1}, respectively). Thus, below-cloud scavenging coefficients of irreversibly scavenged gases such as HNO_3 and HCl are several times those of their corresponding aerosols (Jylhä, 1999a, 1999b). In-cloud scavenging processes of gases and aerosols are hard to compare by this simple estimation procedure, because in-cloud scavenging of aerosol phases involves complexity of cloud dynamical and microphysical processes. Model calculations supported by observational data are necessary to estimate which phases are more efficiently scavenged for determination of net (in-cloud and below-cloud) wet deposition.

The secondary acidification effect was first identified in volcanic SO_2 plumes (Satsumabayashi et al., 2004). Miyakejima volcano, 180 km south of Tokyo, has erupted continuously since July 2000, resulting in considerable SO_2 emissions into the troposphere. One year after the start of emissions measurement in September 2000 (Kazahaya, 2001), SO_2 emissions totaled 9 Tg, equivalent to half the 20 Tg of anthropogenic SO_2 emissions from China in 2000. According to ground-based observations of gases and aerosols at Happo Ridge observatory (1,850 m ASL, 300 km north of Miyakejima volcano), the fraction of gaseous HNO_3 and HCl in the Miyakejima volcanic plume exceeded 95% (September 2000), whereas in the same season the fraction of these gases in contaminated air masses of the Asian continental outflow was approximately 40% (September, 1999). Consequently, the bimonthly mean NO_3^- and Cl^- concentrations in precipitation (net wet deposition) in August and September 2000 at Happo Ridge, after the eruption, increased by 2.7 and 1.9 times, respectively, compared with the same months in 1999, before the eruption.

Extensive studies of the seasonal and diurnal variations in gas–aerosol partitioning of semivolatile components and the mechanisms causing partitioning changes have been conducted (Moya et al., 2001; Lee et al., 2006; Morino et al., 2006). It was confirmed that the partitioning importantly influences surface fluxes of pollutants (Nemitz and Sutton, 2004) and climate (Adams et al., 2001; Schaap et al., 2004). The current study series on secondary acidification provides new evidence that changes in the gas–aerosol partitioning have important environmental impacts.

In section 2, we describe the secondary acidification process in detail. We present the results of our previous study series on secondary acidification due to the Miyakejima volcanic eruption in section 3, based on observational evidence (sect. 3.1) and modeling (sect. 3.2). In section 4, we describe secondary acidification occurring during long-range transport of anthropogenic air pollutants. We conduct an observational analysis (sect. 4.1) to reveal the current status, and perform model studies (sect. 4.2) to analyze possible future scenarios. We summarize our major findings in section 5. Here, we focus mainly on accelerated deposition of nitrate rather than that of chloride, because anthropogenic chloride emissions contain large uncertainty.

2. Secondary acidification process

Secondary acidification is defined as the process by which acid deposition is indirectly accelerated in association with changes in the gas–aerosol partitioning of semi-volatile atmospheric constituents, such as nitric acid, hydrochloric acid, and ammonia, even though emissions of these species and their precursors remain constant.

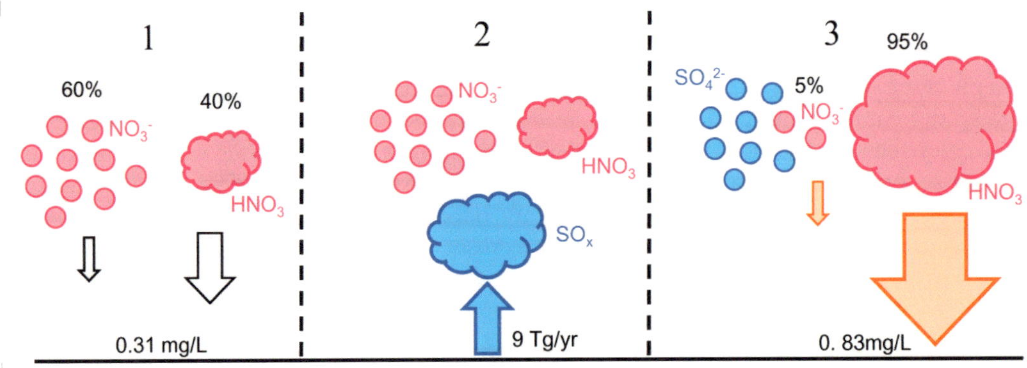

1. Nitric acid is partitioned into HNO_3 gas and NO_3^- aerosol.
   Dry and wet deposition rates of HNO_3 gas are higher than those of NO_3^- aerosol.
2. If SO_2 gas and consequently, H_2SO_4 gas increase, but total (gas + aerosol) nitric acid remain unchanged, then
3. H_2SO_4 quickly forms $(NH_4)_2SO_4$ aerosol, NO_3^- shifts to gas phase, increasing the concentration of HNO_3 gas.
   Consequently, deposition of total nitric acid increases, although total nitric acid concentration remain unchanged.

Fig. 1. Schematic illustration of secondary acidification by nitric acid. Values shown in the figure are those observed during the Miyakejima volcanic eruption event, discussed in section 3.1.

Gas–aerosol equilibrium of semi-volatile inorganic components in solid aerosols	Reaction No.
$NH_3(g) + HNO_3(g) \leftrightarrow NH_4NO_3(s)$	(R1)
$NH_3(g) + HCl(g) \leftrightarrow NH_4Cl(s)$	(R2)
Gas–aerosol equilibrium of semi-volatile inorganic components in liquid aerosols	
$NH_3(g) + HNO_3(g) \leftrightarrow NH_4^+ + NO_3^-$	(R3)
$NH_3(g) + HCl(g) \leftrightarrow NH_4^+ + Cl^-$	(R4)
As sulfuric acid gas increases via photochemical oxidation of SO_2	
$SO_2(g)$ + OH radical (g) $\rightarrow H_2SO_4(g)$	(R5)
$H_2SO_4(g) + NH_3(g) \rightarrow NH_4HSO_4(p)$	(R6)
$H_2SO_4(g) + 2NH_3(g) \rightarrow (NH_4)_2SO_4(p)$	(R7)
As sulfate increases via aqueous-phase oxidation	
[1]$S(IV) + O_3(aq) \rightarrow S(VI) + O_2$	(R8)
$HSO_3^- + H_2O_2(aq) \rightarrow SO_4^{2-} + H_2O$	(R9)
$SO_4^{2-} + 2NH_4^+ \leftrightarrow (NH_4)_2SO_4$	(R10)
In the presence of sea-salt particles	
$2NaCl + H_2SO_4(g) \rightarrow Na2SO_4 + HCl(g)$	(R11)
$NaCl + HNO_3(g) \rightarrow NaNO_3 + HCl(g)$	(R12)
In the presence of calcite-rich dust particles	
$CaCO_3 + H_2SO_4(g) \rightarrow CaSO_4 + H_2O + CO_2(g)$	(R13)
$CaCO_3 + HNO_3(g) \rightarrow Ca(NO_3)_2 + H_2O + CO_2(g)$	(R14)

1. $S(IV) \equiv SO_2 \cdot H_2O$, HSO_3^-, and SO_3^{2-}; $S(VI) \equiv HSO_4^-$ and SO_4^{2-}

Table 1. Chemical reactions describing the changes in gas–aerosol partitioning of semi-volatile inorganic components involved in the secondary acidification process.

Figure 1 illustrates schematically secondary acidification effects of nitric acid caused by increases in SO_2 emissions. The values used in Figure 1 are those measured at the Happo Ridge observatory and on Miyakejima Island, and reflect secondary acidification effects due to the eruption of Miyakejima volcano (see section 3.1 for details). Table 1 summarizes typical chemical reactions between atmospheric constituents involved in the secondary acidification process. Nitric acid is partitioned into HNO_3 gas and NO_3^- aerosol in the atmosphere (Figure 1, panel 1; R1 and R3 in Table 1). Since the partitioning is sensitive to temperature, over East Asia the gas phase is dominant in summer and at lower altitude, whereas the aerosol phase is dominant in winter and at higher altitude (Morino et al., 2006; Hayami et al., 2008; Kajino et al., 2008). This partitioning is also altered by the presence of other inorganic components. Hereafter, for simplicity, we focus on thermodynamic equilibrium in the NH_3–HNO_3–H_2SO_4–H_2O system. An increase in SO_2 emissions (Figure 1, panel 2), is followed by the oxidation of SO_2 [S(IV)] to S(VI), that is, either to H_2SO_4 gas by a gas-phase photochemical reaction (R5), or to SO_4^{2-} by aqueous-phase reactions (R8 and R9) in liquid aerosol or rain droplets. Because the vapor pressure of H_2SO_4 gas is extremely low, ammonium sulfate aerosols form immediately (R6 and R7). In the aqueous phase, SO_4^{2-}, because it is a strong acid, forms an ion pair with NH_4^+ (R10). Because sulfate consumes ammonia in the gas phase, the equilibrium of (R1 and R3) shifts leftward, and, as a result, HNO_3 gas evaporates from the aerosol phase (Figure 1, panel 3).

Wet and dry deposition rates of the highly reactive gaseous HNO_3 are high (Seinfeld and Pandis, 2006). Thus, as the SO_4^{2-} concentration increases, the concentration fraction of HNO_3

increases, with the result that deposition of total nitrate (t-NO_3 = HNO_3 + NO_3^-) is enhanced, even though the total nitrate concentration, as well as that of its precursors (i.e., NO_x), remains unchanged.
In the presence of abundant sea salt or mineral dust particles, however, HNO_3 gas is deposited on particle surfaces, expelling Cl^- and CO_3^-, respectively, into the gas phase (R12 and R14). Na^+ from sea salt and Ca^{2+} from mineral dust particles can also be counterions of SO_4^{2-} (R11 and R13). In such cases, increases in the gas phase fraction of t-NO_3 due to increased SO_4^{2-} and subsequent consumption of NH_3 are suppressed (see also section 4.1 and Kajino et al., 2008).

3. Eruption of Miyakejima volcano and the resulting secondary acidification effects in Japan

The eruption of Miyakejima volcano (Mt. Oyama, 139°32'E, 34°05'N, summit elevation 815 m ASL; Figure 2), 180 km south of Tokyo, Japan, beginning in July 2000 has resulted in the emission of huge amounts of sulfur dioxide. The annual mass of sulfur dioxide emitted was vast (9 Tg yr^{-1}; Kajino et al., 2004), equivalent to half the annual anthropogenic emission from China in 2000 (20 Tg yr^{-1}, Streets et al., 2003). Gases, aerosols, and precipitation have been sampled at the Happo Ridge observatory (137°48'E, 36°41'N, 1,850 m ASL, 330 km north of the volcano; Figure 2) in the central mountainous region of Japan since May 1998, two years before the eruption began (Satsumabayashi et al., 2004). Kajino et al. (2004, 2005) used a chemical transport model to simulate the emission, transport, transformation, and deposition of inorganic compounds such as SO_4^{2-}, NO_3^-, and NH_4^+ of anthropogenic and volcanic origin for the one-year period from September 2000 to August 2001. In this section, we highlight the outcomes of our previous research, focusing on the effects of the volcanic eruption on concentrations and deposition of inorganic compounds over the far East Asian region.

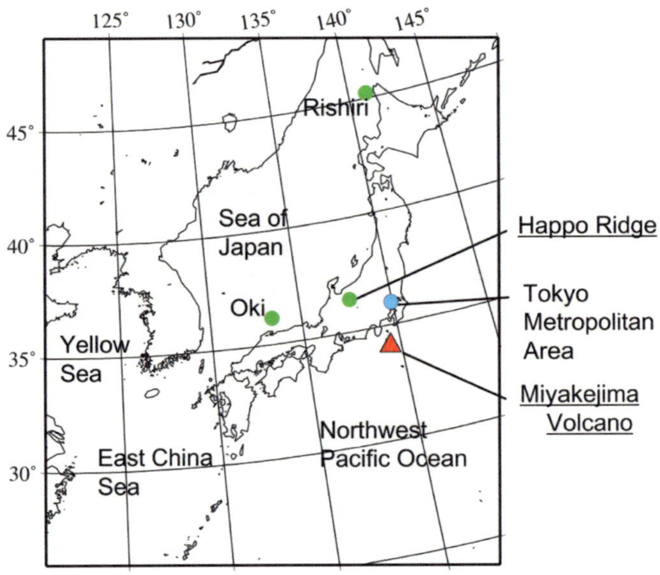

Fig. 2. Map of Japan showing the locations of the Happo Ridge observatory, Miyakejima volcano, the Tokyo Metropolitan Area, and the EANET monitoring stations Oki and Rishiri (see section 4).

3.1 Observational evidence

Temporal variations in smoke height (m) and SO_2 emissions (ton day^{-1}) from Miyakejima volcano (Figure 3) were measured with a correlation spectrometer (COSPEC) by the Japan Meteorological Agency (Kazahaya, 2001). From the start of the observation, total measured SO_2 emissions were 9 Tg yr^{-1}, corresponding to about 70% of the global emissions from volcanoes from the 1970s to 1997 (13 Tg yr^{-1}; Andreas and Kasgnoc, 1998) and to about half the anthropogenic SO_2 emitted from China in 2000 (20 Tg yr^{-1}). The maximum emission, about 82,200 ton day^{-1}, was observed at 10:48 LT on 16 November 2000. This value is equivalent to the anthropogenic emission from all of Asia in 2000 (34.3 Tg yr^{-1}, ~94,000 ton day^{-1}; Streets et al., 2003). The observed smoke height on the same day was only 1,000 m, indicating that almost the entire amount was released into the Planetary Boundary Layer. The emission gradually decreased to about 10,000 ton day^{-1} about 1 year after the onset of eruption. In 2002, the emission was still substantial, at 16.8% of Chinese anthropogenic emissions and 3.8 times Japanese anthropogenic emissions (Kajino et al., 2011). The continuous injection of the volcanic plume containing SO_2 into the Planetary Boundary Layer (i.e., the observed smoke height continued below 2,000 m) necessarily affected surface air quality and environmental acidification over far East Asia substantially.

At Happo Ridge, aerosol samples are collected daily for 3 hours, from 12:00 to 15:00 LT, with a high-volume air sampler. The four-stage filter pack method was used for intensive sampling of gaseous and aerosol inorganic compounds during two weeks in September 1999 and one week in September 2000. Meteorological parameters and hourly concentrations of SO_2, NO_x, O_3, and PM_{10} are monitored automatically. Satsumabayashi et al. (2004) have described the observation methods in detail.

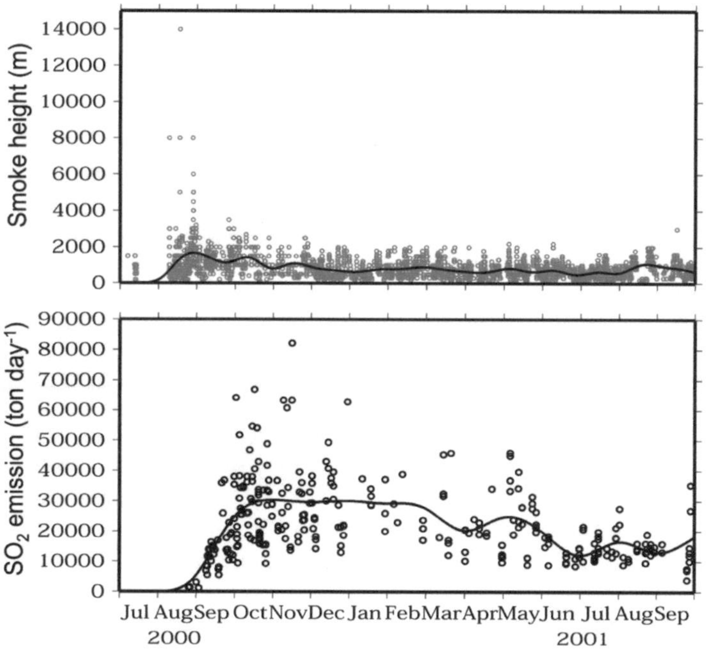

Fig. 3. Time series of observed smoke height (top) and SO_2 emissions (bottom) from Miyakejima volcano. The data were interpolated using a spline function (solid lines) for use as input in the model simulation.

Sampling date and time (LT)	SO_4^{2-} mg m^{-3}	Particle phase fraction	
		Nitrate	Ammonium
	Air mass of Asian continental origin before the eruption (1999)		
13 Sep 12:00–15:00	12.3	0.60	0.78
13 Sep 15:00–18:00	12.1	0.61	0.77
13 Sep 18:00–21:00	8.80	0.57	0.76
13 Sep 21:00–24:00	8.10	0.82	0.72
14 Sep 00:00–03:00	10.7	0.50	0.74
Average	10.4	0.62	0.75
	Air mass directly affected by the volcanic eruption (2000)		
15 Sep 12:00–18:00	32.0	0.00	0.96
15 Sep 18:00–24:00	20.3	0.00	0.94
16 Sep 00:00–06:00	11.0	0.17	0.76
16 Sep 06:00–12:00	6.40	0.00	0.75
Average	17.4	0.04	0.85

Table 2. Gas–aerosol partitioning observed at Happo Ridge before and after the onset of the eruption.

We selected two high sulfate concentration events, from 12:00 LT 13 September to 3:00 LT 14 September 1999, before the onset of the eruption, and from 12:00 LT 15 September to 12:00 LT 16 September 2000, just after the onset of the eruption, and examined SO_4^{2-} concentrations and gas–aerosol partitioning of t-NO_3 and t-NH_4 (= NH_3 + NH_4^+) measured at Happo Ridge (Table 2). Prior to the eruption, in September 1999, the gas–aerosol partitioning of nitrate in the contaminated air mass from the Asian continent tended to favor the aerosol phase: 62% in the aerosol phase versus 38% in the gas phase (Figure 1, panel 1). Similarly, the gas–aerosol partitioning of ammonia also favored the aerosol phase (72% aerosol, 25% gas). In September 2000, two months after the onset of eruption, the gas–aerosol partitioning of nitrate in the air mass from Miyakejima Island was biased almost entirely toward the gas phase (4% aerosol, 96% gas), whereas the aerosol phase fraction of ammonium was higher (85%) than it was before the eruption onset. This result is consistent with thermodynamic equilibrium theory (Table 1).

Table 3 lists the mean bimonthly concentrations of trace chemical components in gases, aerosols, and precipitation measured at Happo Ridge before and after the onset of the eruption. After the eruption, the concentrations of SO_2 gas, SO_4^{2-} aerosol, and SO_4^{2-} in precipitation increased dramatically, by 15, 3, and 6.8 times, respectively, compared with their concentration before the eruption. The concentration of NH_4^+, a major counterion of SO_4^{2-} in aerosols doubled, and it increased in precipitation, by 5 times after the eruption. O_3 and PM_{10} (aerosols smaller than 10μm in diameter) concentrations were slightly higher in September 2000 than before the eruption, but the difference was small compared with the concentration differences in inorganic compounds, indicating that photochemical activity and the total aerosol concentrations were not very different between the period before and that after the eruption began. However, NO_3^- in precipitation increased by 2.7 times after the eruption, whereas aerosol NO_3^- concentrations did not differ between the two periods. Unfortunately, continuous measurement data for HNO_3 gas are not available, so t-NO_3 cannot be determined. Because secondary acidification is defined as an increase in NO_3^- deposition while the t-NO_3 concentration remains unchanged, we cannot prove that the observed increase in bimonthly mean NO_3^- in rainwater was caused by secondary acidification. Nevertheless, the observations are consistent with secondary acidification theory.

	Gas, ppb		Aerosol, µg m^{-3}				Precipitation, mg L^{-1}		
	SO_2	O_3	PM_{10}	SO_4^{2-}	NO_3^-	NH_4^+	SO_4^{2-}	NO_3^-	NH_4^+
Before eruption (Aug. and Sep. 1999)	0.2	38	14	2.2	0.2	0.75	0.25	0.31	0.06
After eruption (Aug. and Sep. 2000)	3.3	46	17	6.5	0.24	1.56	1.70	0.83	0.33

Table 3. Mean bimonthly concentrations of various components of gases, aerosols, and precipitation measured at Happo Ridge before the onset of the eruption and one year later.

It has still not been proved whether wet scavenging of HNO_3 gas or of NO_3^- aerosol is more efficient. Because Henry's law constant of HNO_3 is extremely large and below-cloud scavenging of aerosols is not very efficient, the below-cloud scavenging coefficient of HNO_3 is several times the size of the aerosol scavenging coefficients (Jylhä, 1999a, 1999b). In-cloud aerosol scavenging is much more efficient because the aerosol particles act as cloud condensation nuclei (CCN). HNO_3 can also dissolve in cloud droplets, however, so it is likely that wet scavenging of HNO_3 is more efficient than that of NO_3 aerosols.

Because no methods are available for quantitative measurement of wet scavenging efficiency, numerical models that incorporate cloud dynamical and microphysical processes must be used to determine the wet scavenging efficiency of HNO_3 compared with that of NO_3 aerosols. However, because such models require many assumptions and several parameterizations, especially for modeled gas–aerosol–cloud interaction processes, the numerical answer includes substantial uncertainty. Thus, the results require careful interpretation. In the next section, the simulation results for secondary acidification effects due to the eruption of Miyakejima Volcano are discussed.

3.2 Model simulations

Kajino et al. (2005) investigated secondary acidification due to the Miyakejima eruption over the one-year period from September 2000 to August 2001 by using a chemical transport model (MSSP, Model System for Soluble Particles; Kajino et al., 2004). They concluded that dry and wet deposition of nitrate increased by 0.2–0.8 and 0.5–4 mg m^{-2} day^{-1}, respectively, over far East Asia on average during the year. However, the MSSP model does not include an aerosol dynamics module, and it assumes thermodynamic equilibrium of semi-volatile inorganic compounds. Therefore, we recently developed a more sophisticated aerosol chemical transport and deposition model, and then used the new model to revisit the volcanic eruption study and examine secondary acidification effects of the plume.

The new aerosol chemical transport model used for the simulation is called Regional Air Quality Model 2. In RAQM2, a simple version of a modal-moment aerosol dynamics model (MADMS; Kajino, 2011) was implemented to achieve a completely dynamical (non-equilibrium) solution for gas-to-particle mass transfer over a wide range of aerosol diameters, from 1 nm to greater than 1 µm. MADMS is unique in that it is capable of solving inter-modal coagulation between two modes with very different log-normal size parameters. To consider a variety of atmospheric aerosol properties, including size, chemical composition, and mixing states, a category approach of the EMTACS (Eulerian

Multiscale Tropospheric Aerosol Chemistry and dynamics Simulator) model (Kajino and Kondo, 2011) is utilized. The Advanced Research WRF (Weather Research and Forecasting) model (version 3.1.1; Skamarock et al., 2008) is used to simulate the meteorological field. Lateral and upper boundary concentrations used for the RAQM2 simulation were hourly concentrations of NO_x, O_x, CO, and volatile organic carbons simulated by a global-scale stratospheric and tropospheric chemical climate model (MRI-CCM2; Deushi and Shibata, 2011). Thus, RAQM2 is a so-called one-way nested model of the global chemistry model. The RAQM2 used Streets et al. (2003)'s anthropogenic emission inventory as well as other sources of aerosol emissions such as biogenic sources, open biomass burning, and wind-generated emissions (dust and sea-salt). The model domain is illustrated in Figure 4. There are 100 × 70 horizontal grids, and the resolution is 60 km with a Lambert conformal conic map projection and a reference latitude/longitude of 35°N/115°E. Vertically, there are 28 layers from the ground to 100 hPa in WRF, and 13 layers from the ground to 10 km in RAQM2, with terrain-following coordinates.

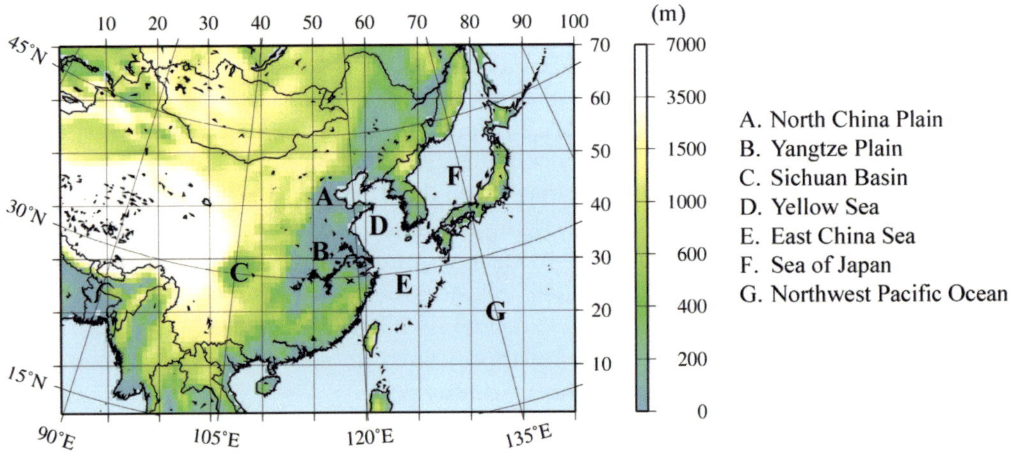

Fig. 4. The WRF and RAQM2 model domain, topography used in the model (color scale, m ASL), and locations of important geographical features.

The simulations were conducted for the month of October 2000. Since wind and precipitation patterns change seasonally, the concentrations and deposition patterns of air pollutants vary substantially during a year. We selected October 2000 for the simulation because during that month volcanic emission was active and the volcanic plume was carried to the Japan archipelago. The Miyakejima volcanic emissions are injected into horizontal grid (X = 87, Y = 39), which corresponds to the location of Miyakejima volcano (Mt. Oyama, 139°32'E, 34°05'N). The SO_2 emission flux and injection height were estimated by spline interpolation of the measured data (solid lines in Figure 3; ~30,000 ton day^{-1}) and uniformly distributed to the vertical grids (Z ≈ 5–7) corresponding to the interval from the summit elevation (815 m ASL) to the measured smoke height (~1,500 m above the crater).

Figure 5 illustrates monthly mean surface concentrations of anthropogenic and volcanic SO_2 and SO_4^{2-} (volcanic data were derived by subtracting the simulation result obtained without including volcanic emissions from that obtained with volcanic emissions), the gas-phase fraction of t-NO_3, and its increase due to increased volcanic SO_4^{2-} in October 2000. The volcanic SO_2 concentration over the area downwind of the volcano (Figure 5b) was

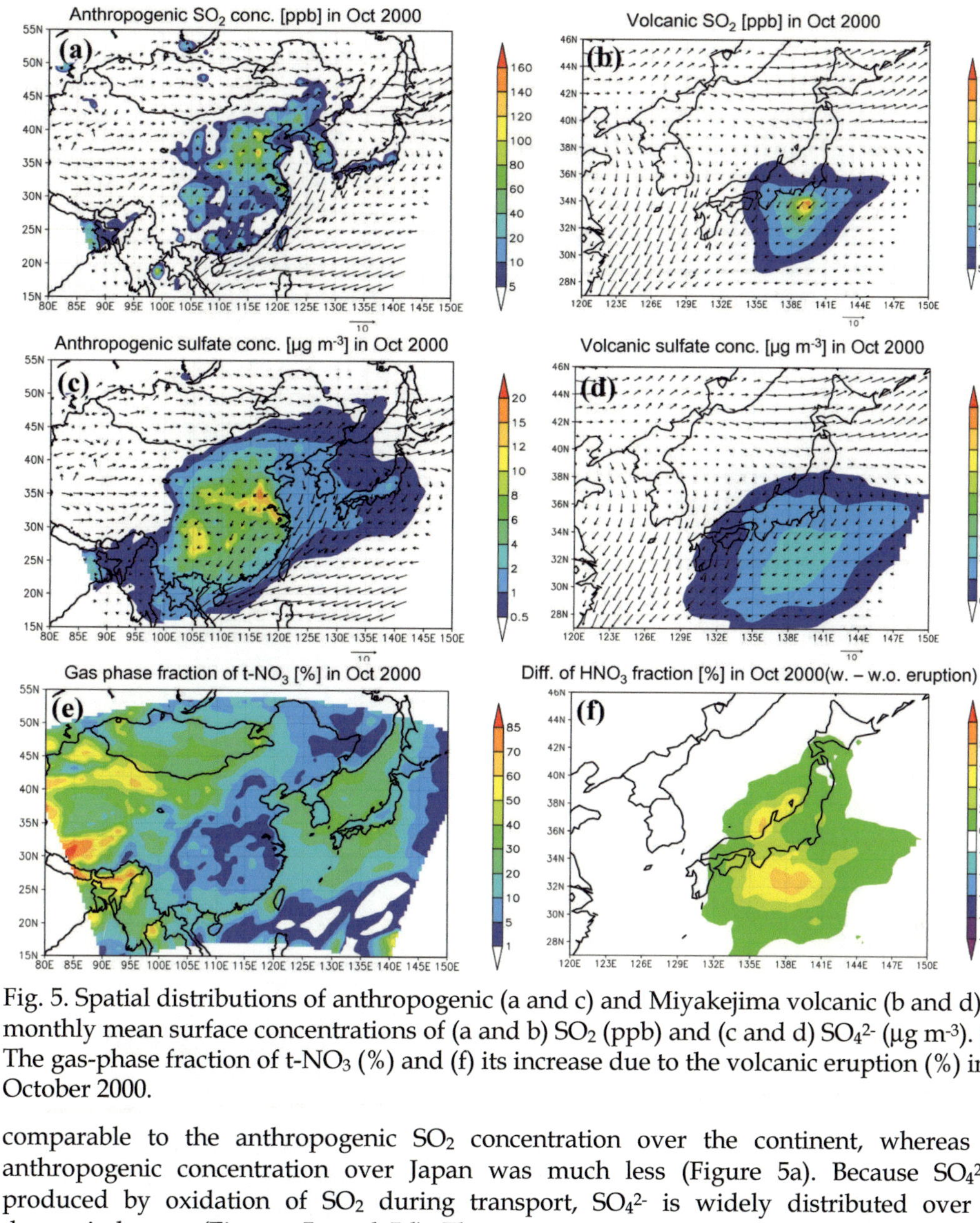

Fig. 5. Spatial distributions of anthropogenic (a and c) and Miyakejima volcanic (b and d) monthly mean surface concentrations of (a and b) SO_2 (ppb) and (c and d) SO_4^{2-} (μg m^{-3}). (e) The gas-phase fraction of t-NO_3 (%) and (f) its increase due to the volcanic eruption (%) in October 2000.

comparable to the anthropogenic SO_2 concentration over the continent, whereas the anthropogenic concentration over Japan was much less (Figure 5a). Because SO_4^{2-} is produced by oxidation of SO_2 during transport, SO_4^{2-} is widely distributed over the downwind areas (Figures 5c and 5d). The maximum concentration of volcanic SO_4^{2-} was smaller than that over the land, probably because photochemical oxidants such as OH radicals, O_3, and H_2O_2 are more abundant over the continent. In central Japan, the SO_4^{2-} concentration was doubled as a result of the volcanic eruption. The gas phase fraction of t-NO_3 was smaller over the continent (1–15%) than over the ocean or the Japan archipelago (20–40%) (Figure 5e), because the surface temperature is higher over the ocean than over the continent in October. These values are consistent with those observed at Happo Ridge. Expulsion of NO_3^- from the aerosol phase occurred as a result of the volcanic SO_2 emission, with an increase of the gas-phase fraction of about 2–5% (Figure 5f). The gas-phase fraction is small compared with the observed fraction at Happo Ridge, where the gas-phase fraction reached 100% during volcanic plume transport events. It is probably because simulated ammonia concentration is overestimated so that it fixed HNO_3 in the aerosol phase and inhibited the expulsion of NO_3^- due to increased sulfate.

Figures 6a-d illustrates the monthly cumulative dry and wet deposition of t-NO_3 in October 2000 and changes due to the volcanic eruption. Figures 6e and 6f show monthly precipitation over East Asia and around Japan, respectively. Secondary acidification in dry deposition, that is, the increase of nitrate in dry deposition, in the region downwind of the Miyakejima volcano, is evident (0.5–3 mg m^{-2}) accounting for at maximum about 10% of the total amount (20–80 mg m^{-2}).

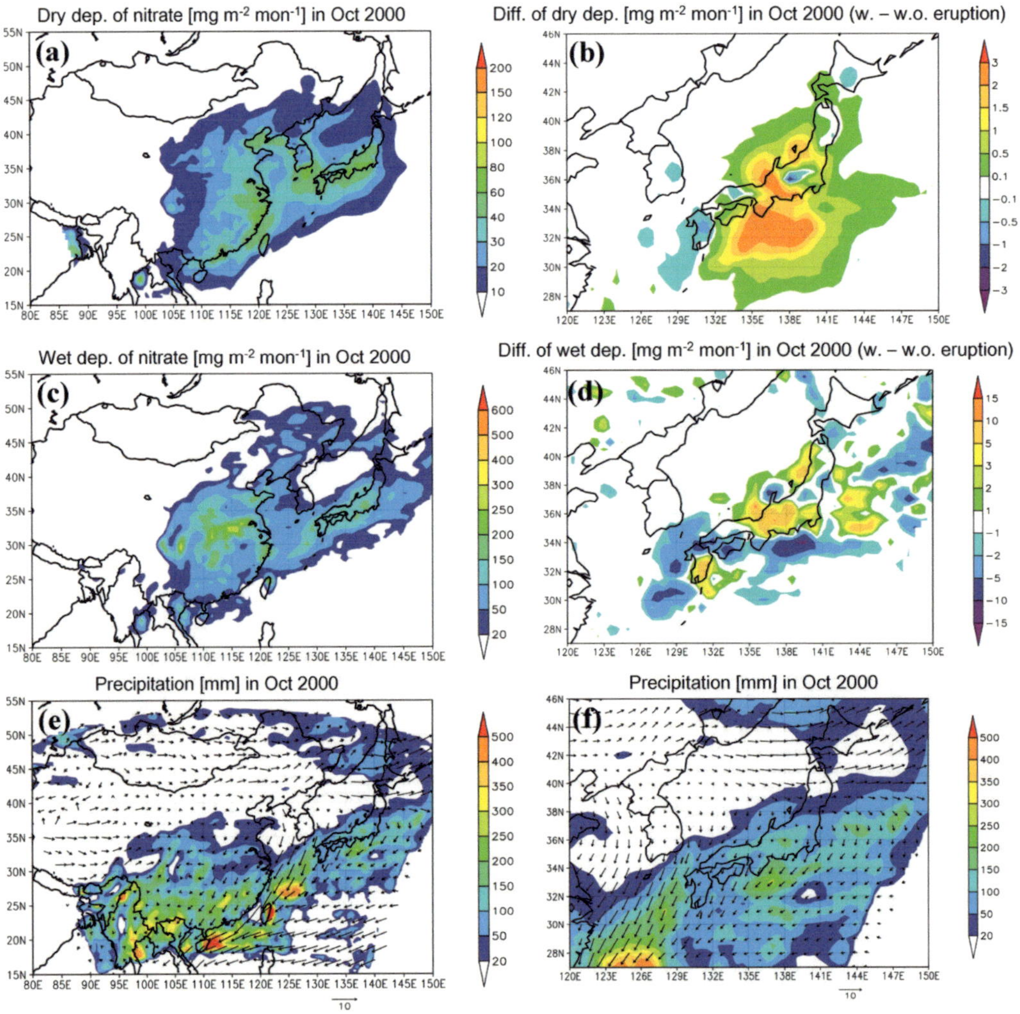

Fig. 6. Spatial distributions of cumulative monthly dry deposition of (a) total nitrate (gas + aerosol) (mg m^{-2}) and (b) the difference due to the eruption. (c and d) The same as (a) and (b) but for wet deposition (mg m^{-2}). (e and f) Monthly cumulative precipitation (mm) in October 2000.

Secondary acidification due to wet deposition was rather complex. An increase in nitrate in wet deposition was simulated over the central Japan land mass (5–10 mg m^{-2}), whereas a decrease of 2–15 (mg m^{-2}) was simulated over the ocean south of the archipelago, where precipitation was heavier. The decrease is due probably to dominance of in-cloud scavenging of aerosols (NO_3^- acting as CCN) over the dissolution of HNO_3 gas, so that the increased fraction of HNO_3 resulted in a decrease in the wet deposition of t-NO_3.The modeled wet deposition increase was smaller than that predicted by MSSP (15–120 mg m^{-2}), which simulated that an increase in the wet deposition of nitrate was comparable to that of sulfate (see Area 1, October 2000 in Table 3 of Kajino et al., 2005).

Even though the RAQM2 model is more sophisticated than the MSSP model, it is hard to judge which model results are closer to reality. MSSP does not solve aerosol processes dynamically. The aerosol size distribution is diagnosed as a simple function of relative humidity, and all particles of inorganic compounds such as SO_4^{2-}, NO_3^-, and NH_4^+ are assumed to act as CCN and to become dissolved in cloud droplets (Kajino et al., 2004). In contrast, the more sophisticated RAQM2 model solves the evolution processes (condensation, evaporation, and coagulation) of the aerosol size distribution dynamically for aerosol diameter ranges from 1 nm to greater than 1 μm. Then, the CCN activity of particles is calculated using the parameterization of Abdul-Razzak and Ghan (2000). This parameterization was developed from cloud parcel model calculations, including aerosol–cloud interaction processes. The critical diameter D_c (above which aerosol particles can act as CCN) is calculated from the predicted aerosol size distribution and chemical composition, the temperature and humidity in the environment, and the updraft velocity. Ironically, however, because it is quite challenging to reproduce atmospheric aerosol size distributions, especially for sub-micron particles, RAQM2 can fail to reproduce the CCN activity of aerosols, which is very sensitive to particle size. Nevertheless, the simulated results of the two models should not be far from reality, as the model results for various concentrations of inorganic components in the air and in precipitation have been evaluated extensively using measurement data (Kajino et al., 2011, and references therein; Kajino and Kondo, 2011).

4. Secondary acidification as a result of long-range transport over East Asia

In section 3, we showed that secondary acidification occurred during an extreme episode. In this section, we investigate secondary acidification under general air pollution conditions. Asia is one of the largest anthropogenic SO_x emitting regions in the world, and because of the rapid economic growth in Asia, emissions there might change drastically in the future. Therefore, understanding the impacts of secondary acidification in regions downwind of large emission sources and how they change depending on future emission changes is important. In section 4.1, we present indications of secondary acidification based on Acid Deposition Monitoring Network in East Asia (EANET) data. In section 4.2, we use a chemical transport model to simulate long-range transport of inorganic air pollutants over the East Asian region. To evaluate the secondary acidification due to future emission changes, we carried out simple sensitivity studies by increasing or decreasing emissions of SO_2 and NH_3, the gaseous counterparts of SO_4^{2-} and NO_3^- aerosols, in model simulations.

4.1 Observational evidence

We found indications of secondary acidification in the EANET monitoring data. EANET began collecting data on a regular basis in 2001, following established guidelines and technical procedures and adopting a quality assurance/quality control program (EANET, Guidelines for Acid Deposition Monitoring in East Asia; available from http://www.eanet.cc/product.html). In 2002, two additional Japanese monitoring stations began to measure concentrations of gas and aerosol components by the filter pack (FP) method, increasing the number of monitoring stations in Japan to the current 10. At these stations, 7- or 14-day cumulative concentrations of gaseous species (HNO_3, HCl, NH_3, SO_2) and aerosol components (SO_4^{2-}, NO_3^-, Cl^-, NH_4^+, Na^+, Mg^{2+}, K^+, Ca^{2+}) are monitored, and 1-day cumulative concentrations in precipitation (SO_4^{2-}, NO_3^-, Cl^-, NH_4^+, Na^+, Mg^{2+}, K^+, Ca^{2+}) are measured by collecting precipitation samples and analyzing them by ion chromatography.

The FP method is known to have several distinct artifact problems. Because particulate NH_4NO_3 and NH_4Cl collected on a filter might volatilize during a 7- or 14-d sampling period, there is a possibility that thermodynamic equilibrium may not have been attained. In addition, high humidity can reduce the concentration of gaseous species because of trapping by condensed water in the filter pack. To evaluate these sampling artifacts in the EANET FP data, we employed a multicomponent gas–aerosol equilibrium model called Simulating Compositions of Atmospheric Particles at Equilibrium (SCAPE; Kim et al., 1993; Meng et al., 1995) to recalculate the gas–aerosol partitioning of t-NO_3 at six selected Japanese EANET monitoring sites and compared the results with the measurement data (Kajino et al., 2008). We found fairly high correlations, low mean errors, and low root mean square errors between the observed and simulated values (see Table 2 of Kajino et al., 2008). We therefore concluded that long-duration FP data from EANET are sufficiently consistent for detecting the secondary acidification effect in our current analysis.

In the present study, we used monitoring data collected by two of the Japanese EANET stations, Rishiri and Oki. As shown in Figure 2, these stations are situated on small islands facing the Sea of Japan, so plumes from the continental outflow are efficiently detected. Rishiri is further north than Oki, so the climate and air quality features differ between these two stations.

Table 4 summarizes the monitoring results from Oki and Rishiri. Non-sea-salt (nss)-SO_4^{2-}, defined as the concentration difference between total [SO_4^{2-}] and sea-salt-originated SO_4^{2-} (= 0.251 × [Na^+]) consists mostly of anthropogenic and partly of volcanic sulfate. Here, the square brackets [ ] denote atmospheric concentration (μg m^{-3}). $C_{nssS/N5}$ is the molar ratio of nss-SO_4^{2-} to the N(V) concentration, namely t-NO_3, where C denotes the atmospheric concentration. P_{gHNO3} is the molar fraction of HNO_3 gas relative to t-NO_3. F_s, defined as,

$$F_s \equiv \frac{S(IV)}{S(IV)+nssS(VI)} = \frac{[S]_{SO_2}}{[S]_{SO_2}+[S]_{nss-SO_4^{2-}}} \tag{1}$$

is the ratio of tetravalent *S* to total (tetravalent plus hexavalent) anthropogenic *S* in the air. Under the assumption that all SO_2 is of non-sea-salt origin, a low F_s value indicates that SO_2 is sufficiently oxidized to SO_4^{2-} during long-range transport. $D_{N5/S6}/C_{N5/S6}$ is the ratio of the N/S molar ratio in precipitation to the atmospheric ratio (where D indicates wet deposition):

$$D_{N5/S6}/C_{N5/S6} \equiv \frac{d[NO_3^-]}{d[SO_4^{2-}]} \times \frac{[SO_4^{2-}]_p/96}{[HNO_3]_g/63+[NO_3^-]_p/62} \tag{2}$$

Here, *d[ ]* denotes the concentration in precipitation (μmol L^{-1}), and the subscripts *g* and *p* denote the gas phase and the particulate phase, respectively. As atmospheric concentrations were measured by the FP method on a 7- or 14-d cycle, the daily measured wet deposition amounts are also averaged over the same sampling period.

Station name	Oki			Rishiri		
Observation period	04/2003 - 03/2005			07/2002 - 03/2005		
Number of samples	77			72		
Longitude/Latitude	133.18/36.70			141.20/45.12		
	Mean concentration ± standard deviation, $\mu g\ m^{-3}$					
$[nss\text{-}SO_4^{2-}]$	3.54	±	2.05	1.86	±	0.95
$[t\text{-}NO_3]$	1.21	±	0.61	0.73	±	0.47
$[t\text{-}NH_4]$	1.26	±	0.77	0.68	±	0.34
[Crustals] [a]	2.64	±	1.58	2.31	±	1.26
	Mean meteorological measurements with standard deviations					
T, °C	14.48	±	7.17	6.50	±	8.72
RH, %	74.10	±	6.16	75.69	±	6.00
	Mean molar ratios with standard deviations					
$[nss\text{-}SO_4^{2-}]/[SO_4^{2-}]$	0.86	±	0.11	0.78	±	0.13
$C_{nssS/N5}$	1.99	±	1.06	1.96	±	1.00
P_{gHNO3}	0.30	±	0.22	0.20	±	0.16
F_s	0.37	±	0.16	0.34	±	0.14
	Correlation coefficient (R)					
T versus P_{gHNO3}	0.63			0.46		
$C_{nssS/N5}$ versus P_{gHNO3} [b]	0.75			-0.03	→	0.41
P_{gHNO3} versus $D_{N5/S6}/C_{N5/S6}$ [b]	0.74			0.17	→	0.70
F_s versus $D_{N5/S6}/C_{N5/S6}$ [b]	-0.45			0.08	→	-0.43

[a]$[Crustals] = [Na^+] + [Mg^{2+}] + [K^+] + [Ca^{2+}]$

[b]The correlation coefficient R at Rishiri improved when it was calculated using only data collected when $[nss\text{-}SO_4^{2-}]/[SO_4^{2-}] > 0.8$ and $T > 0$ °C .

Table 4. Information about Oki and Rishiri stations and monitoring results.

In general, Rishiri and Oki can be characterized as follows. Rishiri is located in northern Japan where temperatures drop to below 0 °C in winter and the marine atmosphere is relatively clean. Thus, the mean anthropogenic indicator $[nss\text{-}SO_4^{2-}]/[SO_4^{2-}]$ was lower than 0.8. As the counterions of NO_3^-, sea-salt-originated crustal concentrations of Na^+ and Mg^{2+} were high (>2 $\mu g\ m^{-3}$), resulting in a low P_{gHNO3} (0.2) value. In contrast, Oki is located in coastal western Japan, where it is influenced by Asian continental outflows, resulting in relatively higher values of $[nss\text{-}SO_4^{2-}]/[SO_4^{2-}]$ (>0.85) and P_{gHNO3} (0.3). Relative humidity (RH) did not differ significantly between the two stations.

Measured P_{gHNO3} was positively correlated with T, as explained in section 2. We found a slightly larger correlation between P_{gHNO3} and $C_{nssS/N5}$ at Oki in western Japan, indicating a marked expulsion of NO_3^- by SO_4^{2-} in the continental outflow. Conversely, at Rishiri the correlation coefficient between P_{gHNO3} and $C_{nssS/N5}$ was near zero, which we attributed to the effects of abundant sea-salt components (R12). P_{gHNO3} and $D_{N5/S6}/C_{N5/S6}$ were positively correlated at Oki. Here, $D_{N5/S6}/C_{N5/S6}$ represents the N(V)/S(VI) molar concentration ratio in precipitation to that in the atmosphere (Eq. 2). Accordingly, a positive correlation between P_{gHNO3} and $D_{N5/S6}/C_{N5/S6}$ indicates a relatively higher wet deposition rate of t-NO_3, compared with that of S(VI), as the gas phase fraction of t-NO_3 increases, that is, secondary acidification. Together with the positive correlation between $C_{nssS/N5}$ and P_{gHNO3}, the expulsion of particulate NO_3^- to the gas phase by anthropogenic SO_4^{2-} results in acceleration of the wet deposition flux of t-NO_3 at Oki.

At Rishiri, where no correlation was detected between $C_{nssS/N5}$ and P_{gHNO3}, analysis on the basis of the $D_{N5/S6}/C_{N5/S6}$ ratio is not applicable. Because of the abundance of cations of sea-salt origin, the correlation coefficient between P_{gHNO3} and $D_{N5/S6}/C_{N5/S6}$ is 0.17. In addition, the secondary acidification effect is usually based on the calculation of the vapor–liquid equilibrium of semivolatile components, but Rishiri station is in northern Japan, where precipitation during winter consists mostly of ice and snow rather than rain. Therefore, we selected data sampled when the mean temperature exceeded 0 °C and [nss-SO_4^{2-}]/[SO_4^{2-}] was greater than a threshold of 0.8, indicating less influence of sea-salt particles. The same analysis using the selected samples resulted in an increase in the correlation coefficient between $C_{nssS/N5}$ and P_{gHNO3} from –0.03 to 0.41, and that between P_{gHNO3} and $D_{N5/S6}/C_{N5/S6}$ increased from 0.17 to 0.70. Secondary acidification is thus evident under the substantial influence of anthropogenic contamination when the influence of sea-salt particles is low and the temperature is sufficiently high for vapor–liquid equilibrium to be established. Neutralization effects by the Asian dust particles (R14) may negate the increase in P_{gHNO3} caused by the increase in SO_4^{2-}. We thus performed a similar analysis using [nss-Ca^{2+}]/[Ca^{2+}], but found no significant effects in the current data sets. Here, [nss-Ca^{2+}] is defined as the difference between total [Ca^{2+}] and sea-salt-originated Ca^{2+} (= 0.038 [Na+]); therefore, higher values of [nss-Ca^{2+}]/[Ca^{2+}] would indicate the occurrence of Asian dust events, which are characterized by abundant calcite.

Figure 7 shows the relationships between $C_{nssS/N5}$ and $P_{g\text{-}HNO3}$ (%), $P_{g\text{-}HNO3}$ (%) and $D_{N5/S6/CN5/S6}$, and F_s and $D_{N5/S6/CN5/S6}$ at Oki and Rishiri. At Oki, the correlations are higher because of its location in western Japan where it is influenced by the pronounced Asian continental outflow. At Rishiri, however, the sea-salt effect is predominant, leading to very weak correlations (open circles in Figs. 7d–f), although they are enhanced significantly by sample selection for T > 0 °C and [nss-SO_4^{2-}]/[SO_4^{2-}] > 0.8 (closed circles).

It is notable that the regression lines obtained from the two stations are very similar despite the ~1,200 km distance separating them. P_{gHNO3} is less than 30% when $C_{nssS/N5}$ is less than 2 at Oki (Fig. 7a), whereas P_{gHNO3} is often greater than 50% under the lower influence of sea-salt particles (closed circles) when $C_{nssS/N5}$ is less than 2 at Rishiri (Fig. 7d). The slope of this regression line for Rishiri is steeper than that for Oki, because the average concentration of t-NH_4, which can fix NO_3^- as ammonium nitrate in the aerosol phase, is low at Rishiri, only half the average concentration at Oki. However, when cations originating from sea salt are abundant at Rishiri (open circles in Figure 7d), P_{gHNO3} is less than 20%, even when $C_{nssS/N5}$ is larger than 4.

The F_s value is an indicator of the distance that pollutants have been transported. Relatively short-distance transport within Japan is indicated by an F_s value greater than approximately 0.6. Conversely, F_s smaller than 0.2 indicates long-range transport from the Asian continent to Japan (Satsumabayashi et al., 2004). At Oki, $D_{N5/S6/CN5/S6}$ was greater than 3 when F_s was smaller than 0.2 (Fig. 7c), indicating that secondary acidification effects were greater at Oki when air pollutants had been transported a long distance. The same trend was observed for the selected samples at Rishiri (closed circles, Fig. 7f).

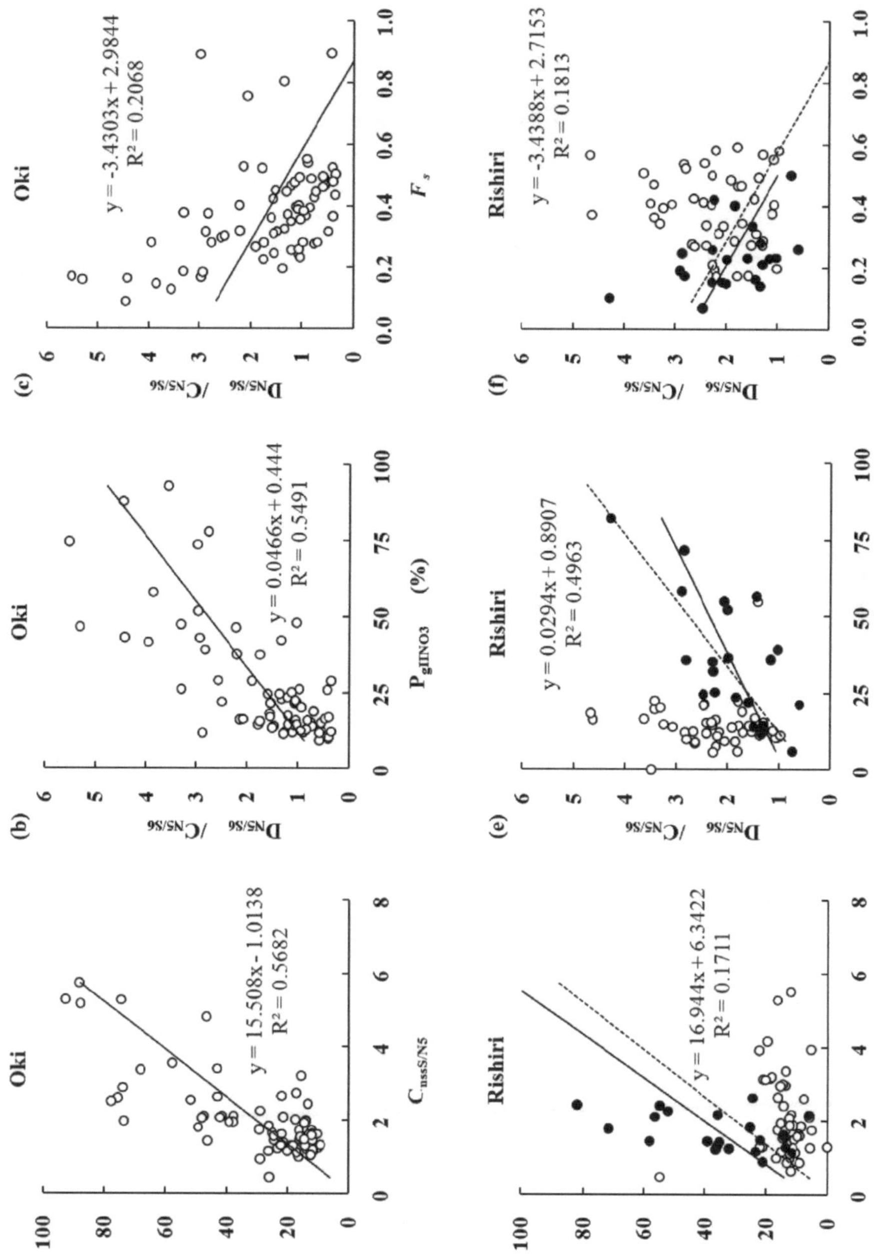

Fig. 7. Scattergrams of (a) $C_{nssS/N5}$ and P_{gHNO3}, (b) P_{gHNO3} and $D_{N5/S6}/C_{N5/S6}$, (c) F_s and $D_{N5/S6}/C_{N5/S6}$ at Oki. (d–f) Same as (a–c) for Rishiri. The open circles in (a–f) show all data. The closed circles in (d–f) show data of samples collected when T > 0 °C and $[nss\text{-}SO_4^{2-}]/[SO_4^{2-}] > 0.8$. The solid regression lines in (d–f) are for all data, and the dashed regressions lines are for only the data shown by the closed circles.

4.2 Model simulations

4.2.1 Future emission scenarios

In all four scenarios (A1, A2, B1, and B2) considered in the Special Report on Emissions Scenarios (SRES) from the Intergovernmental Panel on Climate Change (IPCC) (2000), emission rates are increasing faster in Asia than elsewhere in the world. In these scenarios, SO_x emissions in Asia are expected to peak between 2020 and 2050, becoming twice as high as in 2000. NO_x emissions will continue to increase even when SO_x is no longer increasing, reaching a maximum after 2040 at levels more than twice current levels in the lowest estimation scenario (B1), and in the A2 scenario reaching a level approximately four times the current level in 2100. SRES does not evaluate NH_x emission rates, which have a marked effect on gas-aerosol partitioning, but Klimont et al. (2001) estimated that NH_x emissions will increase to 1.7 times the 1995 level in East Asia by 2030.

Fujino et al. (2002) developed the AIM/Trend model (Asian-Pacific Integrated Model) to assess future environmental loads based on past socio-economic trends. They estimated a change in Asian SO_2 emissions from 30.8 Tg yr^{-1} in 1998 to 21.9–85.1 Tg yr^{-1} in 2032, depending on the socio-economic scenario, Security First (SF), Market First, Sustainability First, or Policy First. The largest emission growth ratio is estimated for the SF scenario, which assumes a world of great disparities where inequality and conflict brought about by socio-economic and environmental stresses will prevail. In that scenario, SO_2 emission in China is predicted to grow to 2.42 times the 2000 level by 2030 (Kajino and Ueda, 2007).

Representative Concentration Pathways (RCPs) are new future emission scenarios that have recently been provided as part of the Coupled Model Intercomparison Project Phase 5 (http://cmip-pcmdi.llnl.gov/cmip5) prepared for the Fifth Assessment Report (AR5) of the IPCC, which is scheduled to be published in 2013. The current version of the RCP database (version 2.0.5) is available at http://www.iiasa.ac.at/web-apps/tnt/RcpDb/. The four RCP future emission scenarios are RCP 3-PD (IMAGE, van Vuuren et al., 2007), RCP 4.5 (MiniCAM; Clarke et al., 2007), RCP 6.0 (AIM; Fujino et al., 2006; Hijioka et al., 2008), and RCP 8.5 (MESSAGE, Riahi et al., 2007). The RCP 2.0.5 emission scenarios for the period until 2050 predict that total Asian SO_2 emissions will increase from 32.4 Tg yr^{-1} in 2000 to as high as 46.3 Tg yr^{-1} (1.43 times; RCP 8.5) in 2020 and then decrease to as little as 12.5 (a 38.6% reduction; RCP 3-PD) in 2050. Total NO_x emissions will increase from 26.3 Tg NO_2 yr^{-1} in 2000 to a high as 48.1 Tg NO_2 yr^{-1} (1.83 times; RCP 8.5) in 2020 and decrease to as little as 25.4 Tg NO_2 yr^{-1} (a 96.6% reduction; RCP 4.5) in 2050. Total Asian NH_3 emissions will increase from 20.9 Tg yr^{-1} in 2000 to as high as 32.7 Tg yr^{-1} (1.56 times; RCP 3-PD), with the minimum predicted value being an increase to 25.6 Tg yr^{-1} (1.22 times; RCP 4.5).

4.2.2 Secondary acidification due to future emissions changes

To investigate secondary acidification effects due to future emissions changes, we performed simple sensitivity studies by simulating increased or decreased emissions of SO_2 and NH_3, the gaseous counterparts of SO_4^{2-} and NO_3^- aerosols (Table 5). The control run (CNTRL) used Regional Emission inventory in ASia (REAS; Ohara et al., 2007) data for 2005 (Kurokawa et al., 2009). The values in Table 5 are the ratio to the CNTRL emissions and were applied uniformly over the whole model domain.

"S2" and "Sh" indicate double and half the SO_2 emission of the CNTRL run, and "NH2" and "NHh" indicates double and half the NH_3 emission of the CNTRL run. We expected secondary acidification effects to be greatest in the S2NHh run, in which SO_2 was increased and the counterpart of NO_3^- was decreased, and that they would be least in the ShNH2 run.

The model (MRI-CCM2-WRF-RAQM2) and the model settings used for these sensitivity studies were identical to those described in section 3.2, except 90 × 60 horizontal grids were used with the reference latitude/longitude being 37°N/123°E (the model domain is not shown as it is not very different from that shown in Figure 4). The simulation was conducted for March 2006. In spring in East Asia, considerable long-range transport occurs because cyclones and anti-cyclones propagating eastward carry contaminated air masses by turn in cycles of about 5 days.

RUN	CNTRL	S2	S2NHh	Sh	ShNH2
SO_2 emission	1	2	2	0.5	0.5
NO_x emission	1	1	1	1	1
NH_3 emission	1	1	0.5	1	2

Table 5. Ratios of emissions to that of CNTRL run used for sensitivity studies to evaluate secondary acidification due to future emission changes.

Figure 8 illustrates the simulated (CNTRL) spatial distributions of the SO_2 and NO_x emission fluxes and monthly mean surface concentrations of SO_4^{2-} and t-NO_3 in March 2006. SO_2 and NO_x emissions peaks are seen in large emission source regions, and SO_4^{2-} and t-NO_3 are transported widely to southward and eastward downwind regions.

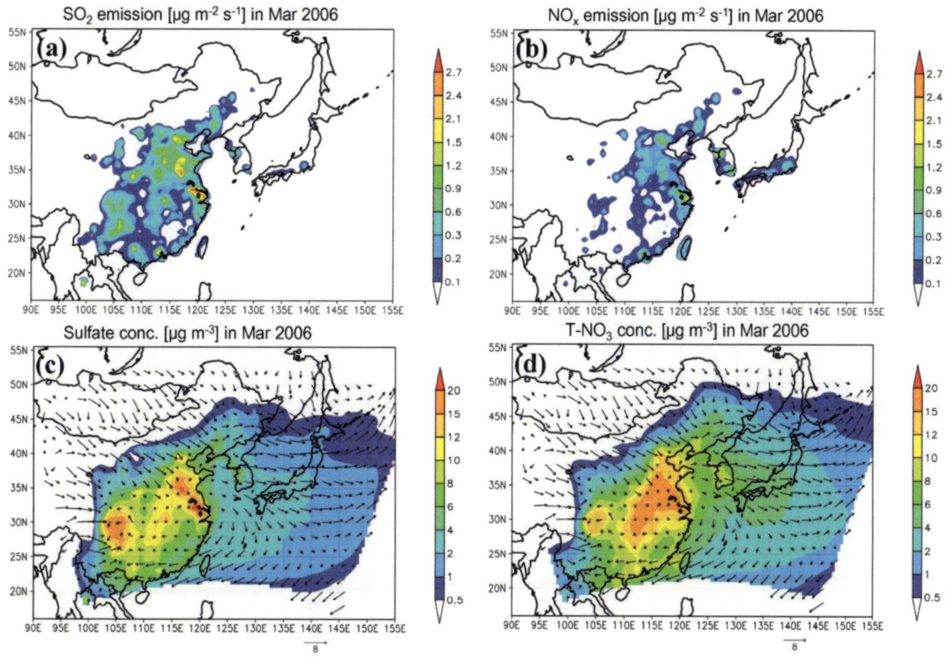

Fig. 8. The simulated (CNTRL) spatial distributions of (a) the SO_2 emission flux (μg m^{-2} s^{-1}), (b) the NO_x emission flux (μg m^{-2} s^{-1}), (c) the monthly mean surface sulfate concentration (μg m^{-3}), and (d) the monthly mean surface total (gas + aerosol) nitrate concentration (μg m^{-3}) in March 2006.

Figure 9 illustrates the simulated (CNTRL) spatial distributions of the gas phase fraction of nitrate, the monthly accumulated precipitation, and the monthly accumulated dry and wet deposition of t-NO_3. The gas phase fraction is larger over the ocean (20–40%) than over the continent (1–30%) because the surface temperature is higher over the ocean in spring. Also because of temperature differences, the gas phase fraction over the land is larger in the south

(5-30%) than in the north (1-5%). The monthly mean surface temperature over the ocean ranges over about 5-20 °C, whereas it ranges from -20 to 0 °C over the northern continent, and from 0 to 15 °C over the southern continent (not shown). In general, the dry deposition amount and the surface concentration are expected to correlate with each other given a relatively constant dry deposition rate. However, the dry deposition amounts are larger over the southern edge of the continent and western Japan, whereas the surface concentrations are larger over the North China Plain, the Sichuan Basin, and the Yangtze Plain. The horizontal distribution of the dry deposition is rather similar to that of the gas phase fraction, because the modeled dry deposition velocities of HNO_3 gas (0.9-2.7 cm s^{-1}) are much larger than those of NO_3^- aerosols (0.02-0.1 cm s^{-1} over the land, 0.2-1 cm s^{-1} over the ocean). The wet deposition amounts are large where both the precipitation and the concentrations are large, and they are about twice to three times the dry deposition amounts.

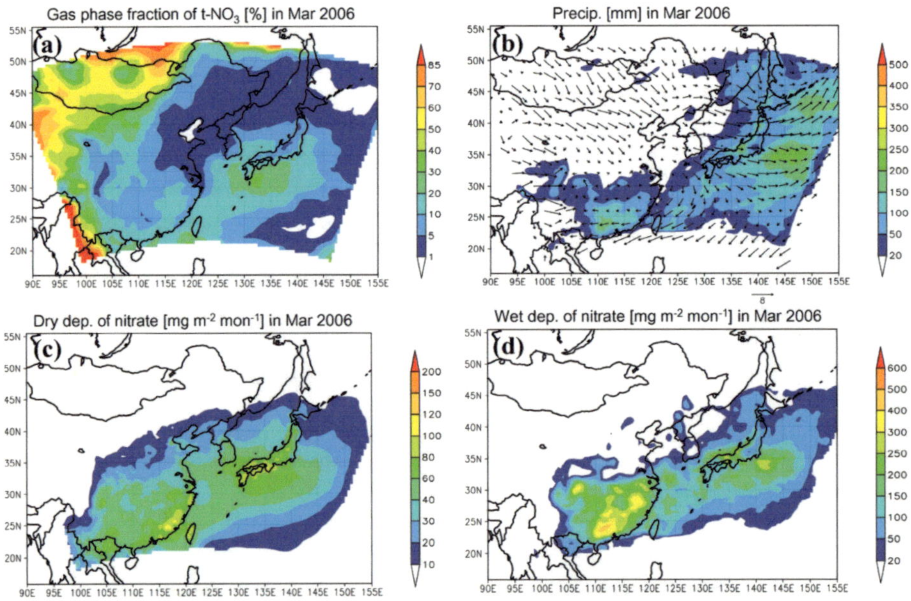

Fig. 9. Spatial distributions of (a) the gas phase fraction of nitrate (%), (b) monthly accumulated precipitation (mm) with surface wind vectors (m s^{-1}), (c) monthly dry deposition amount of nitrate (μg m^{-2} mon^{-1}), and (d) wet deposition of nitrate (μg m^{-2} month) in March 2006.

Figure 10 illustrates the gas phase fraction of nitrate and monthly accumulated total (dry + wet) deposition of t-NO_3 in the CNTRL run and the deviations from the control in the S2 and the S2NHh runs. Doubling SO_2 emissions causes the gas phase fraction to increase by 1-6% over southern China and over the ocean (Figure 10c and d). The increase of the gas phase fraction over northern China is less than 1%, however, because of the low temperatures there. In general, because the East Asian atmosphere is ammonia-rich and is sodium-rich over the ocean, so the expulsion of NO_3^- to the gas phase is not very significant. However, gas phase fraction, of as large as 20% over northern China, is seen, because the counterpart of NO_3^- is decreased substantially. As a result of the increase in the gas phase fraction, the total deposition of nitrate increases by about 5-20 mg m^{-2}, corresponding to about 10% of the total deposition in CNTRL (50-300 mg m^{-2}), when SO_2 emissions double. The increase is larger than 20 mg m^{-2} over wide areas when NH_3 emission is halved, accounting for as much as 50% of the total nitrate of the CNTRL.

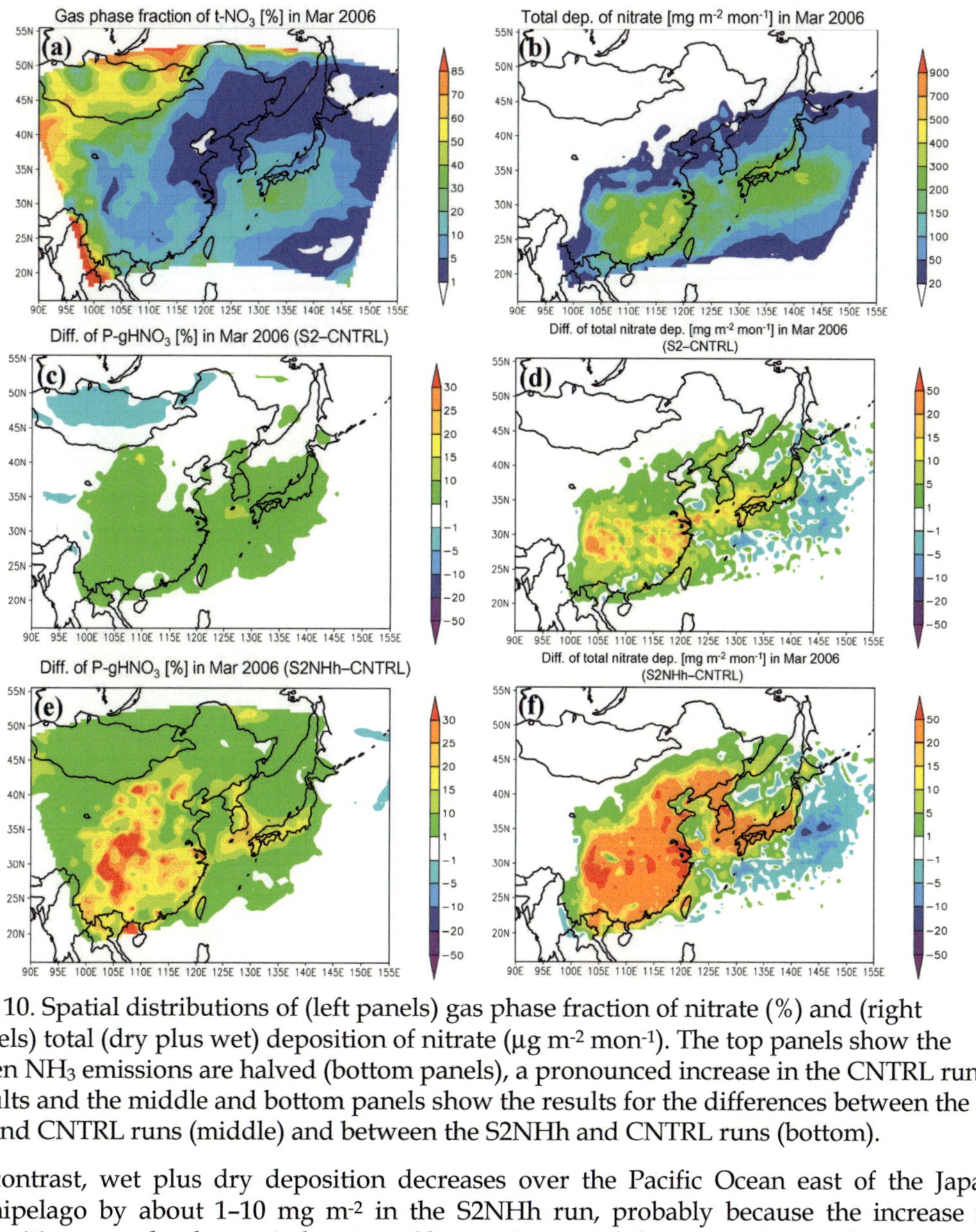

Fig. 10. Spatial distributions of (left panels) gas phase fraction of nitrate (%) and (right panels) total (dry plus wet) deposition of nitrate (μg m^{-2} mon^{-1}). The top panels show the when NH_3 emissions are halved (bottom panels), a pronounced increase in the CNTRL run results and the middle and bottom panels show the results for the differences between the S2 and CNTRL runs (middle) and between the S2NHh and CNTRL runs (bottom).

In contrast, wet plus dry deposition decreases over the Pacific Ocean east of the Japan archipelago by about 1–10 mg m^{-2} in the S2NHh run, probably because the increase in deposition over the downwind regions (the continent and the ocean close to the continent) causes the concentration over the regions further downwind to decrease. Consequently, nitrate deposition also decreases in the regions further downwind.

As discussed before in Sections 3.1 and 3.2, the wet deposition efficiencies of HNO_3 gas and NO_3^- aerosol cannot be directly compared with each other because NO_3^- aerosol particles can act effectively as CCN. When cloud production and NO_3^- aerosol activation are very efficient, secondary acidification may not occur. In contrast, when mature clouds are present and the gravitational fall of rain droplets is dominant, HNO_3 gas is more efficiently captured by water droplets and secondary acidification may occur. The RAQM2 model can show the quantitative results of secondary acidification due to wet deposition, and the simulation results should not differ much from reality because the model results for the concentrations of inorganic components in the air as well as for precipitation have been evaluated extensively with measurement data. However, in the current off-line coupled WRF-RAQM2 framework, processes related to wet deposition, such as aerosol activation, cloud dynamics, and cloud microphysics, are based on many assumptions and various parameterizations. Thus, it is still not possible to determine whether wet scavenging of HNO_3 gas or of NO_3^- aerosol is in reality more efficient.

4.2.3 Adverse effects of an SO_2 emission decrease: a decrease in nitrate deposition downwind may cause an increase in deposition even further downwind

The widespread installation of flue-gas desulfurization (FGD) devices is expected to decrease Asian SO_2 emissions in the future. In China, FGD devices are now being installed in many coal-fired power plants. From 2001 to 2006, FGD penetration increased from 3% to 30%, causing a 15% decrease in the average SO_2 emission factor of coal-fired power plants (Zhang et al., 2009).

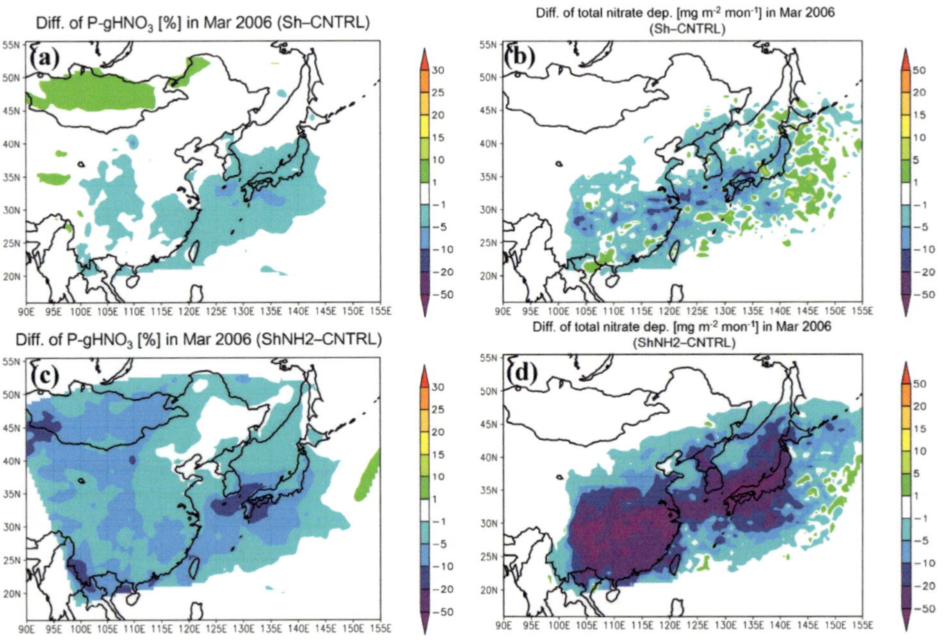

Fig. 11. Spatial distributions of the differences in the gas phase fraction of nitrate (%) (left panels) and total (dry + wet) deposition of nitrate (μg m^{-2} mon^{-1}) (right panels). The upper and lower panels show the results for (Sh - CNTRL) and the (S2NHh - CNTRL) runs, respectively.

As a result of future SO_2 emission decreases, less secondary acidification should occur. However, a decrease in nitrate deposition downwind will also mean that t-NO_3 will be transported longer distances, which may result in increased deposition of t-NO_3 in regions further downwind.

Figure 11 shows changes in the gas phase fraction of nitrate and in total deposition when SO_2 emissions are decreased by half. Both the gas phase fraction of nitrate and total nitrate deposition in downwind regions decrease, by 1–5% and 1–20 mg m^{-2}, respectively (upper panels). When NH_3, the counterpart of NO_3^- in aerosols, is doubled and SO_2 emissions are halved, the gas phase fraction of nitrate decreases substantially over downwind regions, which results in a significant decrease in total nitrate deposition (20–50 mg m^{-2}). In the Sh run, the surface mean t-NO_3 concentration over Pacific coastal regions of Japan increases by 0.5–2% (not shown) and the increase in the total deposition is about 1–5 mg m^{-2} over the same regions (Figure 11b), although the increase is small compared to the total deposition (100–400 mg m^{-2}, Figure 10b).

5. Conclusion

We studied secondary acidification, which is enhanced deposition of NO_3^- caused by an increase in the SO_4^{2-} concentration, using field observation data as well as numerical simulations of a volcanic eruption event and the long-range transport of air pollutants. Because the vapor pressure of H_2SO_4 gas is extremely low, increased SO_4^{2-} expels NO_3^- in the aerosol phase to the gas phase, resulting in an increase in the HNO_3 gas fraction. As wet and dry deposition rates of HNO_3 gas are considered to be more efficient than those of NO_3^- aerosols, the deposition of total nitrate (HNO_3 gas plus NO_3^- aerosols) is consequently enhanced, even though its total concentration remains unchanged.

Secondary acidification was prominent when the Miyakejima Volcano (180 km south of Tokyo) erupted, emitting a huge amount of SO_2 (9 Tg yr^{-1}) into the lower atmosphere (~2000 m ASL). At the Happo Ridge observatory (1850 m ASL, 300 km north of the volcano), the fraction of gaseous HNO_3 increased from 40% before the eruption to 95% after the eruption, and the bimonthly mean NO_3^- concentration in precipitation increased by 2.7 times after the eruption. The numerical simulation using the RAQM2 model predicted that as a result of the volcanic SO_2 emissions, the SO_4^{2-} concentration would double and the gas phase fraction of t-NO_3 would increase from 20–40% to 22–45% per month on average over central Japan, which is downwind of Miyakejima volcano. The increase of dry and wet deposition due to the volcanic emission was about 0.5-3 and 5–10 (mg m^{-2} mon^{-1}), respectively. Wet deposition was decreased in some regions, probably because CCN activation and cloud droplet formation of NO_3^- aerosols is more efficient than dissolution of HNO_3 gas into water droplets.

At the Japanese EANET monitoring station at Oki, we found positive correlations between the following observational parameters:

1. SO_4^{2-} concentration in atmosphere and gas phase fraction of HNO_3
2. The gas phase fraction of HNO_3 and wet deposition rate of total nitrate
3. A long-range transport indicator and the wet deposition rate of total nitrate

These positive correlations indicate that secondary acidification occurs during the long-range transport of air pollutants from the Asian continent to Japan. Secondary acidification is less efficient in the presence of abundant sea-salt particles, because the contained Na^+ reacts with nitrate to form $NaNO_3$, keeping it in the aerosol phase.

We also simulated secondary acidification due to future anthropogenic SO_2 emission changes using the RAQM2 model. If SO_2 emissions double, the gas phase fraction increases 1–6% over southern China and over the ocean, resulting in an increase of about 10% in total nitrate deposition over the region. The Asian atmosphere is generally ammonia-rich, so the expulsion of NO_3^- to the gas phase is not significant. However, if emission of NH_3, as the counterpart of NO_3^-, is decreased by half, along with the doubling of SO_2 emissions, then the expulsion of NO_3^- is significant and total nitrate deposition over the downwind region increases by as much as 50%. Asian SO_2 emissions are likely to decrease in the future because of the installation of flue-gas desulfurization devices and petroleum refineries. As SO_2 emissions decrease, nitrate deposition may also decrease in downwind regions. On the other hand, the decrease in nitrate deposition in downwind regions means that total nitrate will be transported greater distances to regions further downwind.

Our results also indicate that to simulate the concentrations and depositions of t-NO_3 accurately, accurate estimations of emission inventories of SO_2 and NH_3 and of its precursor NO_x are important.

Simulated dry deposition velocities and wet scavenging rates include substantial errors and uncertainties in most numerical models, because those parameters are quite difficult to evaluate from observational data. Therefore, as simulation techniques become more advanced, we should revisit this issue again to update our knowledge about what really happens in the atmosphere.

6. Acknowledgment

We thank Dr. Hikaru Satsumabayashi of Nagano Environmental Conservation Research Institute, Japan, for providing measurement data from Happo Ridge and for engaging in meaningful analysis and discussions.

7. References

Abdul-Razzak, H. & Ghan, S. J. (2000). A parameterization of aerosol activation: 2. Multiple aerosol types, *J. Geophys. Res.*, 105, pp.6837-6844, doi:10.1029/1999JD901161.

Adams, P. J.; Seinfeld, J. H.; Koch, D.; Mickley, L. & Jacob, D. (2001). General circulation model assessment of direct radiative forcing by the sulfate-nitrate-ammonium-water inorganic aerosol system. *J. Geophys. Res.*, Vol. 106, No. D1, pp. 1097-1111.

Andreas, R. J. & Kasgnoc, A. D. (1998). A time-averaged inventory of subaerial volcanic sulfur emission. *J. Geophys. Res.*, Vol.103, No.D19, pp.25,251-25,261.

Brook, J. R.; Di-Giovanni, F.; Cakmak, S., & Meyers, T. P. (1997). Estimation of dry deposition velocity using inferential models and site-specific meteorology - Uncertainty due to siting of meteorological towers. *Atmos. Environ.*, Vol. 31, No. 23, pp. 3911-3919

Clarke, L.; Edmonds, J.; Jacoby, H.; Pitcher, H.; Reilly, J. & Richels, R. (2007). Scenarios of greenhouse gas emissions and atmospheric concentrations. Sub-report 2.1A of Synthesis and Assessment Product 2.1 by the U.S. Climate Change Science Program and the Subcommittee on Global Change Research. Department of Energy, Office of Biological & Environmental Research, Washington, 7 DC., USA, 154 pp.

Deushi, M. & Shibata, K. (2011). Development of an MRI Chemistry-Climate Model ver.2 for the study of tropospheric and stratospheric chemistry, *Papers in Meteor. Geophys.*, in press.

Fujino, J.; Matsui, S.; Matsuoka, Y. & Kainuma, M. (2002). AIM/Trend: Policy Interface, Climate Policy Assessment, Eds. M. Kainuma, Y. Matsuoka and T. Morita, Springer, pp.217-232.

Fujino, J.; Nair, R.; Kainuma, M.; Masui, T. & Matsuoka, Y. (2006). Multi-gas mitigation analysis on stabilization scenarios using AIM global model. Multigas Mitigation and Climate Policy. *The Energy Journal Special Issue.*

Hayami, H.; Sakurai, T.; Han, Z.; Ueda, H.; Carmichael, G.; Streets, D.; Holloway, T.; Wang, Z.; Thongboonchoo, N.; Engardt, M.; Bennet, C.; Fung, C.; Chang, A.; Park, S. U.; Kajino, M.; Sartelet, K.; Matsuda, K. & Amann, M. (2008). MICS-Asia II: Model intercomparison and evaluation of particulate sulfate, nitrate and ammonium. *Atmos. Environ.*, Vol.42, pp.3510-3527.

Hijioka, Y.; Matsuoka, Y.; Nishimoto, H.; Masui, M. & Kainuma, M. (2008). Global GHG emissions scenarios under GHG concentration stabilization targets. *J. Global Environ. Eng.*, Vol.13, pp.97-108.

Intergovernmental Panel on Climate Change (2000), *Special Report on Emissions Scenarios,* edited by N. Nakicenovic, 599 pp., Cambridge Univ. Press., New York.

Jylhä, K. (1999a). Relationship between the scavenging coefficient for pollutants in precipitation and the radar reflectivity factor. Part I: derivation. *J. Appl. Meteorol.*, Vol. 38, pp. 1421-1434

Jylhä, K. (1999b). Relationship between the scavenging coefficient for pollutants in precipitation and the radar reflectivity factor. Part II: Applications. *J. Appl. Meteorol.*, Vol. 38, pp. 1435-1447.

Kajino, M. (2011). MADMS : Modal Aerosol Dynamics model for multiple Modes and fractal Shapes in the free-molecular and near-continuum regimes. *J. Aerosol Sci.*, Vol. 42, No. 4, pp.224-248.

Kajino, M . & Kondo, Y. (2011). EMTACS : Development and regional-scale simulation of a size, chemical, mixing type, and soot shape resolved atmospheric particle model. *J. Geophys. Res.*, Vol. 116, D02303, doi :10.1029/2010JD015030

Kajino, M. & Ueda, H. (2007). Increase in nitrate deposition as a result of sulfur dioxide emission increase in Asia: indirect acidification. *Air Pollution Modeling and its Application XVIII.*, Eds. C. Borrego, E. Renner, Elsevier, ISBN:978-0-444-52987-9, pp. 134-143

Kajino, M.; Ueda, H. ; Satsumabayashi, H ; & An, J. (2004). Impacts of the eruption of Miyakejima Volcano on air quality over far east Asia. *J. Geophys. Res.*, Vol. 109, D21204, doi:10.1029/2004JD004762

Kajino, M.; Ueda, H. ; Satsumabayashi, H ; & Han, Z. (2005). Increase in nitrate and chloride deposition in east Asia due to increased sulfate associated with the eruption of Miyakejima Volcano. *J. Geophys. Res.*, Vol. 110, D18203, doi:10.1029/2005JD005879

Kajino, M ; Ueda, H. & Nakayama, S. (2008). Secondary acidification : Changes in gas-aerosol partitioning of semivolatile nitric acid and enhancement of its deposition due to increased emission and concentration of SOx. *J. Geophys. Res.*, Vol.113, D03302, doi:1029/2007JD008635

Kajino, M.; Ueda, H.; Sato, K. & Sakurai, T. (2010). Spatial distribution of the source-receptor relationship of sulfur in Northeast Asia. *Atmos. Chem. Phys. Discuss.*, Vol.10, pp.30,089-30,127.

Kazahaya, K. (2001). Amount of volcanic gases erupted by Miyakejima Volcano, in *Miyakejima Island Eruption and Wide Area Air Pollution* (in Japanese), Jpn. Soc. For Atmos. Environ., pp. 17-26.

Kim, Y. P.; Seinfeld, J. H. & Saxena, P. (1993). Atmospheric gas-aerosol equilibrium: I. Thermodynamic model, *Aerosol Sci. Technol.*, Vol.19, pp.157-181.

Klimont, Z.; Cofala, J.; Schopp, W.; Amann, M.; Streets, D. G.; Ichikawa, Y. & Fujita, S. (2001). Projections of SO_2, NO_x, NH_3 and VOC emissions in East Asia up to 2030, *Water Air Soil Pollut.*, Vol. 130, pp.193-198.

Kurokawa, J.; Ohara, T.; Uno, I.; Hayasaka, M. & Tanimoto, H. (2009). Influence of meteorological variability on interannual variations of springtime boundary layer ozone over Japan during 1981-2005. *Atmos. Chem. Phys.*, Vol. 9, pp.6287-6304.

Lee, S.; Ghim, Y. S.; Kim, Y. P. & Kim, J. Y. (2006). Estimation of the seasonal variation of particulate nitrate and sensitivity to the emission changes in the greater Seoul area. *Atmos. Environ.*, Vol. 40, pp. 3724-3736.

Meng, Z.; Seinfeld, J. H.; Saxena, P.; Kim, Y. P. (1995). Atmospheric gas-aerosol equilibrium: IV. Thermodynamics of carbonates, *Aerosol Sci. Technol.*, Vol.22, pp.131-154

Morino, Y.; Kondo, Y.; Takegawa, N.; Miyaazaki, Y.; Kita, K.; Komazaki, Y.; Fukuda, M.; Miyakawa, T.; Moteki, N. & Worsnop, D. R. (2006). Partitioning of HNO3 and particulate nitrate over Tokyo: Effects of vertical mixing. *J. Geophys. Res.*, Vol. 111, D15215, doi:10.1029/2005JD006887

Moya, M.; Ansari, A. S. & Pandis, S. N. (2001). Partitioning of nitrate and ammonium between the gas and particulate phases during the 1997 IMADA-AVER study in Mexico City. *Atmos. Environ.*, Vol. 35, pp. 1791-1804.

Nemitz, E. & Sutton, M. A. (2004). Gas-particle interactions above a Dutch heathland: III. Modeling the influence of the NH_3-HNO_3-NH_4NO_3 equilibrium on size-segregated particle fluxes. *Atmos. Chem. Phys.*, Vol. 4, pp. 1025-1045.

Ohara, T.; Akimoto, H.; Kurokawa, J.; Horii, J.; Yamaji, K.; Yan, X. & Hayasaka, T. (2007). An Asian emission inventory of anthropogenic emission sources for the period 1980-2020. *Atmos. Chem. Phys.* Vol. 7, pp. 4419-4444.

Riahi, K.; Gruebler, A. & Nakicenovic, N. (2007). Scenarios of long-term socio-economic and environmental development under climate stabilization. *Technological Forecasting and Social Change*, Vol. 74, No. 7, pp.887-935.

Satsumabayashi, H.; Kawamura, M.; Katsuno, T.; Futaki, K.; Murano, K.; Carmichael, G. R.; Kajino, M.; Horiguchi, M. & Ueda, H. (2004). Effects of Miyake volcanic effluents on airborne particles and precipitation in central Japan. *J. Geophys. Res.*, Vol. 109, D19202, doi: 1029/2003JD004204

Schaap, M.; van Loon, M.; ten Brink, H. M.; Dentener, F. J. & Builtjes, P. J. H. (2004). Secondary inorganic aerosol simulations for Europe with special attention to nitrate. *Atmos. Chem. Phys.*, Vol. 4, pp. 857-874

Seinfeld, J. H. & Pandis, S. N. (2006). *Atmospheric Chemistry and Physics: From Air Pollution to Climate Change, second edition,* Wiley Interscience, New York.

Skamarock, W. C.; Klemp, J. B.; Dudhia, J.; Gill, D. O.; Barker, D. M.; Duda, M. G.; Huang, X. Y.; Wang, W. & Powers, J. G. (2008). A description of the advanced research WRF version 3, *Tech. Note, NCAR/TN~475+STR,* 125 pp. Natl. Cent. Atmos. Res., Boulder, Colo.

Streets, D. G.; Bond, T. C.; Carmichael, G. R.; Fernandes, S. D.; Fu, Q.; He, D.; Klimont, Z.; Nelson, S. M.; Tsai, N. Y.; Wang, M. Q.; Woo, J.-H. & Yarber, K. F. (2003). An inventory of gaseous and primary aerosol emissions in Asia in the year 2000. *J. Geophys. Res.*, Vol.108, No.D21, 8809, doi:10.1029/2002JD003093.

van Vuuren, D. P.; den Elzen, M. G. J.; Lucas, P. L.; Eickhout, B.; Strengers, B. J.; van Ruijven, B.; Wonink, S.; van Houdt, R. (2007) Stabilizing greenhouse gas concentrations at low levels: an assessment of reduction strategies and costs. *Climate Change*, 81, pp.119-159.

Zhang, Q.; Streets, D. G.; Carmichael, G. R.; He, K. B.; Huo, H.; Kannari, A.; Klimont, Z.; Partk, I. S.; Reddy, S.; Fu, J. S.; Chen, D.; Duan, L.; Lei, Y.; Wang, L. T. & Yao, Z. L. (2009). Asian emissions in 2006 for the NASA INTEX-B mission. *Atmos. Chem. Phys.*, Vol. 9, pp.5131-5153.

Part 2

Air Pollution Monitoring and Modelling

3

Gas Sensors for Monitoring Air Pollution

Kwang Soo Yoo
Department of Materials Science and Engineering, University of Seoul, Korea

1. Introduction

The air pollution caused by exhaust gases from automobiles has become a critical issue. In some regions, fossil fuel combustion is a problem as well. The principal gases that cause air pollution from automobiles are nitrogen oxides, NO_x (NO and NO_2), and carbon monoxide (CO). Because NO_x gases with sulfur oxides (SO_x) emitted from coal fired plants cause acid rain and global warming and produce ozone (O_3) that leads to serious metropolitan smog from photochemical reaction, they must be detected and reduced [1-5].

In addition, as greater amounts of oil organic compounds are currently being produced by applied construction materials and households, the number of people who develop various symptoms after moving into a new apartment (*e.g.*, tickle, vertigo, headache, skin trouble) is increasing [6,7]. The principal gases that cause this phenomenon (called "sick-building syndrome") are formaldehyde (HCHO) and volatile organic compounds (VOCs) [8]. Especially, formaldehyde is the most dangerous among indoor pollutants as it could harm all kinds of organisms. Considering these, the allowed concentration of formaldehyde in the Netherlands and Germany is only 0.1 ppm [9,10]. Therefore, gas sensors with excellent reactivity and stability are needed.

The first decade of the 21st century has been labeled by some as the "Sensor Decade." A sensor is a device that converts a physical phenomenon into an electrical signal. As such, sensors represent part of the interface between the physical world and the world of electrical devices, such as computers. In recent years, sensors have received people's attention as one of the important devices in electronic systems and enormous capability for information processing has been developed within the electronics industry. Of all sensors, gas sensors and light sensors have been most actively studied [11-13]. The final goal of gas sensor development is to establish the array technology of multifunctional gas sensors that can monitor air pollution with low cost, and is to fabricate the electronic nose using this technology.

Gas sensors are defined as a device that can substitute for human olfaction, and there are many researches being conducted to monitor air pollution by using these gas sensors. Gas sensors can be classified into semiconductor-type, solid electrolyte-type, electrochemical-type and catalytic combustion-type. Among these, the semiconductor-type gas sensor, the most well-known, is operated by changing its conductivity when it is exposed to gas. The semiconductor-type gas sensor has the advantages of rapid reactivity, efficiency, and gas selectivity when suitable additives are applied to it [14,15]. Sensors made of inorganic materials are the most commonly used, especially ceramics. One reason is that many sensors are used in very severe conditions such as high temperature, reactive or corrosive atmosphere and high humidity, and ceramics are most reliable materials in these conditions. Another reason may be that the microstructure of ceramics can be controlled by process conditions. In general, electrical, mechanical and optical properties of a material are controlled by changing its composition. In ceramics, however, these properties are also controlled by changing its microstructure [13]. The gas-sensing materials for semiconductor-type are SnO_2, WO_3, In_2O_3, perovskite-structure oxides, etc., and the electrolyte for solid electrolyte-type gas sensor is $Na_3Zr_2Si_2PO_{12}$ [1,2,4,16-19].

In this chapter, pollutants and sources of air pollution are briefly explained. Then environmental gas sensors for monitoring air pollution are introduced systematically and the fabrication methods and characteristics of each gas sensor are explained at length with recent research trends.

2. Air pollution [20]

Air pollution is the introduction of chemicals, particulate matter, or biological materials that cause harm or discomfort to humans or other living organisms, or cause damage to the natural environment or built environment, into the atmosphere. The atmosphere is a complex dynamic natural gaseous system that is essential to support life on planet Earth. Stratospheric ozone layer depletion due to air pollution has been recognized as a threat to human health as well as to the Earth's ecosystems. Indoor air pollution and urban air quality are listed as two of the world's worst pollution problems in the 2008 Blacksmith Institute World's Worst Polluted Places report [21].

2.1 Pollutants

A substance in the air that can cause harm to humans and to the environment is known as an air pollutant. Pollutants can be in the form of solid particles, liquid droplets, or gases. They may be natural or man-made [22]. Pollutants can be classified as primary or secondary. Usually, primary pollutants are directly emitted from a process, such as ashes from a volcanic eruption, the NO_x and CO gases from a motor vehicle exhaust or SO_x released from factories. Secondary pollutants are not emitted directly. Rather, they form in the air when primary pollutants react or interact. An important example of a secondary pollutant is ground level ozone - one of the many secondary pollutants that make up photochemical smog. Some pollutants may be both primary and secondary: that is, they are both emitted directly and formed from other primary pollutants. Causes and effects of air pollution are shown in Fig. 1.

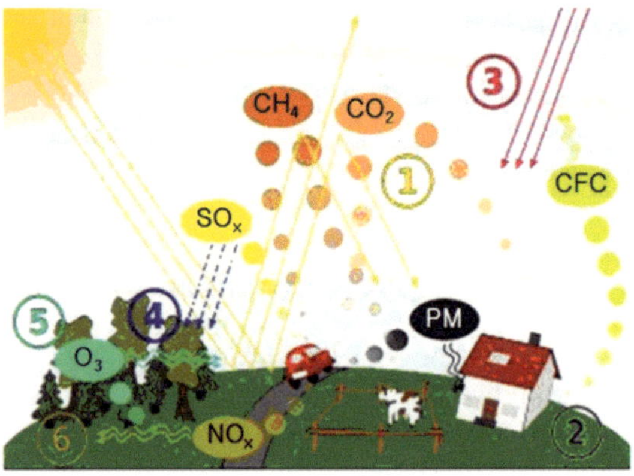

Fig. 1. Schematic drawing, causes and effects of air pollution: (1) greenhouse effect, (2) particulate contamination, (3) increased UV radiation, (4) acid rain, (5) increased ground level ozone concentration, (6) increased levels of nitrogen oxides [20].

2.1.1 Major primary pollutants

- Nitrogen oxides (NO_x): especially nitrogen dioxide (NO_2). NO_2 is emitted from high temperature combustion. Can be seen as the brown haze dome above or plume downwind of cities. This reddish-brown toxic gas has a characteristic sharp, biting odor. NO_2 is one of the most prominent air pollutants.
- Carbon monoxide (CO): CO is a colorless, odorless, non-irritating but very poisonous gas. It is a product by incomplete combustion of fuel such as natural gas, coal or wood. Vehicular exhaust is a major source of carbon monoxide.
- Carbon dioxide (CO_2): CO_2 is a colorless, odorless, non-toxic greenhouse gas associated with ocean acidification, emitted from sources such as combustion, cement production, and respiration.
- Volatile organic compounds (VOCs): VOCs are an important outdoor air pollutant. In this field they are often divided into the separate categories of methane (CH_4) and non-methane (NMVOCs). CH4 is an extremely efficient greenhouse gas which enhances global warming. Other hydrocarbon VOCs are also significant greenhouse gases via their role in creating ozone and in prolonging the life of CH_4 in the atmosphere, although the effect varies depending on local air quality. Within the NMVOCs, the aromatic compounds such as benzene, toluene and xylene are suspected carcinogens and may lead to leukemia through prolonged exposure. 1,3-butadiene is another dangerous compound which is often associated with industrial uses.
- Formaldehyde (HCHO): HCHO is the most dangerous among the indoor pollutants as it could harm all kinds of organisms. As great amounts of oil organic compounds are induced by applied construction materials and households, HCHO and VOCs are produced and cause various symptoms (called "sick-building syndrome") after moving into a new apartment [6-8].
- Ammonia (NH_3): NH_3 is emitted from agricultural processes. It is normally encountered as a gas with a characteristic pungent odor. NH_3 contributes significantly to the nutritional needs of terrestrial organisms by serving as a precursor to foodstuffs and fertilizers. NH_3, either directly or indirectly, is also a building block for the synthesis of many pharmaceuticals. Although in wide use, NH_3 is both caustic and hazardous.
- Sulfur oxides (SO_x): especially sulphur dioxide (SO_2). SO_2 is produced by volcanoes and in various industrial processes. Since coal and petroleum often contain sulphur compounds, their combustion generates SO_2. Further oxidation of SO_2, usually in the presence of a catalyst such as NO_2, forms H_2SO_4, and thus acid rain. This is one of the causes for concern over the environmental impact of the use of these fuels as power sources.
- Particulate matter (PM): Particulates, alternatively referred to as PM or fine particles, are tiny particles of solid or liquid suspended in a gas. In contrast, aerosol refers to particles and the gas together. Some particulates occur naturally, originating from volcanoes, dust storms, forest and grassland fires, living vegetation, and sea spray. Human activities, such as the burning of fossil fuels in vehicles, power plants and various industrial processes also generate significant amounts of aerosols. Averaged over the globe, anthropogenic aerosols - those made by human activities - currently account for about 10 percents of the total amount of aerosols in our atmosphere.

Increased levels of fine particles in the air are linked to health hazards such as heart disease [23], altered lung function and lung cancer.

- Chlorofluorocarbons (CFCs): CFCs are harmful to the ozone layer emitted from products currently banned from use [24,25].
- Persistent free radicals connected to airborne fine particles could cause cardiopulmonary disease.
- Toxic metals, such as lead, cadmium and copper
- Odors such as from garbage, sewage, and industrial processes
- Radioactive pollutants produced by nuclear explosions, war explosives, and natural processes such as the radioactive decay of uranium.

2.1.2 Secondary pollutants

- PM formed from gaseous primary pollutants and compounds in photochemical smog: Smog is a kind of air pollution and the word "smog" means a portmanteau of smoke and fog. Classic smog (London type smog) results from large amounts of coal burning in an area caused by a mixture of smoke and sulfur dioxide. Modern smog (photochemical or Los Angeles type smog) does not usually come from coal but from vehicular and industrial emissions that are acted on in the atmosphere by ultraviolet light from the sun to form secondary pollutants that also combine with the primary emissions to form photochemical smog.
- Ground level ozone (O_3) formed from NO_x and VOCs: O_3 is a key constituent of the troposphere. It is also an important constituent of certain regions of the stratosphere commonly known as the Ozone layer. Photochemical and chemical reactions involving it drive many of the chemical processes that occur in the atmosphere by day and by night. At abnormally high concentrations brought about by human activities (largely the combustion of fossil fuel), it is a pollutant, and a constituent of smog.
- Peroxyacetyl nitrate (PAN) similarly formed from NO_x and VOCs.

2.2 Sources

Sources of air pollution refer to the various locations, activities or factors which are responsible for the releasing of pollutants into the atmosphere. These sources can be classified into two major categories.

2.2.1 Anthropogenic sources (human activity)

- "Stationary Sources" include smoke stacks of power plants, manufacturing facilities (factories) and waste incinerators, as well as furnaces and other types of fuel-burning heating devices.
- "Mobile Sources" include motor vehicles, marine vessels, aircraft and the effect of sound etc.
- Chemicals, dust and controlled burn practices in agriculture and forestry management. Controlled or prescribed burning is a technique sometimes used in forest management, farming, prairie restoration or greenhouse gas abatement. Fire is a natural part of both forest and grassland ecology and controlled fire can be a tool for foresters. Controlled burning stimulates the germination of some desirable forest trees, thus renewing the forest.

- Fumes from paint, hair spray, varnish, aerosol sprays and other solvents.
- Waste deposition in landfills, which generate methane. Methane is not toxic; however, it is highly flammable and may form explosive mixtures with air. Methane is also an asphyxiant and may displace oxygen in an enclosed space. Asphyxia or suffocation may result if the oxygen concentration is reduced to below 19.5% by displacement.
- Military, such as nuclear weapons, toxic gases, germ warfare and rocketry.

2.2.2 Natural sources

- Dust from natural sources, usually large areas of land with little or no vegetation.
- CH_4 gas emitted by the digestion of food by animals, for example, cattle.
- Radon gas from radioactive decay within the Earth's crust. Radon is a colorless, odorless, naturally occurring, radioactive noble gas that is formed from the decay of radium. It is considered to be a health hazard. Radon gas from natural sources can accumulate in buildings, especially in confined areas such as the basement and it is the second most frequent cause of lung cancer, after cigarette smoking.
- Smoke and CO from wildfires.
- Vegetation, in some regions, emits environmentally significant amounts of VOCs on warmer days. These VOCs react with primary anthropogenic pollutants - specifically, NO_x, SO_2, and anthropogenic organic carbon compounds - to produce a seasonal haze of secondary pollutants [26].
- Volcanic activity, which produce sulfur, chlorine, and ash particulates.

3. Environmental gas sensors

A broad definition of environmental monitoring would include all aspects of air and water quality, soil contamination, electromagnetic radiation, noise, even heat release and light source pollution. However, the major environmental gas sensors are to monitor pollution in air, water, and soil as shown in Table 1 [27]. Environmental standard concentration and threshold limit value for six important gases of air pollution are listed in Table 2 [28,29]. Some information about gas sensors on the base of most familiar metal oxides and technological peculiarities of these sensors fabrication, which can be used for such selection, is presented in Tables 3 and 4 [30]. Gas sensors for monitoring principal gases among air pollutants are described in detail by using typical examples here.

	Fixed monitors		Mobile monitors	
	Stationary source	Ambient	Portable	Personal
Air	Industrial emissions, Leaks, Car exhausts, Biochemicals	Air quality	Air quality, Surveys	Gas alarms
Water	Drinking water, Effluent	Water pollution, Intake monitoring	Water pollution, Pollution tracing	Drinking water
Land	Waste disposal		Remediation, Leaks	

Table 1. Classification of Environmental Monitoring Applications [27]

Pollutants	Concentration			Ref.
	Environmental	TLV*	Request of sensors	
NO_x	Below 0.04-0.06 ppm (daily average)	NO_2: 3 ppm, NO: 25 ppm	0.01-0.3 ppm	28
CO_2	-	5000 ppm	200-400 ppm	28
CO	35 ppm† (1 h average)	50 ppm	0.1-10 ppm	28,†29
HCHO	-	1 ppm	-	29
SO_2	Below 0.04 ppm (daily average)	2 ppm	0-2 ppm	28
NH_3	-	25 ppm	-	28
O_3	Below 0.06 ppm (1 h average)	0.1 ppm	0-0.5 ppm	28
CFC**	-	-	20 ppt	28

*TLV: maximum exposure in 8 h period in 40 h work week
**CFC: Chlorofluorocarbon (Freon)

Table 2. Environmental Standard Concentration and Threshold Limit Value (TLV) of Air Pollution

Materials	Advantages	Disadvantages
SnO_2	High sensitivity, Good stability in reducing atmosphere	Low selectivity, Dependence on air humidity
WO_3	Good sensitivity to oxidizing gases, Good thermal stability	Low sensitivity to reducing gases, Dependence on air humidity, Slow recovery process
Ga_2O_3	High stability, Possibility to operate at high temperatures	Low selectivity, Average sensitivity
In_2O_3	High sensitivity to oxidizing gases, Fast response and recovery, Low sensitivity to air humidity	Low stability at low oxygen partial pressure
CTO ($CrTiO_x$)	High stability, Low sensitivity to air humidity	Average sensitivity

Table 3. Main Advantages and Disadvantages of Well-known Metal Oxides for Gas Sensor Applications [30]

Metal oxides	Detection gases	Operating temperature (°C)	Stability	Compatibility with IC fabrication
SnO_2	Reducing gases (CO, H_2, CH_4, etc.)	200-400	Excellent	Imperfect
WO_3	NO_x, O_3, H_2S, SO_2	300-500	Excellent	Low
Ga_2O_3	O_2, CO	600-900	High	Good
In_2O_3	O_3, NO_x	200-400	Moderate	Good
MoO_3	NH_3, NO_2	200-450	Moderate	Moderate
TiO_2	O_2, CO, SO_2	350-800	Enhanced	Moderate
ZnO	CH_4, C_4H_{10}, O_3, NO_x	250-350	Satisfactory	Good
CTO	H_2S, NH_3, CO, volatile organic compounds	300-450	High	Imperfect
Fe_2O_3	Alcohol, CH_4, NO_2	250-450	Low	Moderate

Table 4. Operating Parameters of Solid-state Gas Sensors on the Base of Metal Oxides and Technological Peculiarities of their Fabrication [30]

3.1 NO_x gas sensor

Nitrogen oxide (NO_x) sensing materials reported by several investigators are WO_3, ZnO, SnO_2, In_2O_3, TiO_2, etc. Among these, WO_3 is known as the most promising NO_x gas-sensing material [19,31-39]. These oxides have the advantages of rapid reactivity, efficiency, and gas selectivity when suitable additives are applied to them.

These sensing materials are oxygen-deficient nonstoichiometric compounds. The conductivity of these n-type semiconductors, such as WO_3 and In_2O_3, is estimated based on the electron created by the surplus metal. When sensing materials are exposed to oxidizing gases at temperature ranging from 200°C to 300°C, the concentration of electrons is decreased due to the reaction between the electron and the gas. Consequently, the conductivity decreases and the resistance increases.

As NO_x is also an oxidizing gas, the concentration of electrons is decreased due to the reaction between the electrons in the sensing materials and NO_x gas, as shown in the following equations:

$$NO + 2e^- \longrightarrow \frac{1}{2} N_2 + O^{2-} \tag{1}$$

$$NO_2 + 2e^- \longrightarrow NO + O^{2-} \tag{2}$$

Example [19]:

The powders of various gas-sensing materials were prepared using the solid-state reaction method, starting from the raw materials, WO_3 and In_2O_3. To improve the reactivity and sensitivity of the gas sensors, 0.1-wt% $PdCl_2$ was added as a catalyst. The powders were mixed, dried at 50°C, and then calcined at 1000°C. Thick-film NO_x gas sensors were prepared on alumina substrate. The Pt electrodes were also printed with a silkscreen method before the deposition of the WO_3 and In_2O_3 gas-sensing layer. Schematic diagrams of the sensor are shown in Figure 2. To control the operating temperatures, a printing paste was used to form a Pt heater at the back of the alumina substrate. Pt wires were used as conductuve wires and were attached using silver paste.

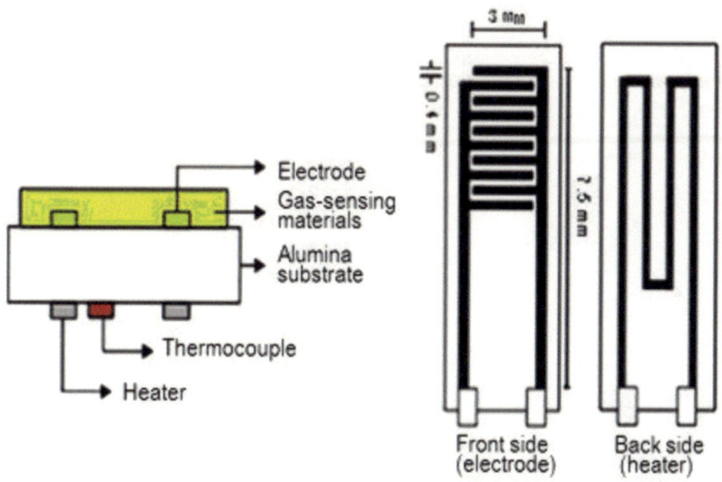

Fig. 2. Schematic diagrams of the gas sensor [19].

The gas-sensing properties were measured in a conventional gas-flow apparatus in the range of 1-5-ppm NO_x by mixing the parent gas (500-ppm NO_x in an N_2 balance) and dry synthetic air. The resistance of the sensor was calculated as:

$$R_s = R_L\left(\frac{V_C}{V_{RL}} - 1\right) \tag{3}$$

where R_s is the resistance of the sensor, R_L is the resistance of the load which was controlled to fix the output voltage to the half of the input voltage because of the change the resistance of the sensor with the change of temperature. V_C is the input voltage and V_{RL} is the output voltage. The sensitivity (S), which refers to the resistance of a sensor that has been exposed to NO_x gas versus the resistance of a sensor that has been exposed to air, was calculated as:

$$S = \left(\frac{R_{gas}}{R_{air}}\right) \tag{4}$$

where R_{gas} is the resistance of the sensor that has been exposed to NO_x gas and R_{air} is the resistance of the sensor that has been exposed to air. In the gas mixtures of NO_x/air, the NO_x concentration varied from 1 ppm to 5 ppm.

As shown in Figures 3 and 4, when the sensors were exposed to NO_x gas, their resistance increased. Below 250°C the resistance of the WO_3 and In_2O_3 were very high, so they could not detect the NO_x gas as there were hardly the resistance change of the WO_3 and In_2O_3. The highest sensitivities of the In_2O_3 to NO_x were at 300°C, as were the highest sensitivities of the WO_3 to NO. The highest sensitivities of the WO_3 to NO_2 were at 250°C, though.

Comparing the sensing property of In_2O_3 with that of WO_3, the sensitivities of In_2O_3 to NO were higher than those of WO_3 to NO, although they were similar. The highest sensitivity (R_{gas}/R_{air}) of In_2O_3 to 5-ppm NO was 10.22 when it was measured at 300°C.

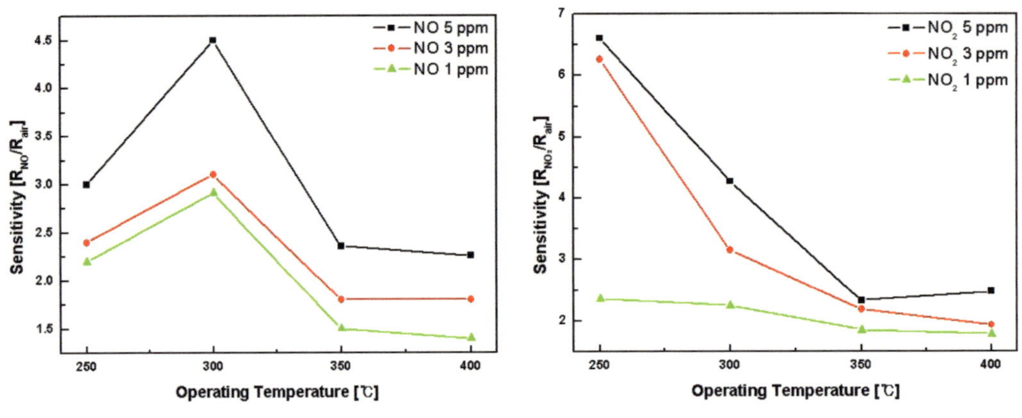

(a) NO gas (b) NO_2 gas

Fig. 3. NO_x Gas-sensing properites of WO_3 [19].

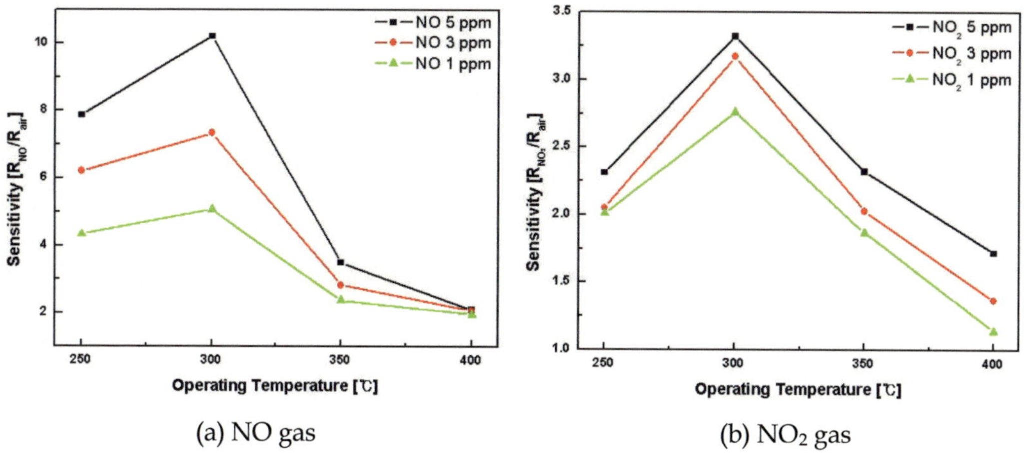

Fig. 4. NO_x Gas-sensing properites of In_2O_3 [19].

3.2 CO_2 gas sensor

Carbon dioxide (CO_2) sensors have been greatly demanded for monitoring or controlling CO_2 in various fields such as combustion process, biology, farming as well as air pollution. So far, many kinds of CO_2 sensors using various materials, such as solid electrolyte, mixed oxide capacitors, polymers with carbonate solution and so on, have been investigated [40-44]. Among them, solid electrolyte-type CO_2 sensors are of particular interest from the viewpoint of low-cost, high-sensitivity, high-selectivity and simple-element structure [45].

Most researches concerning the use of NASICON as active element for gas sensors have been focused on the $Na_{1+x}Zr_2Si_xP_{3-x}O_{12}$ formula, in the composition range of $1.8 < x < 2.4$, because in this range, conductivity shows the largest value [46-48]. A commercial NASICON with a nominal-composition $Na_3Zr_2Si_2PO_{12}$ has been investigated as a CO_2 electrochemical sensor [49,50].

CO_2 sensing properties can be upgraded with auxiliary phases in sensing electrodes, which are binary carbonate systems such as Na_2CO_3-$BaCO_3$, Na_2CO_3-$CaCO_3$, Li_2CO_3-$BaCO_3$, and Li_2CO_3-$CaCO_3$. The binary systems bring about several advantages such as better long-term stability, quick response time, and resistance to water vapor interruption [18,40,51-54]. The device improved in this way has much increased feasibility in practice [55].

Example [18]

The NASICON powder was prepared using the sol-gel method, starting from the solutions of $ZrO(NO_3)_2 \cdot 8H_2O$, $NH_4H_2PO_4$, and $Na_2SiO_3 \cdot 9H_2O$. The solutions were mixed together to form a sol, which was further dehydrated at 80°C to form a gel. The gel was then dried at 120°C for 8 hours to form a fine dry powder, which was then ground and calcined at 750°C to eliminate the organic remains. Afterwards, the calcined material was reground.

The NASICON layer was screen-printed with a paste on the alumina substrate. The Pt electrodes were also screen-printed on the designated regions before and after the deposition of the NASICON layer. The assembly was sintered at 900°C, 1000°C, and 1100°C for 4 hours in air, respectively. After this, a series of auxiliary phases (Na_2CO_3-$CaCO_3$) was screen-printed on the Pt sensing electrode. The schematic diagram of the sensor is shown in Figure 5.

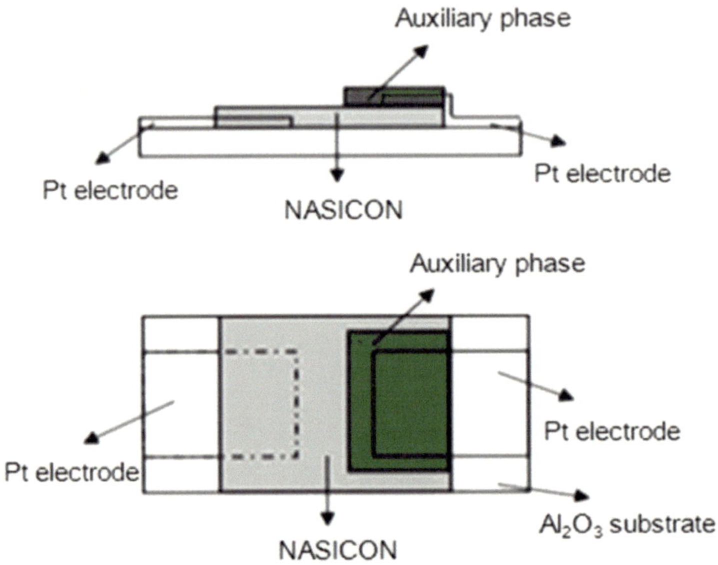

Fig. 5. Schematic diagrams of the CO_2 gas sensor [18].

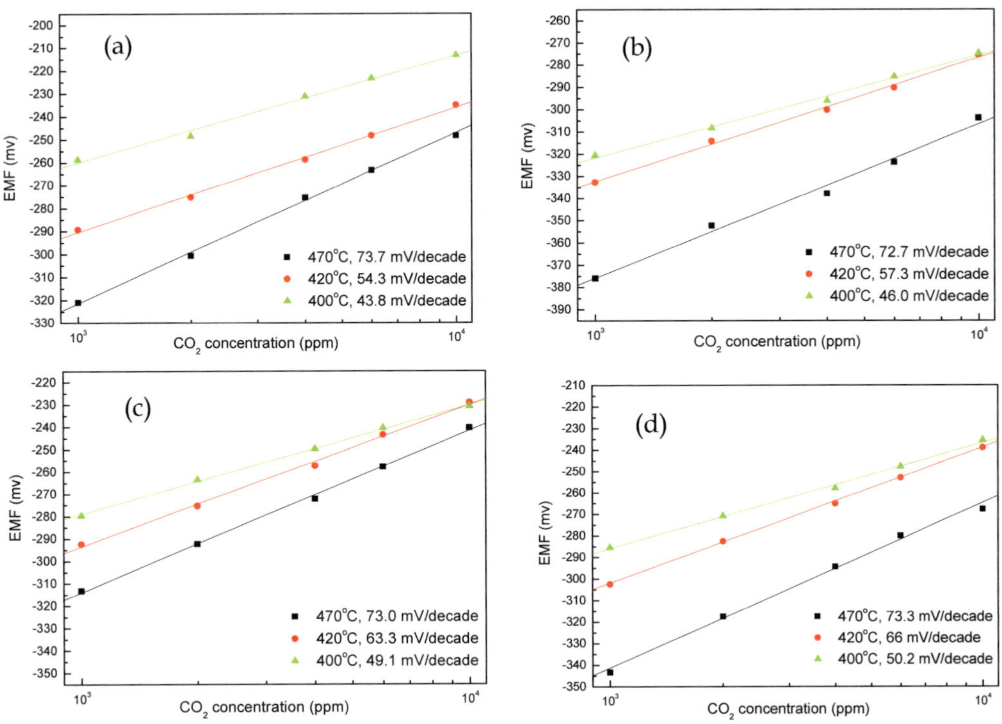

Fig. 6. CO_2 concentration vs. EMF for the CO_2 gas sensors attached with (a) Na_2CO_3-$CaCO_3$ = 1:0, (b) Na_2CO_3-$CaCO_3$ = 1:0.5, (c) Na_2CO_3-$CaCO_3$ = 1:1.5, and (d) Na_2CO_3-$CaCO_3$ = 1:2 [18].

Gas-sensing properties were measured in a conventional gas-flow apparatus by changing the mixing ratio between the parent gas (4% CO_2 in an N_2 balance) and dry synthetic air. The operating temperature was controlled by monitoring the applied voltage and current using the power supply. The sensors were exposed to the flow (100 cm^3/min) of the required sample gases. The gas mixtures of CO_2/air with the CO_2 concentration varied from 1,000 to 10,000 ppm.

Four types of sensors were fabricated from NASICON as a solid electrolyte. A series of Na_2CO_3-$CaCO_3$ mixtures at the molar ratio range of 1:0-1:2 was attached to the sensing electrode. Figure 6 shows the EMF response to CO_2 as a function of the CO_2 concentration at various temperatures. The EMF variation for each sensor at 470°C agreed well with the theoretical value of 74.0 mV/decade, based on a two-electron electrochemical reaction. As the temperature decreased, however, the slope tended to deviate from the ideal. Quite noticeably, the deviation could be suppressed very effectively with Na_2CO_3-$CaCO_3$ (1:2), which allowed 50.2 mV/decade to be kept at temperatures as low as approximately 400°C. An increase in the amount of $CaCO_3$ at the auxiliary phase is fairly effective for keeping the theoretical value at lower temperatures, whereas an adverse effect occurred when the $CaCO_3$ content was insufficient. The mechanism behind such improvements is not yet well understood, though. It requires further research.

3.3 HCHO gas sensor

Formaldehyde (HCHO) is an achromatic toxic gas and has a stimulating scent. When exposed to HCHO gas even just for a short time, a person may develop headache and vertigo, and when exposed to it for a long time, a person may develop asthma and other lung diseases. When exposed to high concentrations of HCHO, a person may develop pneumonia or edema of the lungs [9]. Considering these, the allowed concentrations of formaldehyde in Korea, Denmark, the Netherlands, and Germany are only 2 ppm, 0.2 ppm, 0.1 ppm, and 0.1 ppm, respectively [10]. Therefore, gas sensors with excellent reactivity and stability are needed. In view of the above, numerous attempts are being made to reduce the amount of HCHO in the air. Few studies have been conducted, however, on the detection and the measurement of the amount of HCHO gas in the air by using ceramic gas sensors.

HCHO sensing materials are perovskite-structure oxides (ABO_3) as the semiconductor type. ABO_3-type materials have the advantage of high stability. The sensitivity and selectivity of these kinds of sensors can be controlled by selecting suitable A and B atoms or through chemical doping with $A_{1-x}A_xB_{1-y}B_yO_3$ materials [56].

$La_{1-x}Sr_xFeO_3$ ceramics are ABO_3 perovskite materials. They are nonstochiometric compounds and p-type semiconductors whose conductivity is estimated through the holes created by the surplus oxygen therein. Substitution at the A-site of an element with a different valence (*e.g.*, the replacement of La^{3+} by Sr^{2+}) leads to the formation of oxygen vacancies and high-valence cations at the B-site, which results in a significant change in the catalytic activity [57-60]. When these sensing materials are exposed to reducing gases like CO, CH_4, and HCHO, their conductivity decreases, and their resistance increases because of the chemical surface reactions between the reducing gas and the surplus oxygen [61-63].

Example [17]

$La_{1-x}Sr_xFeO_3$ powders (x = 0, 0.2, 0.5) were prepared through the conventional solid-state reaction method, starting from raw materials of La_2O_3, SrO, and Fe_2O_3. The mixed powders were dried and were calcined at 1000°C.

The $La_{1-x}Sr_xFeO_3$ sensing layers were silkscreen-printed on the alumina substrate. The Pt electrodes were also silkscreen-printed on the designated regions before the deposition of the $La_{1-x}Sr_xFeO_3$ layer. Schematic diagrams of the sensor are shown in Figure 2.

The gas-sensing properties were measured in a conventional gas-flow apparatus by mixing the parent gas (10 to 50 ppm HCHO in N_2 balance) and dry synthetic air. The resistance of the sensor was calculated by using eq. (3). The gas sensitivity, which refers to the resistance of a sensor that has been exposed to HCHO gas versus the resistance of a sensor that has been exposed to air, was calculated as eq. (4). To confirm the selectivity of the sensors, the gas-sensitivities for CO_2, N_2, and C_3H_8 were also measured. The operating temperature was controlled by monitoring the voltage and current applied by using a power supply. The sensors were exposed to a flow (200 cm^3/min) of the required sample gases. Gas mixtures of HCHO/air with the HCHO concentration varying from 10 ppm to 50 ppm were used.

As HCHO gas is a reducing gas, free electrons are released due to the reaction between the surplus oxygen in the sensing materials and the gas [62], as shown in the following equation:

$$HCHO_{(gas)} + O^{2-}_{(ads)} = CO_{2(ads)} + H_2O_{(ads)} + 2e^- \qquad (5)$$

The sensing properties were improved by increasing the number of active sites of oxygen through the replacement of La with Sr. As shown in Figs. 7 to 9, when the sensors were exposed to HCHO gas, their resistance increased. As the reaction yield of the sensing material $La_{0.8}Sr_{0.2}FeO_3$ to the gas and the surplus oxygen increased, its sensing property was improved by increasing the resistance rather than the sensing property of $LaFeO_3$. The highest sensitivity (R_{gas}/R_{air}) of $La_{0.8}Sr_{0.2}FeO_3$ in 50 ppm was 14.7 when it was measured at 150°C. The sensing property of $La_{0.5}Sr_{0.5}FeO_3$ declined, however, when the amount of surplus oxygen was decreased, despite the fact that the number of active sites of oxygen increased. The reason is assumed to be related to the microstructure of the sensor.

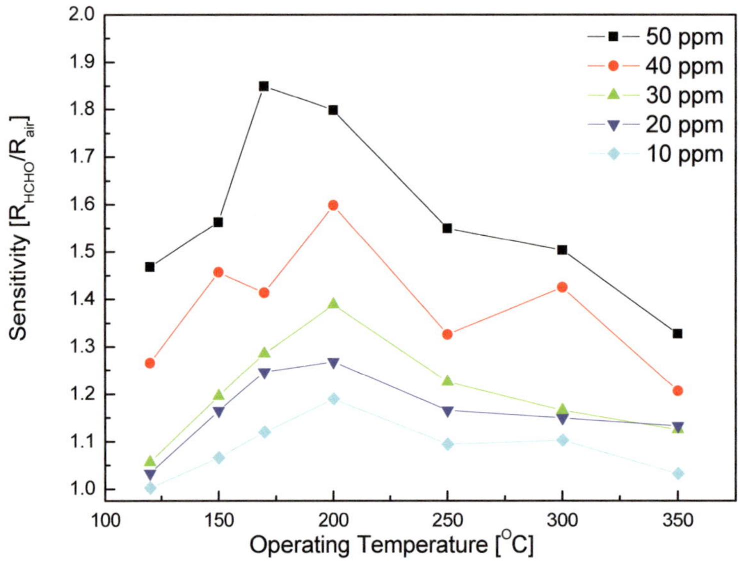

Fig. 7. HCHO Gas-sensing properties of $LaFeO_3$ [17].

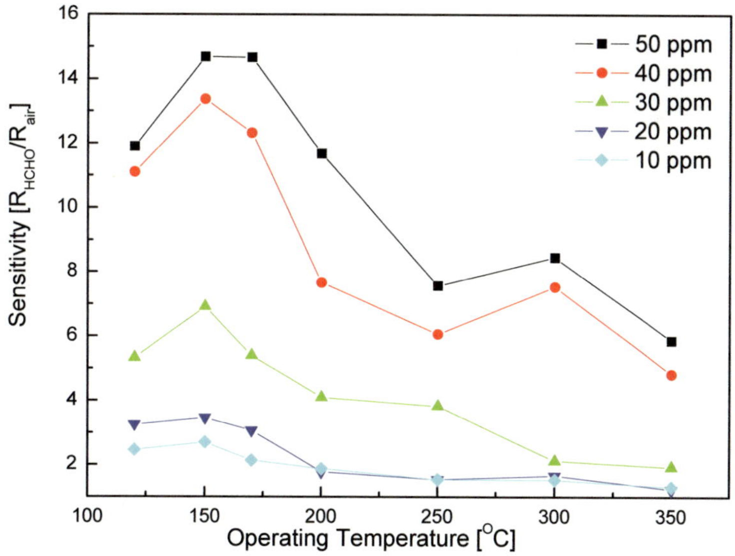

Fig. 8. HCHO Gas-sensing properties of $La_{0.8}Sr_{0.2}FeO_3$ [17].

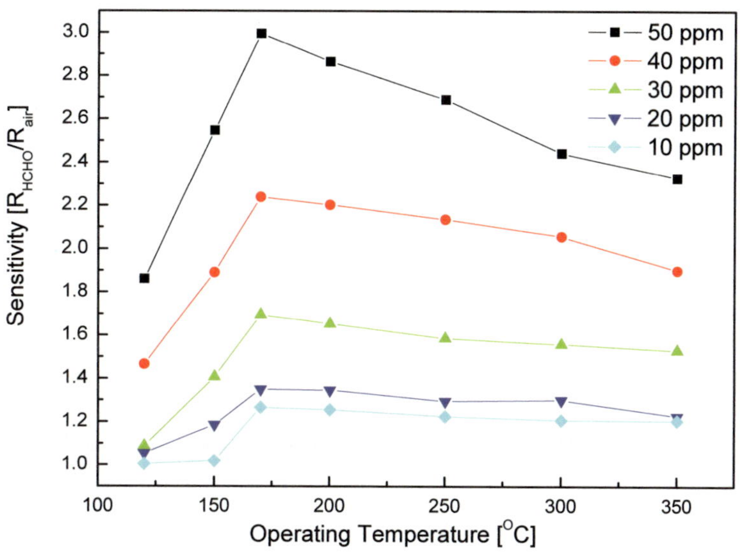

Fig. 9. HCHO Gas-sensing properties of $La_{0.5}Sr_{0.5}FeO_3$ [17].

Considering the selectivity of the sensors, as shown in Table 5, the gas-sensitivity for HCHO gas was higher than those for other gases. As HCHO gas has a very strong reducing property, its sensitivity is over 2.5 because of the reaction between the surplus oxygen in the sensing materials and HCHO gas. On the other hand, other gases do not react to sensing materials, so their sensitivities were near 1. In particular, the $La_{0.8}Sr_{0.2}FeO_3$ sensor could selectively detect HCHO gas.

	$R_{3\%CO_2} / R_{air}$	$R_{2000ppmC_3H_8} / R_{air}$	$R_{50ppmHCHO} / R_{air}$
$LaFeO_3$	1.03	1.00	1.80
$La_{0.8}Sr_{0.2}FeO_3$	0.89	1.07	14.7
$La_{0.5}Sr_{0.5}FeO_3$	0.80	0.95	2.50

Table 5. Gas Selectivity of the Sensors Measured at 150℃ [17]

3.4 Other gas sensors

3.4.1 CO gas sensor

Carbon monoxide (CO) is a colorless, odorless, and tasteless gas which is slightly lighter than air. Because the development of CO gas sensors was urgent to avoid gas poisoning caused by imperfect combustion of kerosine or gas in a heater, many commercial SnO_2-based sensor devices have been realized by several investigators since 1980's. These gas sensors often operate at high temperature up to 400°C, in order for high sensitivity.

Recently, in order to decrease the operating temperature, catalysts such as Pt, Pd, or Au [64] are added, and metal oxides (*e.g.* WO_3, In_2O_3 [65], MoO_3 [66], V_2O_5 [67]) are doped into the SnO_2 matrix. Especially, mixed oxides, normally tailored by doping metal cations into an oxide matrix, have attracted a great deal of interest in applications from catalysis to gas-sensing [67].

The electrochemical CO gas sensor is also useful for a fire alarm. If a sensor could detect CO in concentrations of 50-100 ppm, it could become a more useful fire detector than the smoke sensor [68].

3.4.2 NH_3 gas sensor

Ammonia (NH_3) is extensively used in preparing fertilizers, pharmaceuticals, surfactants, and colorants, with a global production. It presents many hazards to both humans and environment. Detection of NH_3 is required in many applications, including leak-detection in air-conditioning systems as well as in sensing of trace amounts of ambient NH_3 in air for environmental analysis, breath analysis for medical diagnoses, animal housing, and more [69].

Recently, various NH_3 gas sensors based on different sensing mechanisms have been developed. For example, the WO_3 nanofibers showed rapid response and recovery characteristics to NH_3, and gas-sensing mechanism was explained in terms of surface resistivity and barrier height model [70,71]. It was reported that polypyrrole (PPy)/$ZnSnO_3$ nanocomposites also exhibited a higher response to NH_3 gas [72], and by combining the merits of a chitosan polymer and a porous Si photonic crystal, the optical sensor showed high sensitivity, selectivity, and stability [69].

3.4.3 Others

Hydrogen sulfide (H_2S) is a colorless, very poisonous, and flammable gas with the characteristic foul odor of rotten eggs at concentrations up to 100 ppm. An ultrahigh sensitive H_2S gas sensor was developed utilizing Ag-doped SnO_2 thin film on the alumina substrate [73]. This Ag-SnO_2 nanocomposite showed excellent sensing properties upon exposure to H_2S as low as 1 ppm at 70°C. Cuong *et. al.* [74] reported a solution-processed gas sensor based on vertically aligned ZnO nanorods on a chemically converted grapheme film. This sensor effectively detected 2 ppm of H_2S in oxygen at room temperature.

In addition, the sulfur dioxide (SO_2) gas sensor using an alkali metal sulfate-based solid electrolyte [75] and ozone (O_3) gas sensor of In_2O_3 thin-film type [76] were developed. Recently, gas sensor array for monitoring the perceived car-cabin air quality was reported [34,77]. The technological process in microelectromechanical system (MEMS) metal oxide gas sensors in terms of stability and reproducibility has promoted the technology for mass market applications. Tille [34] suggested an automotive air quality gas sensor using micro-structured silicon technology as shown in Figure 10. The metallization and the gas-sensing layer were electrically isolated from the heating layer by a passivation. Reducing gases (*e.g.* CO, C_xH_y) result in an increase in conductivity and oxidizing gases (*e.g.* NO_2) produce a reduction in the conductivity of the metal oxide. For detection of various gases, several sensor elements such as SnO_2, ZnO, or WO_3 could be combined.

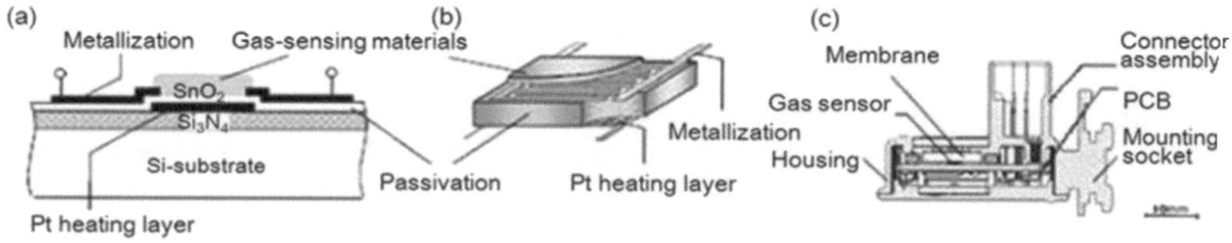

Fig. 10. Schematic illustration of a micro-structured metal oxide gas sensor (a) cross section; (b) metallization as inter-digital structure, and heating layer as platinum meander structure; (c) cross section of a typical automotive air quality sensor with embedded metal oxide gas sensor [34].

In the future, smart sensors with high sensitivity, good reliability, and rapid response by using MEMS technology and advanced signal processing should be developed.

4. References

[1] H. Kawasaki, T. Ueda, Y. Suda, and T. Ohshima, *Sensors and Actuators B*, vol. 100, p. 266, 2004.
[2] U. Guth and J. Zosel, *Ionics*, vol. 10, p. 366, 2004.
[3] K. S. Yoo, T. S. Kim, and H. J. Jung, *J. Kor. Ceram. Soc.*, vol. 32, p. 1369, 1995.
[4] D. L. West, F. C. Montgomery, and T. R. Armstrong, *Sensors and Actuators B*, vol. 106, p. 758, 2005.
[5] Y. S. Yoon, T. S. Kim, and W. K. Choi, *J. Kor. Ceram. Soc.*, vol. 41, p. 97, 2004.
[6] H. S. Kang, S. W. Kim, and Y. J. Cho, *J. Kor. Furni. Soc.*, vol. 18, p. 91, 2007.
[7] J. H. Jang and Y. S. Lee, *J. Archi. Ins. Kor.*, vol. 24, p. 299, 2004.
[8] H. J. An, C. H. Cheong, H. J. Kim, and Y. G. Lee, *J. Archi. Ins. Kor.*, vol. 25, p. 51, 2005.
[9] J. Y. Park and M. S. Jung, *J. Soc. Health Edu. Promo.*, vol. 1, p. 260, 1996.
[10] J. W. Seo, *J. Kor. Air-Condi. Refri.*, vol.31, p. 13, 2002.
[11] Japanese R&D Trend Analysis Report No. 6: Ceramic Sensors, KRI International, Inc., Tokyo, 1989.
[12] J. S. Wilson, Sensor Technology Handbook, Elsevier, New York, 2005.
[13] S. Y. Yurish and M. T. S. R. Gomes, Smart Sensors and MEMS, Kluwer Academic Publishers, Dordrecht, 2004.
[14] D. D. Lee, *Ceramist*, vol. 4, p. 57, 2001.

[15] T. S. Kim, Y. B. Kim, K. S. Yoo, K. S. Sung, and H. J. Jung, *J. Kor. Ceram. Soc.*, vol. 34, p. 387, 1997.
[16] T. G. Nenov and S. P. Yordanov, Ceramic Sensors: Technology and Applications, Technomic Publishing Company, Inc., Lancaster, 1996.
[17]] M. W. Son, J. B. Choi, H. J. Kim, K. S. Yoo, and S. D. Kim, *J. Kor. Phys. Soc.*, vol. 54, p. 1072, 2009.
[18] H. B. Shim, J. H. Kang, J. W. Choi, and K. S. Yoo, *J. Electroceram.*, vol. 17, p. 971, 2006.
[19] M. W. Son, J. B. Choi, H. I. Hwang, and K. S. Yoo, *J. Kor. Sensors Soc.*, vol. 18, p. 263, 2009.
[20] Wikipedia, http://en.wikipedia.org/wiki/Air_Pollution, 2011.
[21] "Reports", WorstPolluted.org., http://www.worstpolluted.org/. Retrieved 2010-08-29 by Wikipedia.
[22] "EPA: Air Pollutants", http://www.epa.gov/ebtpages/airairpollutants.html. Retrieved 2010-08-29 by Wikipedia.
[23] Evidence Growing of Air Pollution's Link to Heart Disease, Death // American Heart Association, http//www.newsroom.heart.org/index.php?s=43&item=1029. Retrieved 2010-05-10 by Wikipedia.
[24] "Newly Detected Air Pollutant Mimics Damaging Effects of Cigarette Smoke", http://www.physorg.com/pdf138201201.pdf. Retrieved 2010-08-29 by Wikipedia.
[25] "Infant Inhalation of Ultrafine Air Pollution Linked to Adult Lung Disease", http://www.sciencedaily.com/releases/2009/07/090722123751.htm. Retrieved 2010-08-29 by Wikipedia.
[26] A. H. Goldstein, D. K. Charles, L. H. Colette, and Y. F. Inez, *Proc. National Academy of Sci.*, 2009. http://www.pnas.org/content/106/22/8835.full. Retrieved 2010-12-05 by Wikipedia.
[27] K. W. Jones, in: W. Göpel, J. Hesse, and J. N. Zemel (Ed.), Sensors, vol. 8, Micro- and Nano Sensor Technology/ Trends in Sensor Markets, p. 451, VCH Verlagsgesellschaft mbH, Weinheim, 1995.
[28] K. Colbow and K. L. Colbow, in: W. Göpel, J. Hesse, and J. N. Zemel (Ed.), Sensors, vol. 3, Chemical and Biochemical Sensors Part II, p. 969, VCH Verlagsgesellschaft mbH, Weinheim, 1992.
[29] BITMART, http://www.bitmart.kr/board/view.asp?tb=board_2&num=49&page=1&st =&sc= &sn=, Seoul, Korea, 2009.
[30] G. Korotcenkov, *Mater. Sci. & Eng. B*, vol. 139, p. 1, 2007.
[31] F. Mitsugi, E. Hiraiwa, T. Ikegami, and K. Ebihara, *Surface and Coatings Technology*, vol. 169, p. 553, 2003.
[32] H. Kawasaki, J. Namba, K. Iwatsuji, Y. Suda, K. Wada, K. Ebihara, and T. Ohshima, *Applied Surface Sci.*, vol. 197, p. 547, 2002.
[33] C.-Y. lin, Y.-Y. Fang, C.-W. Lin, J. J. Tunney, and K.-C. Ho, *Sensors and Actuators B*, vol. 146, p. 28, 2010.
[34] T. Tille, *Procedia Eng.* (*Proc. Eurosensors XXIV, Linz, Austria*), vol. 5, p. 5, 2010.
[35] L. Bissi, M. Cicioni, P. Placidi, S. Zampolli, I. Elmi, and A. Scorzoni, *IEEE Transactions on Instrumentation and Measurement*, vol. 60, p. 282, 2011.
[36] A. Serra, M. Re, M. Palmisano, M. V. Antisari, E. Filippo, A. Buccolieri, and D. Manno, , *Sensors and Actuators B*, vol. 145, p. 794, 2010.
[37] P.-G. Su and T.-T. Pan, *Mater. Chem. and Phys.*, vol. 125, p. 351, 2011.

[38] U. Lange, V. M. Mirsky, *Analytica Chimica Acta*, vol. 687, p. 7, 2011.
[39] J. D. Fowler, M. J. Allen, V. C. Tung, Y. Yang, R. B. Kaner, and B. H. Weiller, *ACS Nano*, vol. 3, p. 301, 2009.
[40] T. Kida, Y. Miyachi, K. Shimanoe, and N. Yamazoe, *Sensors and Actuators B*, vol. 80, p. 28, 2001.
[41] Y. Miyachi, G. Sakai, K. Shimanoe, and N. Yamazoe, *Sensors and Actuators B*, vol. 93, p. 250, 2003.
[42] H. J. Kim, H. B. Shim, J. W. Choi, and K. S. Yoo, *Proc. 10th Asian Conf. on Solid State Ionics*, Kandy, Sri Lanka, p. 849, 2006.
[43] H. J. Kim, J. W. Choi, S. D. Kim, and K. S. Yoo, *Mater. Sci. Forum*, vols. 544-545, p. 925, 2007.
[44] J. J. Lai, H. F. Liang, Z. L. Peng, X. Yi, and X. F. Zhai, *J. Phys.: Conf. Series (3rd Internaional Photonics & OptoElectronics Meetings)*, vol. 276, p. 012129, 2011.
[45] F. Qiu, L. Sun, X. Li, M. Hirata, H. Suo, and B. Xu, *Sensors and Actuators B*, vol. 45, p. 233, 1997.
[46] J. P. Boilot, P. Salanié, G. Desplanches, and D. Le Potier, *Mater. Res. Bull.*, vol. 14, p. 1469, 1979.
[47] D. H. H. Quon, T. A. Wheat, and W. Nesbitt, *Mater. Res. Bull.*, vol. 15, p. 1533, 1980.
[48] G. Desplanches, M. Rigal, and A. Wicker, *Am. Ceram. Soc. Bull.*, vol. 59, p. 546, 1980.
[49] N. Miura, S. Yao, Y. Shimizu, and N. Yamazoe, *Sensors and Actuators B*, vol. 9, p. 165, 1992.
[50] Y. Sadaoka, Y. Sakai, M. Matsumoto, and T. Manabe, *J. Mater. Sci.*, vol. 28, p. 5783, 1993.
[51] S. Yao, Y. Shimizu, N. Miura, and N. Yamazoe, *Chem. Lett.*, vol. 1990, p. 2033, 1990.
[52] N. Miura, S. Yao, Y. Shimizu, and N. Yamazoe, *J. Electrochem. Soc.*, vol. 139, p. 1384, 1992.
[53] S. Yao, Y. Shimizu, N. Miura, and N. Yamazoe, *Jpn. J. Appl. Phys.*, vol. 31, p. L197, 1992.
[54] Y. Shimizu, and N. Yamashita, *Sensors and Actuators B*, vol. 64, p. 102, 2000.
[55] T. Kida, H. Kawate, K. Shimanoe, N. Miura, and N. Yamazoe, *Solid State Ionics*, vol. 136, p. 647, 2000.
[56] S. Zhao, J. K. O. Sin, B. Xu, M. Zhao, Z. Peng, and H. Cai., *Sensors and Actuators B*, vol. 64, p. 83, 2000.
[57] V. Lantto, S. Saukko, N. N. Toan, L. F. Reyes, and C. G. Granqvist, *J. Electroceram.*, vol. 13, p. 721, 1992.
[58] I. Waernhus, N. Sakai, H. Yokokawa, T. Grande, M. Einarsrud, and K. Wiik, *Solid State Ionics*, vol. 178, p. 907, 2007.
[59] M. Popa, J. Frantti, and M. Kakihana, *Solid State Ionics*, vols. 154 – 155, p. 437, 2002.
[60] H. K. Hong, B. H. Kim, Y. I. Cheon, and Y. K. Sung, *J. Kor. Elect. Eng.*, vol. 13, p. 372, 1990.
[61] N. N. Toan, S. Saukko, and V. Lantoo, *Physica B*, vol. 327, p. 297, 2003.
[62] Z. Zhong, K. Chen, Y. Ji, and Q. Yan, *Appl. Catalysis A : General* , vol. 156, p. 29, 1997.
[63] S.-D. kim, B.-J. kim, J.-H. Yoon, and J.-S. Kim, *J. Kor. Phys. Soc.*, vol. 51, p. 2069, 2007.
[64] B. Bahrami, A. Khodadadi, M. Kazemeini, and Y. Mortazavi, *Sensors and Actuators B*, vol. 133, p. 352, 2008.
[65] M. W. Son, J. B. Choi, H. I. Hwang, and K. S. Yoo, *J. Kor. Sensors Soc.*, vol. 18, p. 263, 2009.

[66] Z. A. Ansari, S. G. Ansari, T. Ko, and J.-H. Oh, , *Sensors and Actuators B*, vol. 87, p. 105, 2002.
[67] C.-T. Wang and M.-T. Chen, *Sensors and Actuators B*, vol. 150, p. 360, 2010.
[68] T. Fujioka, S. Kusanagi, N. Yamaga, Y. Watabe, K. Doi, T. Inoue, T. Hatai, K. Sato, A. Takemoto, and D. Kouzeki, in: M. Aizawa (ed.), Chemical Sensor Technology, vol. 5, p. 65, Kodansha Ltd., Tokyo, 1994.
[69] Y. Shang, X. Wang, E. Xu, C. Tong, and J. Wu, *Analytica Chimica Acta*, vol. 685, p. 58, 2011.
[70] J.-Y. Leng, X.-J. Xu, N. Lv, H.-T. Fan, and T. Zhang, *J. Colloid & Interface Sci.*, vol. 356, p. 54, 2011.
[71] N. V. Hieu, V. V. Quang, N. D. Hoa, and D. Kim, *Current Appl. Phys.*, vol. 11, p. 657, 2011.
[72] P. Song, Q. Wang, and Z. Yang, *Mater. Letters*, vol. 65, p. 430, 2011.
[73] J. Gong, Q. Chen, M.-R. Lian, N.-C. Liu, R. G. Stevenson, and F. Adami, *Sensors and Actuators B*, vol. 114, p. 32, 2006.
[74] T. V. Cuong, V. H. Pham, J. S. Chung, E. W. Shin, D. H. Yoo, S. H. Hahn, J. S. Huh, G. H. Rue, E. J. Kim, S. H. Hur, and P. A. Kohl, *Mater. Letters*, vol. 64, p. 2479, 2010.
[75] G.-Y. Adachi and N. Imanaka, in: N. Yamazoe (ed.), Chemical Sensor Technology, vol. 3, p. 131, Kodansha Ltd., Tokyo, 1991.
[76] T. Takada, in: T. Seiyama (ed.), Chemical Sensor Technology, vol. 2, p. 59, Kodansha Ltd., Tokyo, 1989.
[77] M. Blaschke, T. Tille, P. Robertson, S. Maier, U. Weimar, and H. Ulmer, *IEEE Sensors J.*, vol. 6, p. 1298, 2006.

4

Development of Low-Cost Network of Sensors for Extensive In-Situ and Continuous Atmospheric CO2 Monitoring

Kuo-Ying Wang[1], Hui-Chen Chien[2] and Jia-Lin Wang[3]

[1]*Department of Atmospheric Sciences, National Central University,*
[2]*Environmental Protection Administration,*
[3]*Department of Chemistry, National Central University,*
Taiwan

1. Introduction

Extensive and dedicated measurements of carbon dioxide concentrations in the atmosphere are increasingly recognized as a necessary step in verifying anthropogenic carbon dioxide emissions and as necessary methods to support international climate agreements (Marquis & Tans, 2008; NRC, 2010; Tollefson, 2010). The successful launch of the Greenhouse Gas Observing Satellite (GOSAT) on 23 Jan 2009 by Japan's Aerospace Exploration Agency (Heimann, 2009), followed by a not successful launch of Orbiting Carbon Observatory (OCO) on 24 Feb 2009 (Brumfiel, 2009; Kintisch, 2009) all vindicate the importance of extensive and accurate carbon dioxide measurements as a necessary step in global carbon emission verification (Haag, 2007; Normile, 2009; Tollefson & Brumfiel, 2009). We note that a replacement to the OCO is now actively in plan in NASA (Hand, 2009). Other satellite instruments such as Aqua AIRS (Chahine et al., 2006), and SCIAMARCHY (Barkley et al., 2006) have also provided retrieved CO2 concentration in the vertical column.

In Europe, an ongoing new research infrastructure called Integrated Carbon Observing System (ICOS) is dedicated to establish and harmonize a network of atmospheric greenhouse sites (http://www.icos-infrastructure.eu). A list of present-day carbon dioxide monitoring sites whose standard gases have traceability to the World Meteorological Organization (WMO) standard is reported in WDCGG (2007).

In addition to these satellite remote sensing measurements and land-based in-situ measurements, carbon dioxides also been measured from in-service commercial aircrafts such as CONTRAIL (Matsueda & Inoue, 1996; Machida et al., 2008) and the planned flights of IAGOS (Volz-Thomas et al., 2007), research aircraft such as the HIPPO (http://www.ucar.edu/news/releases/2009/hippovisuals.shtml), and in-service container cargo ships (Watson et al., 2009;).

Given the important status of carbon dioxide in affecting earth's climate, however, detailed measurements of carbon dioxide close to areas with heavy industrial emissions and intense anthropogenic activities are relatively rare (Tollefson, 2010). This is in a sharp comparison with other intensively observed air pollutants such as ozone, carbon monoxide, nitrogen oxides, sulfur dioxide, and suspended particles. Since detailed measurements of carbon dioxides close to anthropogenic areas where carbon dioxide is being relentlessly emitted into the atmosphere are required to estimate its annual emission inventories (NRC, 2010), more portable and flexible measurements but in the meantime accurate and traceable to WMO standards are needed to significantly increase carbon dioxide measurements where carbon dioxide been emitted. Burns et al. (2009) described a portable trace-gas measuring system to measure carbon dioxide. In this work we develop a GFC-based measurement system for extensive carbon dioxide measurements that are traceable to the WMO NOAA CO2 standards.

2. Method

In this work we use a fast-response high-precision CO2 analyzer as the core for our CO2 measurements. The analyzer, EC9820T, was made by ECOTECH, Australia (ECOTECH, 2007). The EC9820T was built based on the principle of gas filter correlation (GFC) and the nondispersive infrared (IR) absorption of CO2 near 4.5 microns which is used to determine the presence of the CO2.

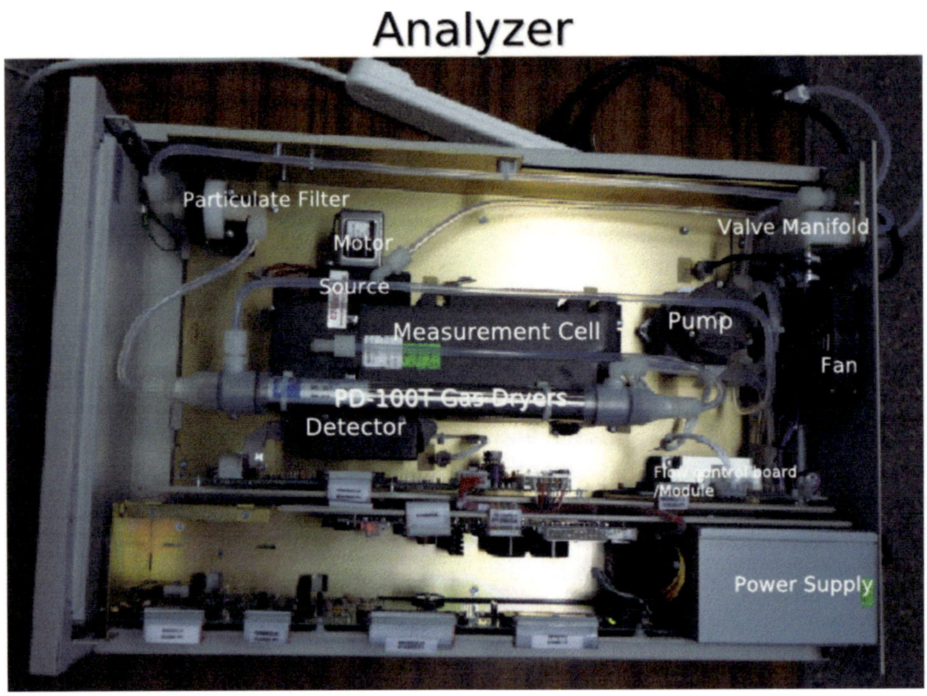

Fig. 1. A top view of the EC9820 CO2 analyzer.

Fig. 1 shows a photo of the top view of the CO2 analyzer used in this work. The analyzer comprises three basic components: the sample flow components (valve manifold, particulate filter, pump, Teflon tubes, dryer, etc), the optical measurement components (motor, IR sources, measurement cell, IR detector), and computer control component (microprocessor boards located at the lower half of the unit, power supply, and fan).

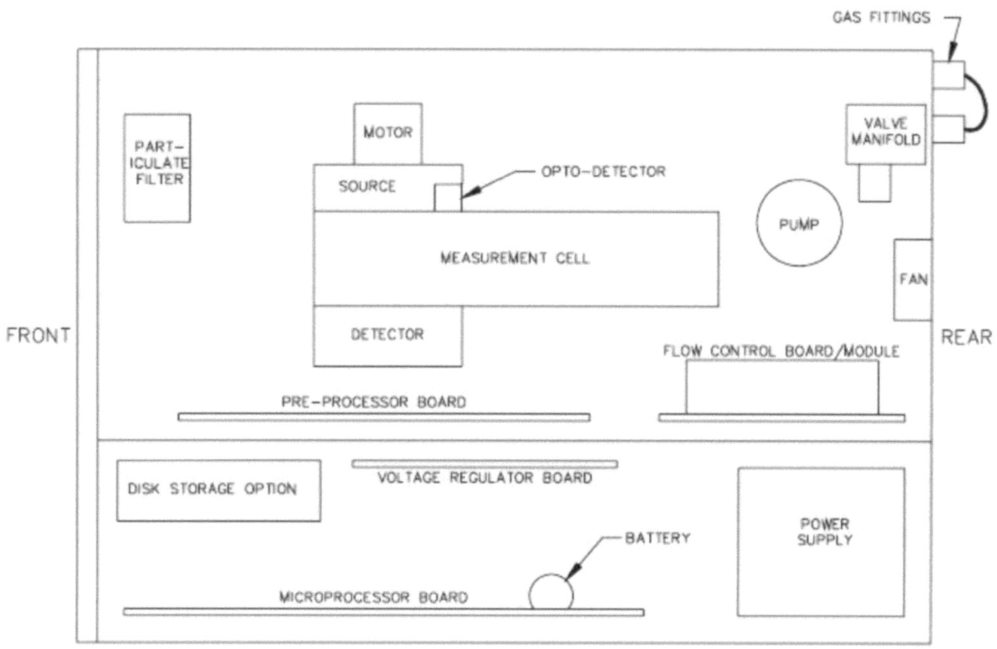

Fig. 2. A schematic diagram showing the major component of the EC9820 CO2 analyzer (ECOTECH, 2007).

The exact locations of these components are shown in more details in Fig. 2. The FRONT presents a mini-terminal like operational interface where the operations and calibrations of the analyzer can be done from this region. The REAR indicates area where sample flow tubes (including ZERO CO2 airs, span gases, exhaust, and purged air which provide zero CO2 air to the chamber that houses gas correlation wheel) are connected with the analyzer.
The optical components are the locations where the GFC principle is in action and measuring the atmospheric CO2 concentrations. The measured results are stored in the onboard computer storages. The measurement cycles, and the control of manifold valves where different airs (zero CO2 air, span CO2 airs, and sample air) are entirely controlled by the onboard microprocessor unit. The analyzer analyzes CO2 concentrations, using the GFC principle, and stores the analyzed (measured) results in the onboard computer storage area. This distinctive capability makes the analyzer a self contained unit which is characteristically suitable to conduct portable and accurate CO2 measurements that are traceable to WMO NOAA standards. The independent of the analyzer from the need of an additional data logger makes the entire operation understandable and sustainable.
Fig. 3 is a flow chart showing a typical loop for sample air measurement. The sample air is sucked in from the SAMPLE IN (on the left) by the the PUMP which connected to SAMPLE EXHAUST (on the right). The pump maintains the sample flow rates at 1 liter per minute. The sample air first passes filter paper where filters our suspended particles in the sample before entering the measurement cell. On the top, the MOTOR drives the rotation of gas filter wheel, which is illuminated with the broadband IR sources (more details of the operational principle of IR sources and gas filter wheel will be discussed later). The DETECTOR detects the concentrations of CO2 in the sample air, and the electrical signals are sent to preprocessor and micro processor boards to determine and store the measured results.

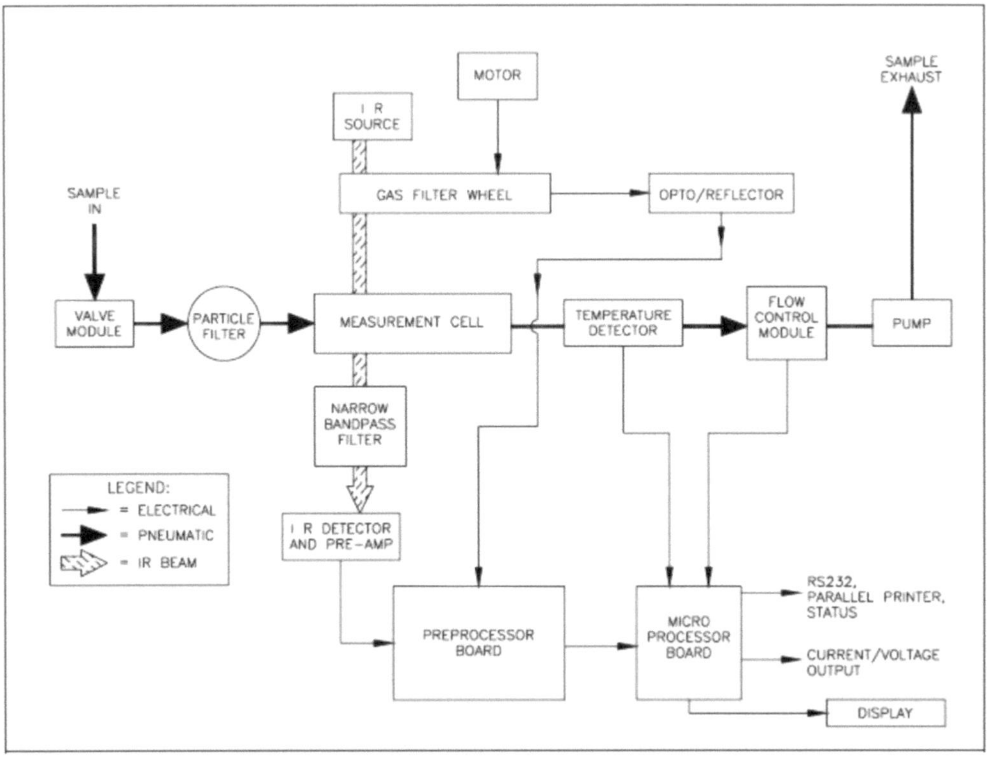

Fig. 3. A flow-chart diagram for the EC9820 CO2 analyzer (ECOTECH, 2007).

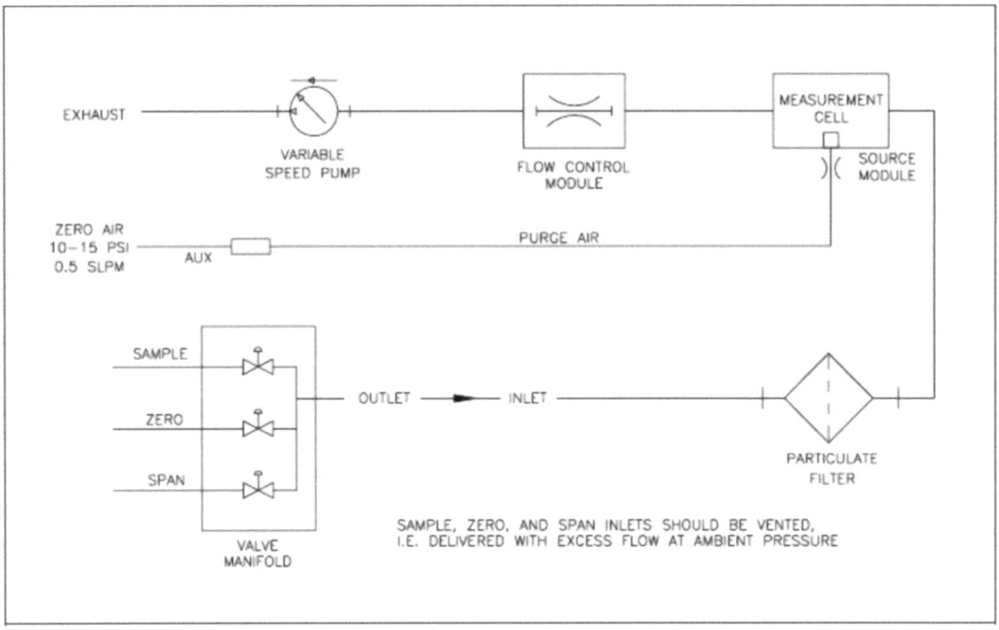

Fig. 4. A pneumatic diagram for EC9820 CO2 analyzer (ECOTECH, 2007).

Fig. 4 shows a pneumatic diagram of the CO2 analyzer. The externally given zero CO2 air, span gases, and sample airs are input to the analyzer through the electronic valve manifold. The span gases normally comprise of two working standards which are calibrated against WMO NOAA CO2 standards provided by NOAA ESRL CCL. All inlet airs pass through a particulate filter to remove suspended particle in the air. The inlet air then enters the measurement cell where GFC principle used to measure CO2 levels. Additional zero CO2 air is provided through auxiliary (AUX) inlet at a flow rate of0.5 liter per minute. The purpose of this purge air is to fill the chamber that houses gas correlation wheel and the IR source with zero CO2 air therefore the interference of CO2 between IR source and gas correlation wheel can be removed.

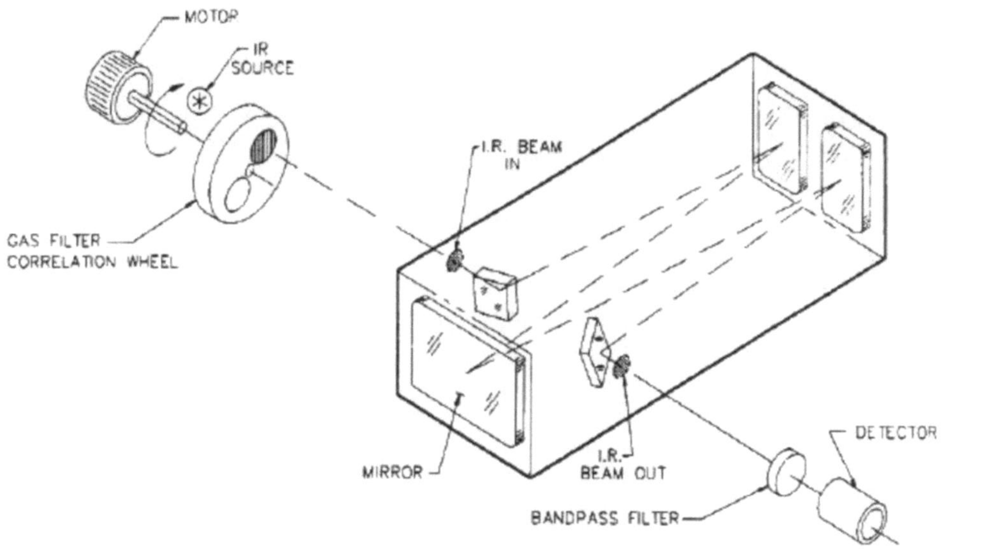

Fig. 5. A schematic diagrame showing the optical component of the analyzer (ECOTECH, 2007)

More detailed structure of GFC principle used in measuring CO2 levels is shown in Fig. 5. From the left-most part is the motor, which rotates the gas filter correlation wheel. Between the motor and the wheel is a broadband IR sources that constantly emit IR sources to the two small chambers that enclose pure CO2 and N2 airs, respectively (Fig. 6). When the IR sources pass CO2 chamber, the IR centered at 4.5 microns will be absorbed and removed while the rest IR spectrums pass CO2 chamber and enter the measurement cell. On the other hand, when the IR sources pass N2 chambers, nothing will be absorbed by the N2 chamber and all IR sources enter the measurement cell where the absorption at 4.5 microns will be occurred due to the CO2 in the measurement cell. The IR sources then pass a narrow band pass filter that allow near 4.5 microns the leave the measurement cell and to be detected by the IR detector on the right-most part.

GFC-based technology has been extensively used for providing CO measurements in the atmosphere (Dickerson & Delany, 1988; Doddridge et al., 1994; Doddridge et al., 1998; Gerbig et al., 1999; Novelli, 1999; Chen and Xu, 2004; Wong et al, 2007; Zellweger et al., 2009).

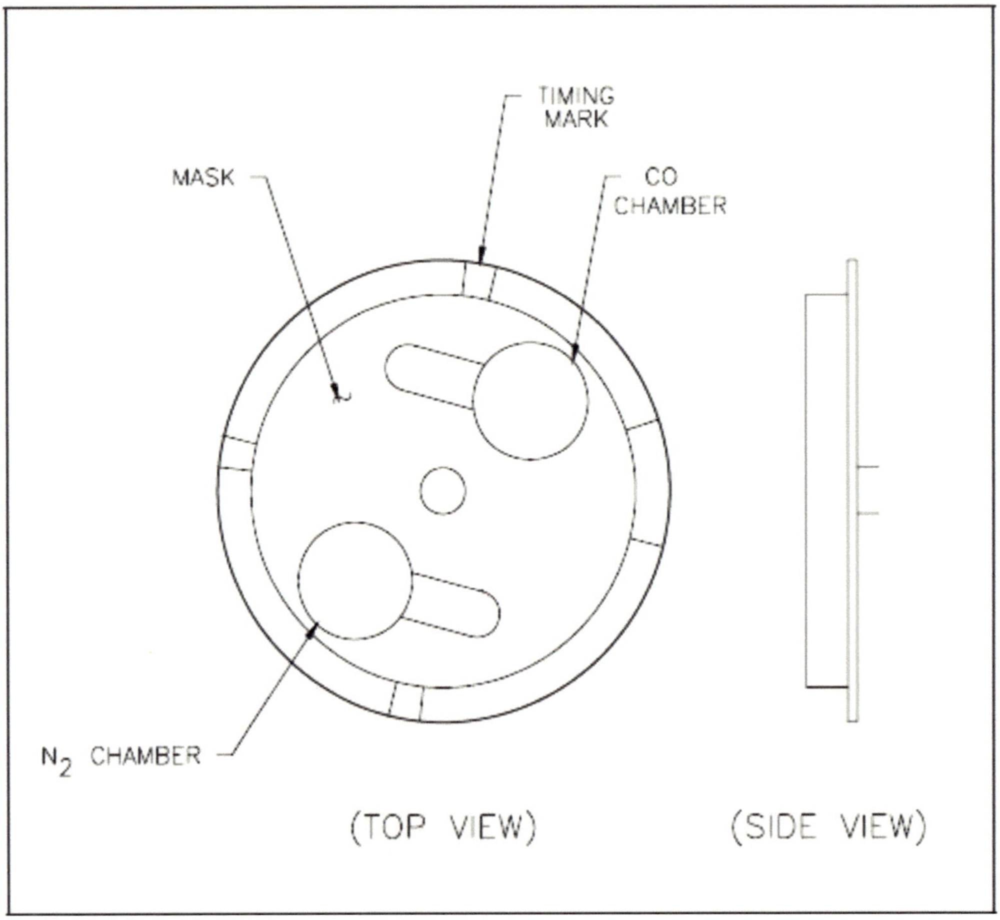

Fig. 6. A schematic diagram showing the top and side views of the gas correlation wheel used in the optical component of the analyzer (ECOTECH, 2007).

3. Results

3.1 Constant SPAN test

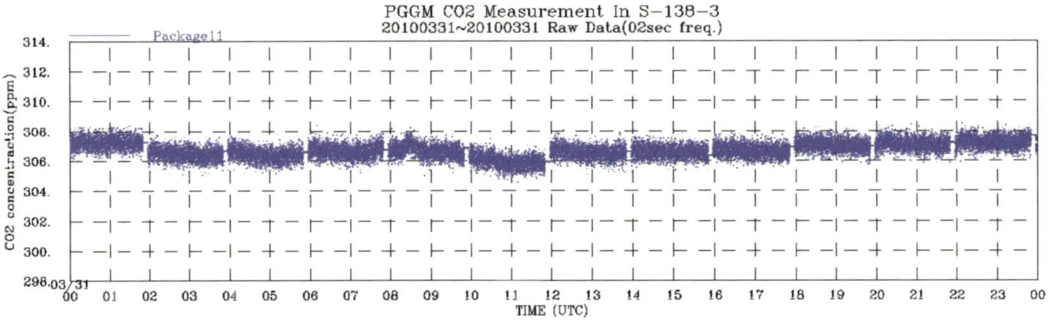

Fig. 8. A constant span test for a CO2 analyzer in the laboratory.

A total of eleven EC9820 CO2 analyzers have been installed since June 2009 for atmospheric CO2 measuring. Each CO2 analyzer was tested in the laboratory before start taking measurements. Fig. 8 shows a constant span test that run continuously for 48 hours using a given working standard. This test was run with 2-hour background frequency (the white gas seen in the data). The output frequency is 2 seconds. The results show that the instrument is very stable from this continuous span test. The constant span tests, and later added constant zero CO2 tests, are good ways to rigorously test if an analyzer is stable and fit for making measurements.

3.2 Inter comparisons between analyzers

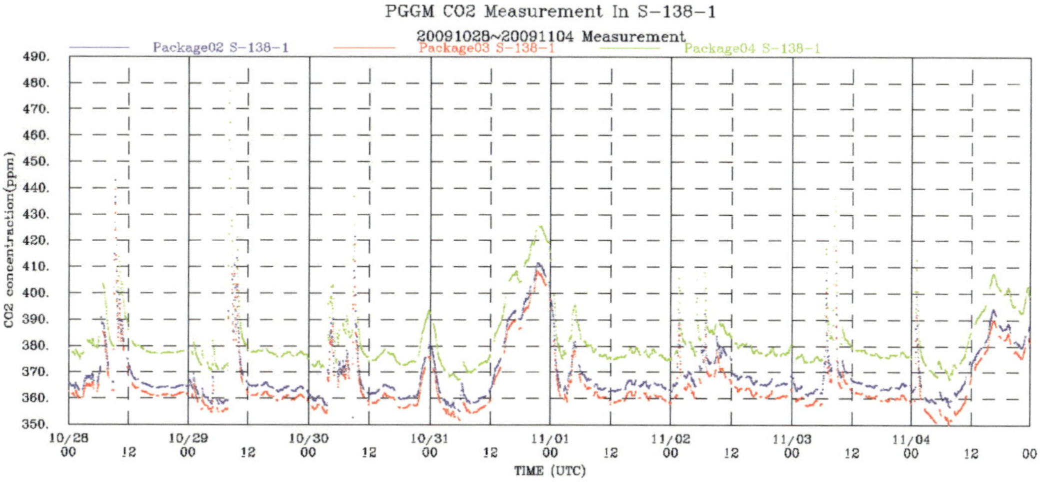

Fig. 9. Inter comparisons of CO2 sample measurements (raw data, in the units of ppm) between three analyzers for the period from 28 October to 4 November 2009.

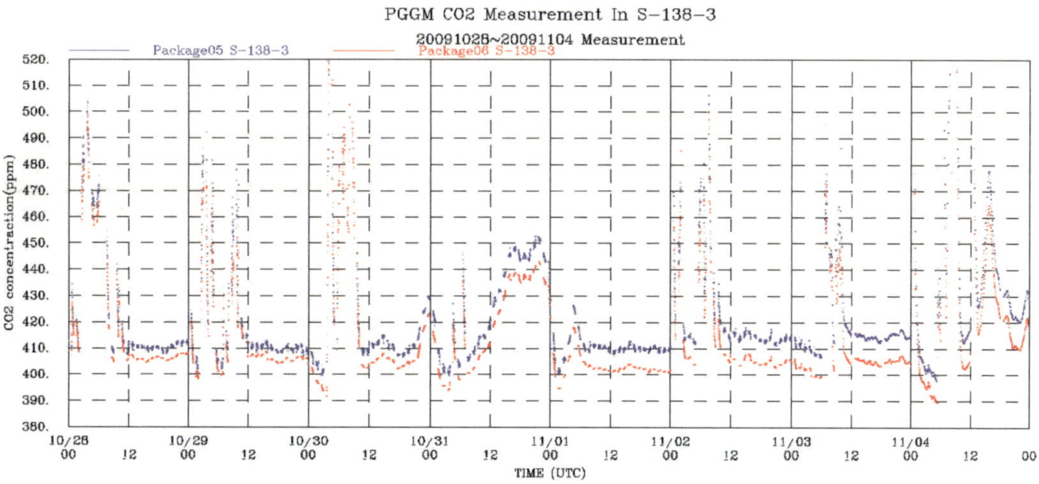

Fig. 10. Intercomparisons of CO2 sample measurements (raw data, in the units of ppm) between two analyzers for the period from 28 October to 4 November 2009.

In addition to the constant span tests and zero tests, analyzers were continuously tested against each other in the laboratory. Fig. 9 shows a time-series plot of a test run between three CO2 analyzers from 28 October 4 November 2009. The results shown here are raw data (un-calibrated data), which shows great consistency in these analyzers. Fig. 10 shows another test results from inter comparisons of two analyzers in a second laboratory. The occasional short bursts of high CO2 concentrations close to 500 ppm were resulted from the researchers making routine maintenance and download of data from the analyzers. Fig. 10 very nicely show that measurements are consistent with each other; and the effect of human presence can be quickly response in the measured data.

3.3 Inter comparisons between GFC and CRDS analyzers

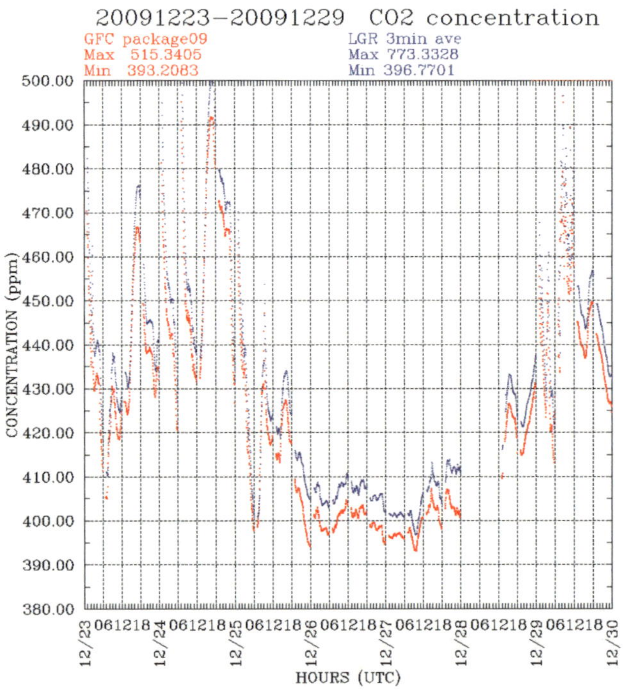

Fig. 11. Inter comparisons of CO2 measurements between GFC and CRDS-based analyzers for the period from 25 November to 8 December 2009

During the period of late 2009 and early 2010, we have had a chance to have an analyzer based on cavity ring down spectroscopy (CRDS) principle running in our laboratory. This analyzer was built by Los Gatos Research Inc., USA(van der Lann et al., 2009). Fig. 11 shows a comparison of time-series plot for the period from 25 November to 8 December 2009. These results indicate a good consistency between the measurements (correlation coefficient= 0.99).

3.4 CO2 measurements in campus

Fig. 12 shows a time-series plot of CO2 measurements at two sites in the campus of National Central University (NCU) for the period from 13 to 21 February 2010. These two sites are separated by 400 m. One analyzer has its sample inlet located at 15 meter height (blue curve), while the other analyzer has its sample inlet located at 35 meter height (red curve). Both measurements were conducted with two working standards that are traceable to WMO NOAA standards. Fig. 12 shows that the measurements are very consistent with each other. The period from 24 to 19 February 2010 is the Chinese New Year in Taiwan, and the measured CO2 concentrations are basically very close to 400 ppm and with little variations. However, after the long holiday was over, the return of working people clearly impacted CO2 levels as shown in the days on 20 and 21 February 2010. After 22 February 2010, Fig. 13 shows variations of CO2 during the normal human activity. The CO2 concentrations vary between 400 and above 470 ppm. The sharp contrast between the long holiday period (Fig. 12) and normal working days (Fig. 13) clearly shows the impact of anthropogenic activity on the atmospheric CO2 concentrations.

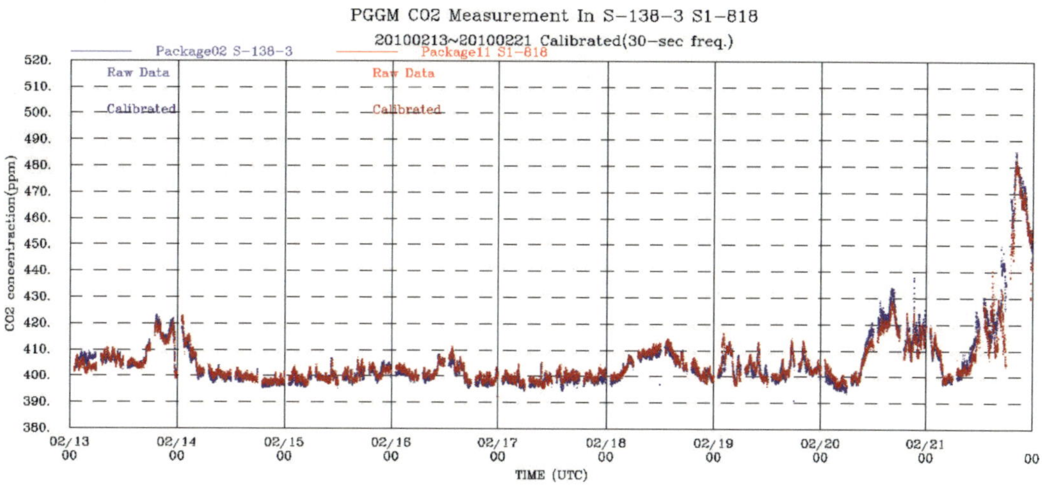

Fig. 12. A time-series plot shown CO2 measurements at two site in the campus of National Central University (NCU) for the period from 13 to 21 February 2010.

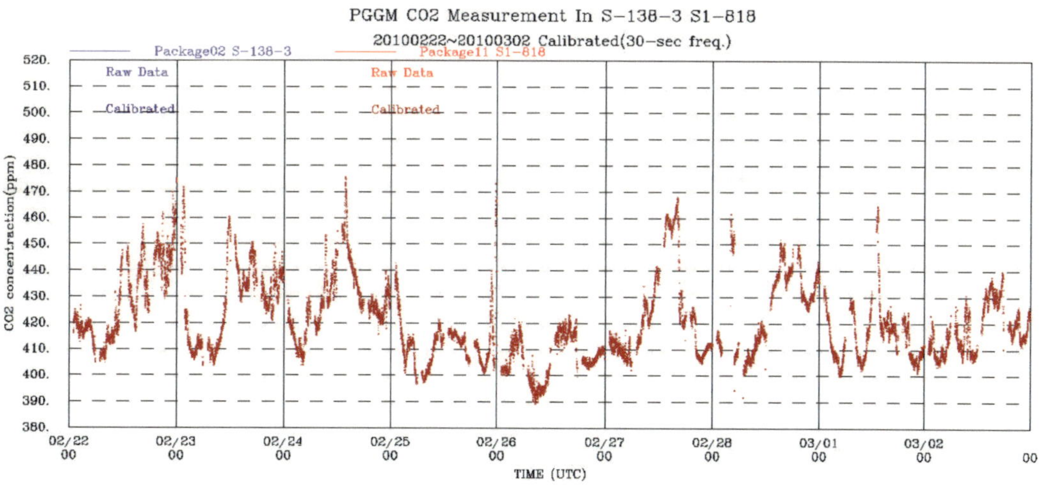

Fig. 13. CO2 measurements at a site in the campus for the period from 22 February to 2 March 2010.

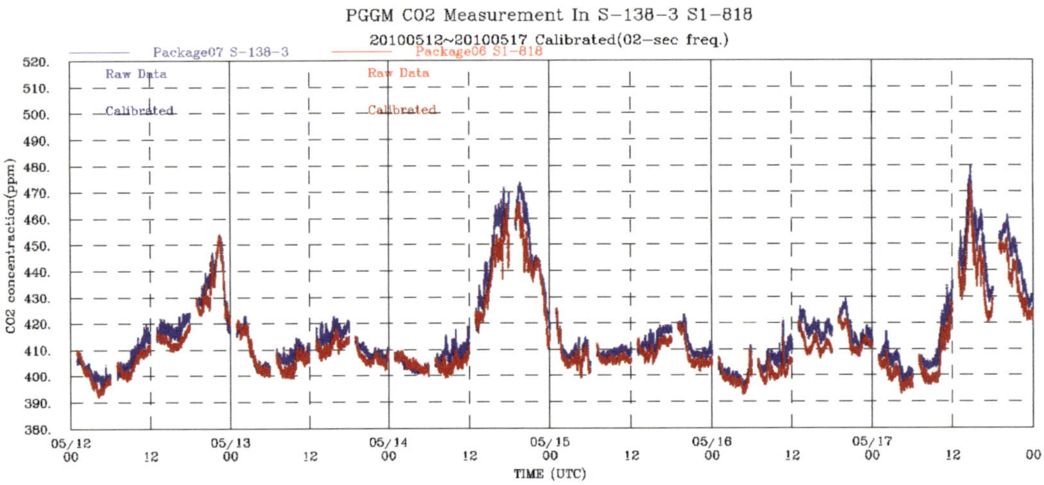

Fig. 14. Time-series plots of CO2 measurements (in the units of ppm) at two sites (blue curve indicates results from 15-m height sample inlet, red curves indicates results from 35-m height sample inlet) in the NCU campus for the period from 12 to 17 May 2010.

Fig. 14 shows another comparison of continuous CO2 measurements at two sites in the NCU campus. Again, we observe the great consistency of CO2 measurements. These results vindicate that our methodology can be consistently applied for a long period, and the use of the GFC-based analyzer with the working standards traceable to WMO NOAA standards ensure that our measurements meet WMO requirements (WDCGC, 2007).

3.5 Indoor CO2 measurements

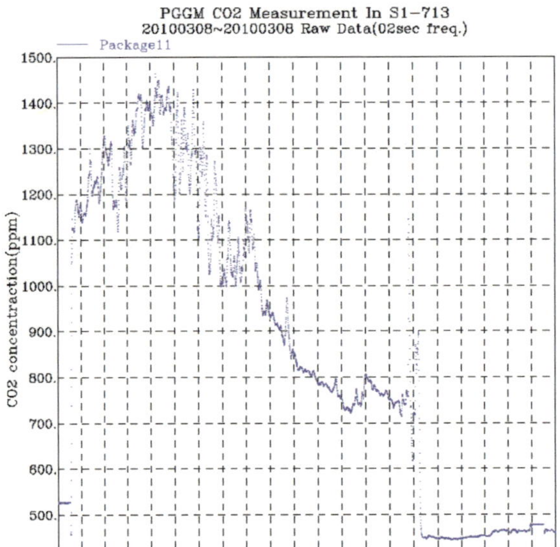

Fig. 15. Indoor CO2 measurements taken in a classroom with 30 students inside on 8 Mar 2010.

In addition to outdoor CO2 measurements shown before, we have also conducted indoor CO2 measurements to understand the variations of CO2 inside a room. Fig. 15 shows a time-series plot of CO2 measurements in a classroom with 30 students at NCU campus on 8 Mar 2010. The build up of the CO2 from 14:30 to 15:20 local time was due to the close of both doors of the classroom. The reduction from about 15:20 to 17:00 was due to the open of a door for ventilation (the air was too stuffy). The effect of ventilation in removing indoor accumulation of CO2 is clearly seen in these results. The measurements also nicely contrast indoor CO2 concentrations with those coming for ambient air, which was taken after 17:00 local time.

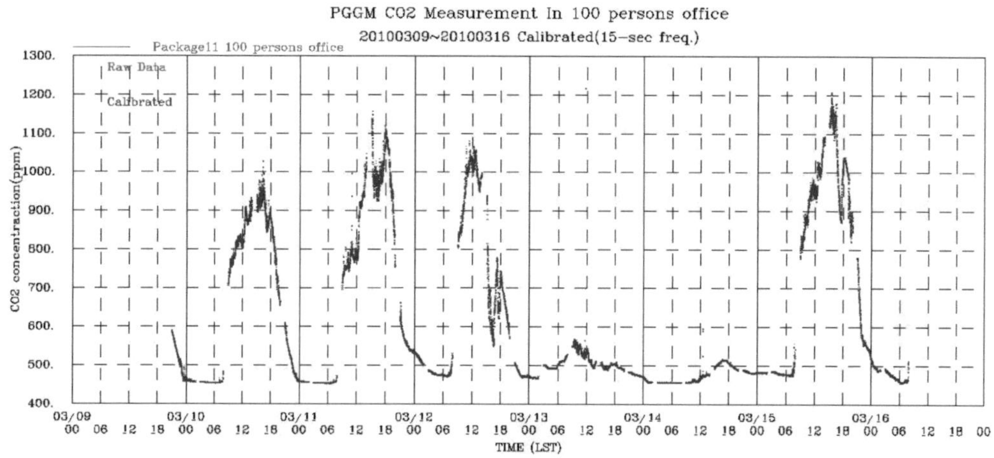

Fig. 16. Indoor CO2 measurements taken in a 100-people working office for the period from 9 Mar to 16 Mar 2010.

Fig. 16 shows another indoor CO2 measurement taken in a 100-people working office for the period from 9 to 16 Mar 2010. Here we see clearly the daily accumulation of CO2 inside the office from about 08:00 local time to peak at about 13:00-15:00 in the early afternoon. The reductions after 15:00 are due to the gradual leaving of people from the office and the accumulated effect of ventilation to counter the CO2 accumulation inside the office. The smooth CO2 concentrations inside the office during the weekend (13-14 March) are clearly seen.

3.6 CO2 measurements onboard a in-service container ship

One of the main motivations to develop GFC-based CO2 measurements is to conduct global CO2 measurements over the Pacific regions. We have vigorously tested this idea since June 2009. Fig. 17 shows a service route from a container ship called EVER DECENT during its service for the period from 22 January to 26 Mar 2010. The detailed operations and installation of the GFC analyzer and CO2 working standards will be reported in a separate work. Fig. 18 shows results from this cruise. The measurements show that CO2 levels are close to 400 ppm when the measurements were taken over the marine atmospheric boundary layer. However, the CO2 levels increase sharply as soon as the ship approached a port. This plot very nicely shows that anthropogenic activity as the key sources for atmospheric CO2 levels. More results on the ship-based measurements will be presented in separate publications this year.

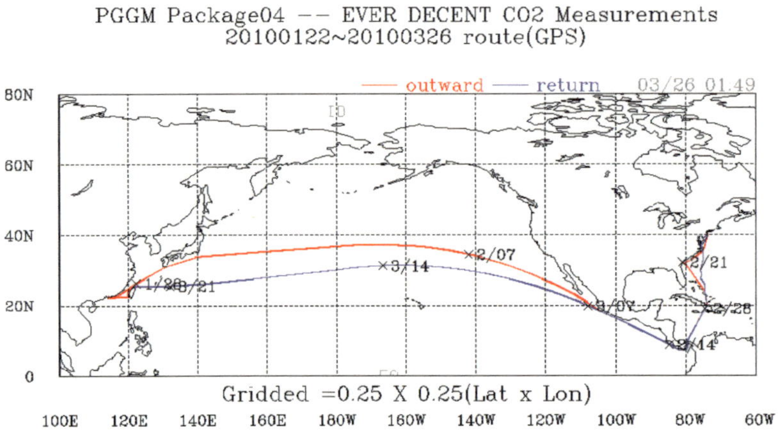

Fig. 17. A service route for EVER DECENT for the period from 22 Janunary to 26 Mar 2010.

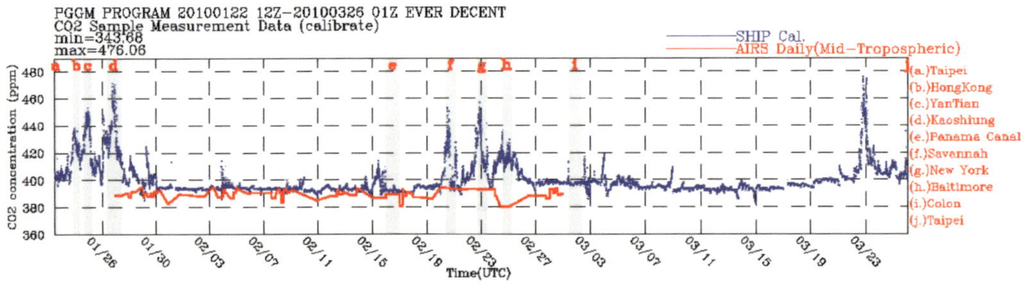

Fig. 18. A time-series plot of a ship-based measurement made by EVER DECENT

4. Summary

In this work we demonstrate the development of a GFC-based technology for making continuous in-situ atmospheric CO2 measurements for climate policy decision makers. All GFC-based analyzers were rigorously tested in the laboratory before went to the field for measurements. All CO2 measurements were made with two CO2 working standards that are traceable to the WMO NOAA CO2 standards. All CO2 working standards were routinely calibrated again six bottles of WMO NOAA CO2 standards.

Great effort has been taken to ensure that the atmospheric CO2 measurements follow the WMO standards (WDCGC, 2007). We have developed a series of tests to verify the measurements, including constant span test, constant zero tests, inter comparisons between GFC analyzers, and inter comparisons between GFC and a CRDS analyzer. We show some results from the land-based measurements taken in the NCU campus, indoor measurements, and a ship-based measurement. These results indicate encouraging results that can significantly increase our understanding of atmospheric CO2 distribution.

5. Acknowledgements

We are very grateful to National Science Council, and Environmental Protection Administration, Taiwan, for funding this work. We benefit tremendously from the help, comments, and continuous discussions from Neil Harris and Andrew Robinson of

Cambridge University, UK; Andreas Volz-Thomas of Forschungszentrum, Jülich, Germany; Jean-Pierre Cammas, CNRS, France; and Charlie Jia-Lin Wang, National Central University, Taiwan; and T. Machida, NIES, Japan. We thank ECOTECH, Australia, for their collaborations to upgrade original EC9820 to a version that meet our standards. We thank Leder Corporation, Taiwan, for their engineering assistance.

6. References

Barkley, M.P.;Monks, P.S.&Engelen, R.J. (2006).*Comparisons of SCIAMARCHY and AIRS CO2 measurements over North America during the summer and autumnof 2003*, Geophys. Res. Lett., 33, L20805, doi:10.1029/2006GL026807.

Brumfiel, G. (2009). Satellite to monitor carbon sinks sinks.*Nature*, doi:10.1038/4571067b.

Chahine, M. et al. (2006). AIRS: Improving weather forecasting and providing new data on greenhouse gases, Bull. Amer. Meteor. Soc.,87, pp.911-926.

Chen, C.&Xu, Z. (2005).*Experiment determination of after-operating gas composition in a transverse-flow CW high power CO2 laser,* Optics Laser Tech.,37, pp.293-297.

Dickerson, R.R.&DelanyA.C. (1988).*Modification of a commercial gas filter correlation CO detector for enhanced sensibility,*Atmos,J.. Oceanic Tech.,5,pp.424-431.

Doddridge, B.G.;Dirmeyer,P.A.;Merrill,J.T.;Oltmas,S.J. &Dickerson, R.R. (1994).*Interannual variability over the eastern North Atlantic Ocean: Chemical and meteorological evidence for tropical influence on regional-scale transport in the extratropics,* Geophys., J.;Res., 99, No. D11,22,pp.923-935.

Doddridge, B.G. et al. (1998), *Gound-based and airborne observations of carbon monoxide during NASA Measurements of Air Pollution From Satellite (MAPS) missions SRL-1 and SRL-2,* Geophys., J.;Res., 103, No. D15, pp.19,305-19,316.

ECOTECH (2007), EC9820 Carbon Dioxide Analyser Service Manual Revision: B, ECOTECH environmental monitoring, *Australia.*

Gerbig, C.;Schmitgen,S.;Kley,D.;Volz-Thomas,A. &Dewey,K. (1999), *An improved fast-response vacuum-UV resonance fluorescence CO instrument,* Geophys,J.. Res., 104, No. D1, pp.1699-1704.

Haag, A. (2007), The crucial measurement, *Nature*, pp.785-787.

Hand, E. (2009), Budget win for climate probe, *Nature*, pp.462, 832.

Heimann, M. (2009), Searching out the sinks, *Nature Geoscience*, pp.2-4.

Kintisch, E. (2009), Loss of carbon observatory highlights gaps in data, *Science,* 323, pp.1276-1277.

Machida, T. et al. (2008), *Worldwide measurements of atmospheric CO2 and other trace gas species using commercial airlines,*Atmos,J.. Oceanic Technol., 25, pp.1744-1754.

Marquis, M.&Tans,P. (2008), Carbon crucible, *Science,* 320, pp.460-461.

Matsueda, H.&InoueH.Y. (1996), *Measurements of atmospheric CO2 and CH4 using a commercial airliner from 1993 to 1994,* Atmos. Environ., 30, pp.1647-1655.

Normile, D. (2009), Tracking CO2's comings and goings from space, *Science,*pp.323, 325.

Novelli, P.C. (1999), *CO in the atmosphere: measurement techniques and related issues,* Chemosphere: Global Change Sci., 1, pp.115-126.

NRC (2010), Verifying greenhouse gas emissions: Methods to support international climate agreements, Committee on Methods for Estimating Greenhouse Gas Emissions, *National Research Council*, pp.144., The National Academies.

Parrish, D.D.;Holloway,J.S. &Fehsenfeld, F.C. (1994), *Routine, continuous measurement of carbon monoxide with parts per billion precision*, Environ. Sci. Tech., 28, pp.1615-1618.

Tollefson, J.&Brumfiel,G. (2009), Climate researchers in a spin after satellite loss, *Nature*, doi:10.1038/news.2009.pp.124.

Tollefson, J. (2010), Greenhouse-gas numbers up in the air, *Nature*, 465, pp.18-19.

Van der Laan, S.;Neubert,R.E.M. &Meijer,H.A.J. (2009), *A single gas chromogratograph for accurate atmospheric mixing ratio measurements of CO2, CH4, N2O, SF6, and CO*, Atmos. Meas. Tech., 2, pp.549-559.

Volz-Thomas, A. et al. (2007), *In-service aircraft for global observations – the future*,IGACtivities, 37, pp.18-22.

Watson, A.J. et al. (2009), Tracking the variable North Atlantic sink for atmospheric CO2, *Science*, 326, pp.1391-1393.

WDCGC (2007), *WMO WDCGC Data Summary*, WDCGC No. 31, *GAW Data*, Vol.iV-Greenhouse Gases and Other Atmospheric Gases, Japan Meteorological Agency in co-operation with World Meteorological Organization.

Wong, H.L.A.;Wang,T.;Ding, A.;BlakeD.R. &Nam,J.C. (2007), *Impact of Asian continental outflow on the concentrations of O3, CO, NMHC and halocarbons on Jeju Island, South Korea during March 2005*, Atmos. Environ., 41, pp.2933-2944.

Zellweger, C.;Klausen,H.J.;Steinbacher,M.;Vollmer,M.&Buchmann,B. (2009).*Inter-comparison of four different carbon monoxide measurement techniques and evaluation of the long-term carbon monoxide time series of Jungfraujoch*, Atmos. Chem. Phys., 9, pp.3491-3503.

5

Aerosol Optical Thickness and PM10 Study by Using a Handheld Spectroradiometer Over Penang

Tan Fuyi, Lim Hwee San, Mohd. Zubir Mat Jafri and K. Abdullah
School of Physics, Universiti Sains Malaysia, Malaysia

1. Introduction

Knowledge of the parameters that are related to the optical properties of atmospheric aerosols is essential for the determination of their climatic effects, development of techniques for remote sensing of aerosol from space or the necessary correction of atmospheric effects in satellite imagery (Sanchez, et al., 1998). The AOT values were derived from the measured transmittance. AOT, τ, is a measure of aerosol loading in the atmosphere (Christopher, et al., 2002). Generally, a higher AOT value indicates higher column of aerosol loading and therefore low visibility (Wang and Christopher, 2003).

Atmospheric components (aerosol and molecules) scatter and absorb solar radiation. This study investigated the use of a handheld spectroradiometer for the retrieval of atmospheric optical thickness (AOT) values over Penang Island. A set of spectroradiometer measured transmittance data was used to derive the aerosol optical thickness (AOT) in the atmosphere. The transmittance values were measured in spectral region from 350nm to 1050nm at the earth surface. The locations were determined with a Global Positioning System (GPS) receiver. We selected the values at 550nm for the present analysis. This study introduces an economical and a simple technique for retrieval of AOT for remote sensing application. The transmittance values were measured around the PenangIsland. Particulate matters of size less than 10 micron (PM10) were collected simultaneously with the acquisition of the transmittance measurements. The results of the analysis indicated that the AOT values were linearly correlated with the PM10 readings.

The objective of this study was to explore the relationship between aerosol optical thickness (AOT) derived from the spectroradiometer transmittance measurements and the correlation of particulate matter less than 10 micron (PM10). PM10 values were colleted simultaneously with the transmittance measurements. This study showed that there was a good correlation between the derived AOT and the measured PM10. In this study, we present results from ground-based measurements for PM10 mapping.

Many researchers have conducted satellite monitoring of the AOT [Husar, et al., (1997) and Liu, et al., (2002), Jiang, et al., (2007) and Guo, et al., (2009)]. In this study, we present a method for air quality mapping from ground-based measurements. Typically, ground based measurements were made with a device of sun photometer. In this study, our purpose is to generate AOT and PM10 maps over Penang Island, Malaysia by using a handheld spectroradiometer. The transmittance values were measured using a handheld spectraradiometer. The sensitivity of this type of spectroradiometer is between the spectral wavelengths from 350 nm to 1050 nm. The wells know Beer-Lambert law was used in this study to retrieve AOT values from the transmittance value measurements. AOT and PM10 maps were generated using Kriging interpolation techniquebased on the measured data. This study indicates that the spectroradiometer is a useful tool for retrieval of AOT and PM 10 values in Penang, Malaysia

2. Study area

The study area is the Penang Island, Malaysia, located within latitudes 5° 9′ N to 5° 33′ N and longitudes 100° 09′ E to 100° 30′ E. The map of the study area is shown in Figure 1. Penang Island is located in equatorial region and enjoys a warm equatorial weather the whole year. Therefore, it is impossible to get the 100 % cloud free satellite image over Penang Island. But, the satellite image chosen is less than 10 % of cloud coverage over the study area. Penang Island located on the northwest coast of Peninsular Malaysia.

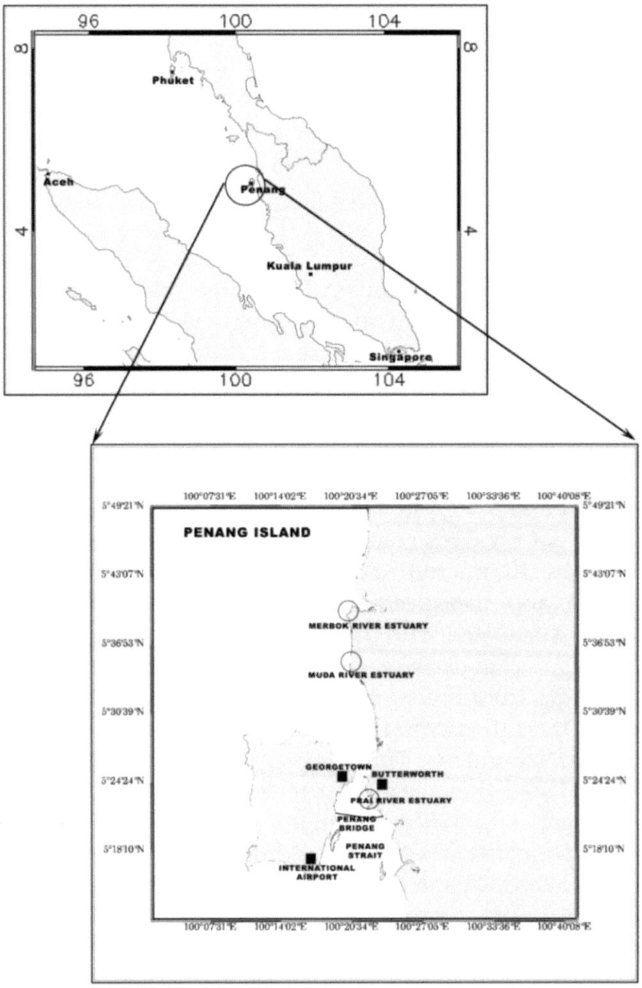

Fig. 1. The study area

Penang is one of the 13 states of the Malaysia and the second smallest state in Malaysia after Perlis. The state is geographically divided into two different entities - Penang Island (or "Pulau Pinang" in Malay Language) and a portion of mainland called "SeberangPerai" in Malay Language. Penang Island is an island of 293 square kilometres located in the Straits of Malacca and "SeberangPerai" is a narrow hinterland of 753 square kilometres (Tan, et al., 2010). The island and the mainland are linked by the 13.5 km long Penang Bridge and ferry.
Penang Island is predominantly hilly terrain, the highest point being Western Hill (part of Penang Hill) at 830 metres above sea level. The terrain consists of coastal plains, hills and mountains. The coastal plains are narrow, the most extensive of which is in the northeast which forms a triangular promontory where George Town, the state capital, is situated. The topography of "SeberangPerai" is mostly flat. Butterworth, the main town in "SeberangPerai", lies along the "Perai" River estuary and faces George Town at a distance of 3 km (2 miles) across the channel to the east (Tan, et al., 2010).
The Penang Island climate is tropical, and it is hot and humid throughout the year. with the average mean daily temperature of about 27°C and mean daily maximum and minimum temperature ranging between 31.4°C and 23.5°C respectively. However, the individual extremes are 35.7°C and 23.5°C respectively. The mean daily humidity varies between 60.9% and 96.8%. The average annual rainfall is about 267 cm and can be as high as 624 cm (Tan, et al., 2010).

3. Data acquisition

The corresponding PM10 measurements were collected at the several selected locations around the Penang Island between 9.00 a.m. to 3.00 p.m. A handheld spectroradiometer was used to collect the sky transmittance data over Penang Island, Malaysia campus on 24th December 2008 from 40 stations in the USM campus (Fig. 2). The AOT was calculated based on the Beer-Lambert-Bouguer law. Table 1 shows the specification of the ASD handheld spectroradiometer used in this study.

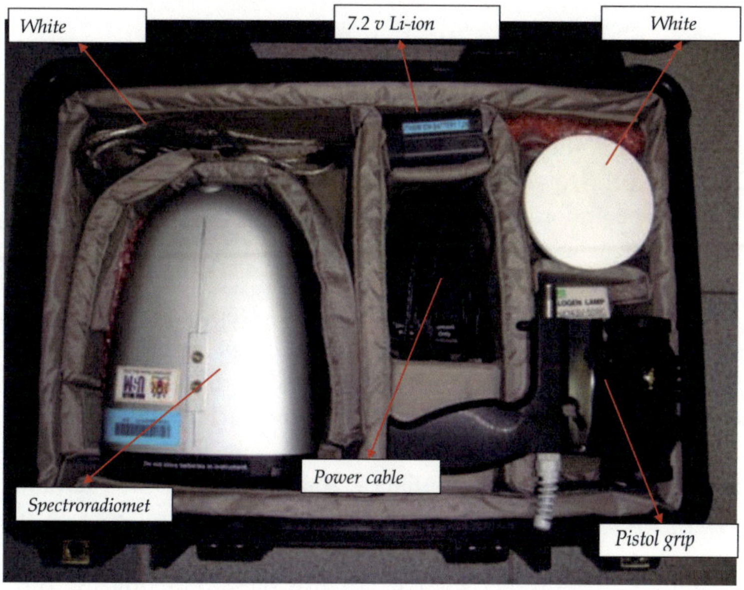

Fig. 2. TheASD handheld Spectroradiometer

FieldSpec® HandHeld (325-1075nm)	A field portable, lightweight, battery powered spectroradiometer.
Spectral Range:	325 - 1075 nm.
Sampling Interval :	1.6 nm.
Spectral Resolution:	3.5 nm at 700 nm.
Noise Equivalent Delta Radiance (NeDL):	5.0 x 10^-9 W/cm^2/nm/sr @ 700 nm.
Weight:	1.2 kg including battery pack.
Size:	22 x 16 x 8.2 cm.

Table 1. Specification of the ASD handheld Spectroradiometer

4. Methodology

The transmittance data over PenangIsland on 24th December 2008 was used in this study. The observation site is located in the northern region of the Peninsular Malaysia. The spectroradiometer measure the atmospheric transmittance from 350nm to 1050nm. The data that have been used in this study were collected at 40 locations in Penang, Malaysia.

The AOT is calculated based on the Beer-Lambert-Bouguer law. The AOT is related to the transmission by the expression

$$T_{d\lambda} = e^{-\frac{\tau_\lambda}{u_s}} \quad \text{(Vermote, et al., 1997)} \tag{1}$$

where

$T_{d\lambda}$ = transmittance for direct irradiance at wavelength, λ

u_s = cosines (θ), θ is the zenith angle

AOT values were obtained after performing the sequence of the following calculations:

1. First, we measured the total solar irradiance. This was done by measuring the radiance reflected from a Spectralon panel placed perpendicular to the direction of the Sun. The measured radiance was converted into irradiance by multiplying by Π and then divided by the reflectance of the Spectralon panel.
2. Second, we measured the diffuse irradiance. This step was performed with the same Spectralon panel maintained in the same orientation as in step 1. The panel was shaded from direct sunlight using a disk of black painted cardboard mounted on a stick. The size of the disk and distance to the panel should be such that the shadow of the disk on the panel is just sufficient to fully shade the area viewed by the FieldSpec HH. As in step 1, we measured the reflected radiance. Again, the measured radiance is converted to irradiance by multiplying by Π and then dividing by the reflectance of the Spectralon panel.
3. Third, we computed the direct solar irradiance by subtracting the diffuse irradiance from the measured total solar irradiance.

4. Fourth, we computed the top-of-atmosphere (TOA) solar irradiance values corresponding to the wavelengths of the spectra measured by the FieldSpec HH. In this study, Wehrli 1985 AM0 Spectrum was chosen to calculate the TOA spectra and interpolate it to the FieldSpec HH wavelengths, multiplying by cosines solar angle and then correct for the exact Earth-Sun distance factor, D, given by Spencer, 1971 as

$$\begin{aligned} D &= 1.00011 + 0.034221\cos\phi \\ &+0.00128\sin\phi + 0.000719\cos 2\phi \\ &+0.000077\sin 2\phi \end{aligned} \tag{2}$$

The day angle, ϕ, in radians is represented by

$$\phi = 2\pi(d-1)/365 \tag{3}$$

where d is the day number of a year (1-365)

5. Finally, we computed the atmospheric transmission by dividing the direct solar irradiance computed in step 3 by the top of atmosphere values calculated in the fourth step. This was done by first exporting the FieldSpec HH spectrum to a text file, importing the text file into a program like MS Excel and then performing the calculations in Excel. Then the AOT values were computed using Equation (1).

5. Data analysis and results

The transmittance values were measured using a handheld spectraradiometer. The sensitivity of this type of spectroradiometer is between the spectral wavelengths from 350 nm to 1050 nm. We selected 6 spectral wavelengths in this study centred at 400.5 nm, 500.5 nm, 600.5 nm, 700 nm, 800 nm and 900 nm. These selected 6 bands are based on the Wehrli 1985 AM0 Spectrum. Then the transmittance values were used to derive the AOT values using Equation (1). AOT maps were created using Kriging interpolation technique for estimating aerosol values to be associated to their distribution patterns (Figure 3).

In this study, we used PM10 as air quality parameter measurements over Penang Island. The relationship between AOT and PM10 was investigated in this study and we discovered that there was a linear relationship between PM10 and AOT. The linear correlation coefficient (R) was greater than 0.8 (Figure 3). Several studies have indicated linear relationship between these parameter [Christopher and Wang, (2003) and Wang and Christopher, (2003)]. We can clearly see from Figure 4 that PM10 values was increasing as the AOT values increasing. Therefore AOT is a useful information for retrieval of PM 10 values.

From our experimental data set, we can see the PM10 increases as the AOT increases. This mean that as the AOT values increases these is an increase of the concentration of PM10. Therefore, we obtained a positive correlation between AOT and PM10. From the AOT and PM10 maps, we can see that the high concentration of the AOT and PM10 values was located near the Bayan Lepas and Prai industrial areas (Black circle) and Georgetown urban areas (Red circle) (Figures 3 and 5). This was due to the high density of population area and industrial area respectively. The air pollution caused by industry has even more grave effects than vehicle exhaust fumes

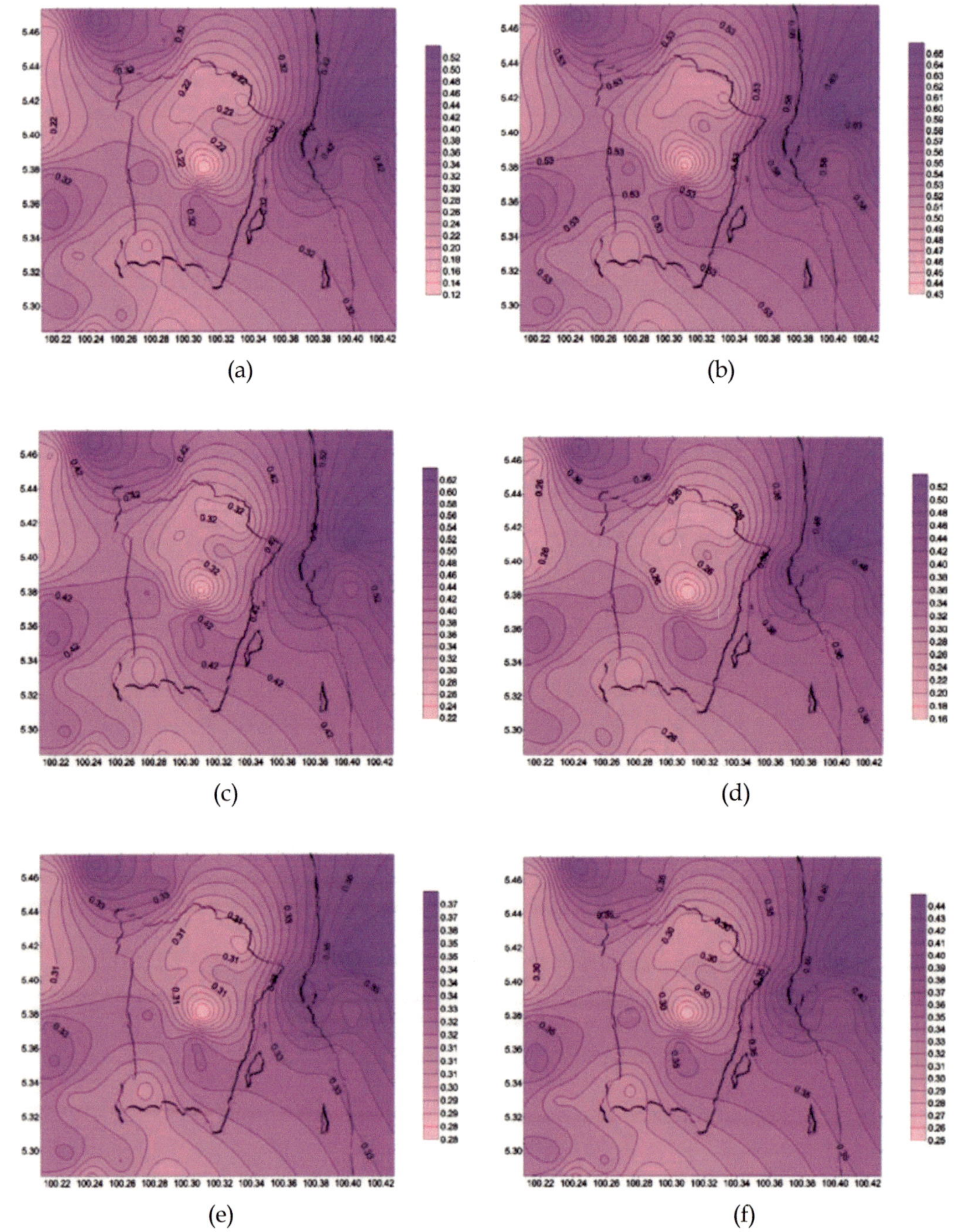

Fig. 3. AOT map for (a) 400.5 nm, (b) 500.5 nm, (c) 600.5 nm, (d) 700 nm, (e) 800 nm and (f) 900 nm

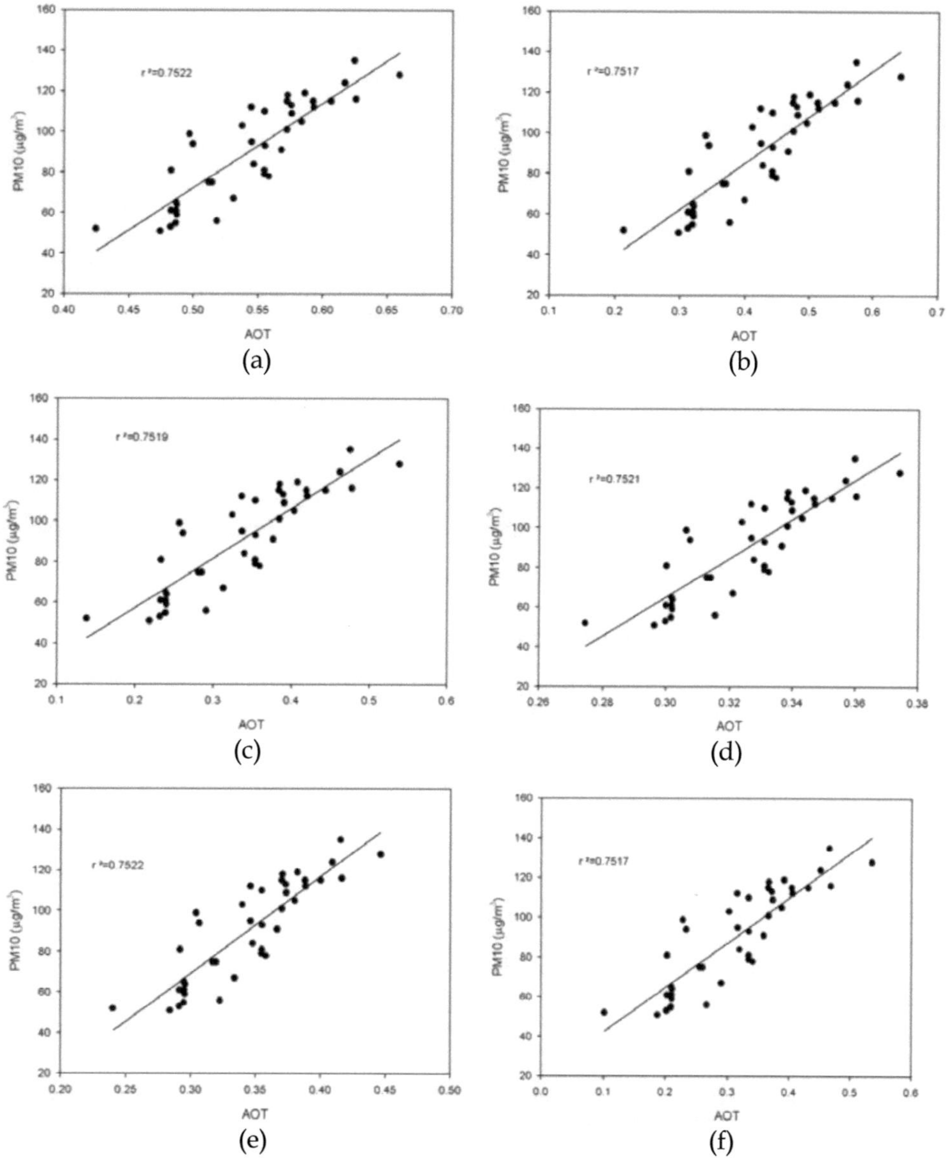

Fig. 4. A linear correlation between AOT and PM10 (µg/m3) values for (a) 400.5 nm, (b) 500.5 nm, (c) 600.5 nm, (d) 700 nm, (e) 800 nm and (f) 900 nm

6. Conclusion

AOT retrieval was produced by the handheld spectroradiometer from the atmospheric transmittance measurements. The results showed that there was a good agreement between AOT and PM10. The spectroradiometer retrieved AOT data had a good positive correlation with the PM10 concentration. This indicates that air quality can be retrieved from the spectroradiometer transmittance measurements. The interpolated AOT map can be used to discern air quality categories such as good, moderate and unhealthy to a relatively high degree of confidence. A linear relationship between PM10 and AOT was found by this study. Findings can be used to analyze the air pollution levels over Penang Island.

7. Acknowledgements

This project was carried out using the USM short term grants and Science Fund. The authors gratefully acknowledge the financial support from the RU grant, Relationship Between Heavy Rain, Flash Floods And Strong Winds In Malaysia, Southeast Asia During Tropical Cyclones, account number: 1001/PFIZIK/811152. We would like to thank the technical staff who participated in this project. Thanks are extended to USM for support and encouragement. Special thanks are extended to Jet Propulsion Laboratory, California Institute of Technology, USA for providing free online ASTER DEM data used in this study through GDEM WIST url that follows. https://wist.echo.nasa.gov/wist-bin/api/ims.cgi?mode=MAINSRCH&JS=1. Thanks are extended to USM for support and encouragement.

8. References

Christopher, S. A., Zhang, J., Holben, B. N. and Yang, S. K., 2002, GOES-8 and NOAA-14 AVHRR retrieval of smoke aerosol optical thickness during SCAR-B, int. j. remote sensing, 23 (22), 4931-4944

Guo, J. P., Zhang, X. Y., Che, H. Z., Gong, S. L., An, X., Cao, C. X., Guang, J., Zhang, H., Wang, Y. Q., Zhang, X. C., Xue, M. and Li, X. L., 2009, Correlation between PM concentrations and aerosol optical depth in eastern China, Atmospheric Environment, 43(37), p.p. 5876-5886

Husar R. B., M. Prospero and L. Stowe, 1997, Characterization of tropospheric aerosols over the oceans with the NOAA advanced very high resolution radiometer optical thickness operational product, Journal Geophysics. Research, 102, 16889-16909

Jiang, X., Liu, Y., Yu, B. and Jiang, M., 2010, Comparison of MISR aerosol optical thickness with AERONET measurements in Beijing metropolitan area, Remote Sensing of Environment, 107(1-2), 15, p.p. Pages 45-53

Liu, G. R., Chen, A. J., Lin, T. H. and Kuo, T. H., 2002, Applying SPOT data to estimate the aerosol optical depth and air quality, Environmental modeling& software, 17, 3-9

Sanchez Oliveros, C., Olmo Reyes, F. J. danAlados-Arboledas, L., 1998, Determination of aerosol optical thickness from measurements of spectral sky radiance. Journal of aerosol Science, 10, p.p. 1199-1211

Tan, K. C., Lim, H. S., MatJafri, M. Z. and Abdullah, K., 2010, Landsat data to evaluate urban expansion and determine land use/land cover changes in Penang Island, Malaysia, Springer, Environmental Earth Sciences, 60(7), p. p.1509–1521, ISSN: 1866-6280 (Print) 1866-6299 (Online), Available online: http://www.springerlink.com/content/d00w220673408052/. Digital Object Identifier: 10.1007/s12665-009-0286-z

Vermote, E., Tanre, D., Deuze, J. L., Herman, M. and Morcrette, J. J., 1997, Second Simulation of the satellite signal in the solar spectrum (6S), [Online] available: http://www.geog.tamu.edu/klein/geog661/handouts/6s/6smanv2.0_P1.pdf

Wang, J. and Christopher, S. A., 2003, Intercomparison between satellite-derived aerosol optical thickness and PM2.5 mass: Implications for air quality studies, Geophysics Research Letters, 30 (21)

Wehrli 1985 AM0 Spectrum, [Online] available: http://rredc.nrel.gov/solar/spectra/am0/wehrli1985.new.html

6

Remote Sensing of PM2.5 Over Penang Island from Satellite Measurements

Sim Chong Keat, Lim Hwee San,
Khiruddin Abdullah and Mohd. Zubir Mat Jafri
School of Physics, Universiti Sains Malaysia, Malaysia

1. Introduction

Satellite remote sensing technique has provided new datasets for monitoring air quality at regional and urban area. Air pollution in Asian cities has grown with the progressing industrialization and urbanization. Air quality standards often refer to respirable suspended particulate matter (PM), being aerosols with a diameter smaller than 2.5 μm (PM2.5). The generated algorithm was based on the aerosol optical depth. Remote sensed data provide comprehensive geospatial information on air quality with both qualitative imagery and quantitative data, such as particulate matters of size less than 2.5 micron (PM2.5).
Remote sensing has been widely used for environmental application such as for air quality and water quality studies. But the atmosphere affects satellite images of the Earth's surface in the solar spectrum. So, the signal observed by the satellite sensor is the sum of the effects from the ground and atmosphere. Surface reflectance is a key to the retrieval of atmospheric components from remotely sensed data. The accuracy of aerosol remote sensing techniques heavily depends on the information on the ground reflectance (Kokhanovskya, et al., 2005).
Environmental pollution is coeval with the appearance of humans. When Homo sapiens first lighted fire, its smoke provided the first medium of environmental pollution. The burning of fuels for heating and cooking has contributed to the air pollution of inner spaces (Makra and Brimblecombe, 2004). Air pollution is one of the most important environmental problems, which concentrates mostly in cities. Generally, human activities induce monotonous accumulation of pollutants [Makra, et al., (2001a) and Makra, et al., (2001b)]. Sources of exposure to particulate air pollution are many, Indoor sources, including cooking, unvented heating appliances and pets often make important contributions to exposures. Road traffic, however, generally provides the major source of ambient particulate pollution (Gulliver and Briggs, 2004).
We used a DustTrak Aerosol Monitor 8520 to collect in situ data. The particulate matters of size less than 2.5 micron data were collected simultaneously during the satellite Landsat overpass the study area. An empirical relationship between PM2.5 derived from the SPOT using regression technique is explored. A new algorithm was developed base on the aerosol characteristics in the atmosphere. We were obtained the atmospheric reflectance values by subtracting surface reflectance from the amount of reflectance measured from the satellite. The satellite recorded reflectance is the sum of the surface reflectance and atmospheric reflectance. The surface refleatance values were retrieved using ATCOR3 in the PCI Geomatica 10.3 image

processing software. The atmospheric reflectance values were later used for PM2.5 mapping using the calibrated algorithm. The efficiency of the developed algorithm, in comparison to other forms of algorithm, will be investigated in this study. Based on the values of the correlation coefficient and root-mean-square deviation, the proposed algorithm is considered superior. It is found that a linear relationship reveal a good results with PM2.5 measurements where the R value exceeding 0.89. The calibrated algorithm will be used to generate the air quality maps over Penang Island, Malaysia. The finding obtained by this study indicates that the SPOT data can be used to retrieved air quality information for remotely sensed data.

2. Remote sensing

Remote sensing is a technique for collecting information about the earth without touching the surface using sensors placed on a platform at a distance from it. The major applications of remote sensing include environmental pollution, urban planning, and earth management. We have to understanding the basic concept of electromagnetic waves well enough to apply remote sensing techniques in our studies. We classify electromagnetic energy by its wavelength. This electromagnetic radiation give an energy source to illuminate the target except the sensed energy is being emitted by the target.

There are two types of remote sensors: active and passive. Passive remote sensors detect reflected energy from the sun back to the sensor; they do not emit energy itself. But active sensors can emit energy or provide its own source of energy and detect the reflected energy back from the target. There are two types of remotely sensed data: airborne and space-borne. Airborne images are captured using sensors placed on aircraft platform while space-borne images are captured using sensors placed on a satellite platform. Remotely sensed data began with the traditional black and white aerial photography and followed by colour photography. However hyperspectral airborne or space-borne images are readily available nowadays. We use both the multispectral airborne and space-borne digital images in our studies. The major advantage of using remote sensing data is that we can produce the final output as maps of the Earth's surface phenomena by studies.

3. Study area

The study area is the Penang Island, Malaysia, located within latitudes 5° 9′ N to 5° 33′ N and longitudes 100° 09′ E to 100° 30′ E. The map of the study area is shown in Figure 1. Penang Island is located in equatorial region and enjoys a warm equatorial weather the whole year. Therefore, it is impossible to get the 100 % cloud free satellite image over Penang Island. But, the satellite image chosen is less than 10 % of cloud coverage over the study area. Penang Island located on the northwest coast of Peninsular Malaysia.

Penang is one of the 13 states of the Malaysia and the second smallest state in Malaysia after Perlis. The state is geographically divided into two different entities - Penang Island (or "Pulau Pinang" in Malay Language) and a portion of mainland called "Seberang Perai" in Malay Language. Penang Island is an island of 293 square kilometres located in the Straits of Malacca and "Seberang Perai" is a narrow hinterland of 753 square kilometres (Tan, et al., 2010). The island and the mainland are linked by the 13.5 km long Penang Bridge and ferry.

Penang Island is predominantly hilly terrain, the highest point being Western Hill (part of Penang Hill) at 830 metres above sea level. The terrain consists of coastal plains, hills and mountains. The coastal plains are narrow, the most extensive of which is in the northeast which forms a triangular promontory where George Town, the state capital, is situated. The topography of "Seberang Perai" is mostly flat. Butterworth, the main town in "Seberang Perai", lies along the "Perai" River estuary and faces George Town at a distance of 3 km (2 miles) across the channel to the east (Tan, et al., 2010).

The Penang Island climate is tropical, and it is hot and humid throughout the year. with the average mean daily temperature of about 27°C and mean daily maximum and minimum temperature ranging between 31.4°C and 23.5°C respectively. However, the individual extremes are 35.7°C and 23.5°C respectively. The mean daily humidity varies between 60.9% and 96.8%. The average annual rainfall is about 267 cm and can be as high as 624 cm (Tan, et al., 2010).

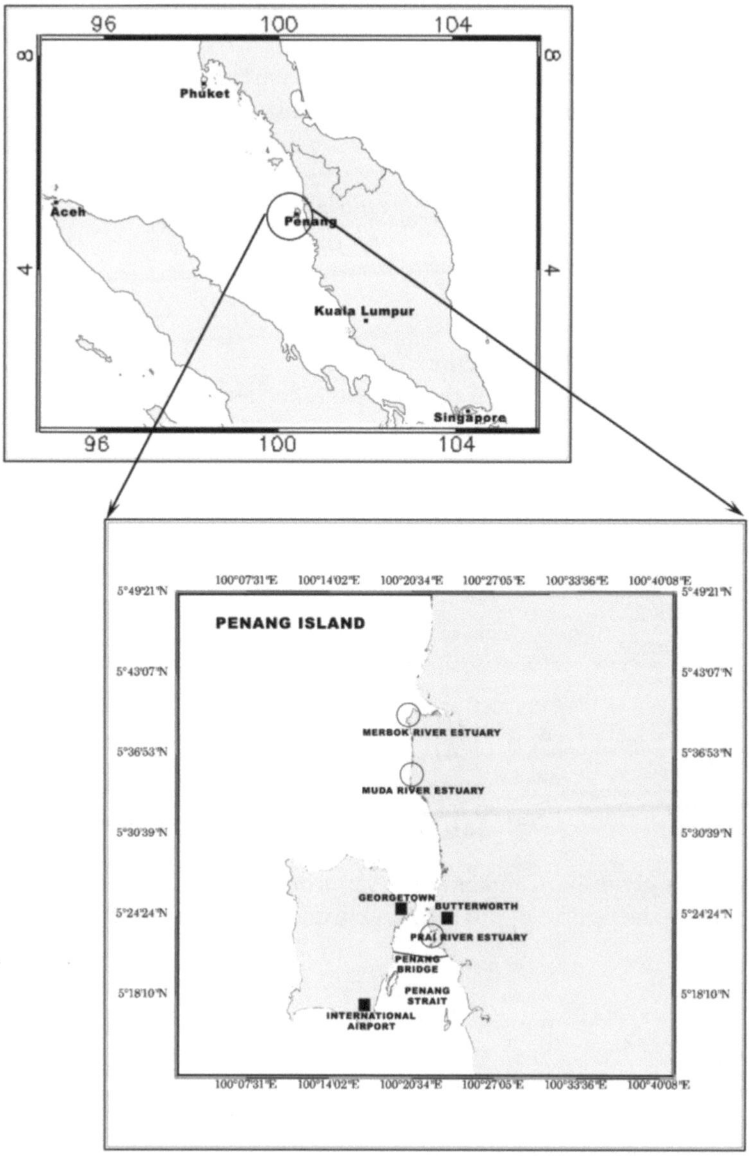

Fig. 1. The study area

4. Algorithm model

The atmospheric reflectance due to molecule, R_r, is given by (Liu, et al., 1996)

$$R_r = \frac{\tau_r P_r(\Theta)}{4\mu_s \mu_v} \tag{1}$$

where
τ_r = Rayleigh optical thickness
$P_r(\Theta)$ = Rayleigh scattering phase function
μ_v = Cosine of viewing angle
μ_s = Cosine of solar zenith angle
We assume that the atmospheric reflectance due to particle, Ra, is also linear with the τa [King, et al., (1999) and Fukushima, et al., (2000)]. This assumption is valid because Liu, et al., (1996) also found the linear relationship between both aerosol and molecule scattering.

$$R_a = \frac{\tau_a P_a(\Theta)}{4\mu_s \mu_v} \tag{2}$$

where
τ_a = Aerosol optical thickness
$P_a(\Theta)$ = Aerosol scattering phase function
Atmospheric reflectance is the sum of the particle reflectance and molecule reflectance, R_{atm}, (Vermote, et al., (1997).

$$R_{atm} = R_a + R_r \tag{3}$$

where
R_{atm} = atmospheric reflectance
R_a = particle reflectance
R_r = molecule reflectance

$$\begin{aligned} R_{atm} &= \left[\frac{\tau_a P_a(\Theta)}{4\mu_s \mu_v} + \frac{\tau_r P_r(\Theta)}{4\mu_s \mu_v} \right] \\ R_{atm} &= \frac{1}{4\mu_s \mu_v} \left[\tau_a P_a(\Theta) + \tau_r P_r(\Theta) \right] \end{aligned} \tag{4}$$

The optical depth is given by Camagni and Sandroni, (1983), as in equation (5). From the equation, we rewrite the optical depth for particle and molecule as equation (6)

$$\tau = \sigma \rho s \tag{5}$$

where
τ = optical depth
σ = absorption
s = finite path

$$\tau = \tau_a + \tau_r \qquad \text{(Camagni and Sandroni, 1983)}$$

$$\tau_r = \sigma_r \rho_r s \quad (6a)$$

$$\tau_p = \sigma_p \rho_p s \quad (6b)$$

Equations (6) are substituted into equation (4). The result was extended to a three bands algorithm as equation (7) Form the equation; we found that PM was linearly related to the reflectance for band 1 and band 2. This algorithm was generated based on the linear relationship between τ and reflectance. Retalis et al., (2003), also found that PM was linearly related to τ and the correlation coefficient for linear was better that exponential in their study (overall). This means that reflectance was linear with PM. In order to simplify the data processing, the air quality concentration was used in our analysis instead of using density, ρ, values.

$$R_{atm} = \frac{1}{4\mu_s\mu_v}\left[\sigma_a\rho_a s P_a(\Theta) + \sigma_r\rho_r s P_r(\Theta)\right]$$

$$R_{atm} = \frac{s}{4\mu_s\mu_v}\left[\sigma_a\rho_a P_a(\Theta) + \sigma_r\rho_r P_r(\Theta)\right]$$

$$R_{atm}(\lambda_1) = \frac{s}{4\mu_s\mu_v}\left[\sigma_a(\lambda_1)PP_a(\Theta,\lambda_1) + \sigma_r(\lambda_1)GP_r(\Theta,\lambda_1)\right]$$

$$R_{atm}(\lambda_2) = \frac{s}{4\mu_s\mu_v}\left[\sigma_a(\lambda_2)PP_a(\Theta,\lambda_2) + \sigma_r(\lambda_2)GP_r(\Theta,\lambda_2)\right]$$

$$P = a_0 R_{atm}(\lambda_1) + a_1 R_{atm}(\lambda_2) \quad (7)$$

where
P = Particle concentration (PM2.5)
G = Molecule concentration
R_{atm} = Atmospheric reflectance, i = 1 and 2 are the band number
a_j = algorithm coefficients, j = 0, 1, 2, … are then empirically determined.

5. Data analysis and results

Remote sensing satellite detectors exhibit linear response to incoming radiance, whether from the Earth's surface radiance or internal calibration sources. This response is quantized into 8-bit values that represent brightness values commonly called Digital Numbers (DN). To convert the calibrated digital numbers to at-aperture radiance, rescaling gains and biases are created from the known dynamic range limits of the instrument.

The satellite image was rectified using the second order polynomial coordinated transformation using nearest neighbor method to relate ground control points (GCP) in the map to their equivalent row and column positions in the TM scene. Overall, both satellite images achieved the RMS errors less than 0.5 pixels in this study (Vicente-Serrano et al. 2008; Schroeder et al. 2006; Kabbara et al. 2008).

The digital numbers (DN values) corresponding to the ground truth sample locations were extracted from all the images. The DN for window size of 3 by 3 was used because the data set produced higher correlation coefficient and lower RMS value. The DN values were converted into radiance values and later reflectance values.

It should be noted that the relfectance values at the top of atmospheric is the sum of the surface reflectance and atmospheric relfectance. The signals measured in each of these visible bands represent a combination of surface and atmospheric effects, usually in different proportions depending on the condition of the atmosphere. Therefore, it is required to determine the surface contribution from the total reflectance received at the sensor. In this study, we extracted the surface reflectance from mid-infrared band because the surface reflectances at various bands across the solar spectrum are correlated to each other to some extended. The surface reflectances of dark targets in the blue and red bands were estimated using the measurements in the mid-infrared band (Quaidrari and Vermote, 1999). Over a simple black target, the observed atmospheric reflectance, is the sum of reflectance of aerosols and Rayleigh contributions (Equation 8). This simplification, however, is not valid at short wavelengths (less than 0.45 pm) or large sun and view zenith angles (Vermote and Roger, 1996). In this study, a simple form of the equation was used in this study (Equation 9). This equation also used by other research in their study (Poop, 2004).

$$Rs - TRr = Ratm \tag{8}$$

$$Rs - Rr = Ratm \tag{9}$$

where:
Rs = reflectance recorded by satellite sensor
Rr = reflectance from surface references
Ratm = reflectance from atmospheric components (aerosols and molecules)
T = transmittance

It should be noted that the relfectance values at the top of atmospheric was the sum of the surface reflectance and atmosphere relfectance. In this study, we used ATCOR3 image correction software in the PCI Geomatica 10.3 image processing software for creating a surface reflectance image. Usually, the absolute radiometric correction methods available only correct for atmospheric effects of satellite imagery (Janzen et al. 2006). Therefore, they may suitable in certain atmospheric correction in flat or terrain area. But, the algorithm proposed cannot reduce the topographic effects. Hence, in mountainous terrain, these methods may bring some errors while apply the algorithms to eliminate the atmospheric effects (Tan, et al., 2010).

In order to remove atmospheric and topographic effects efficiently, a method has been developed and implement in satellite imagery over mountainous area. The algorithm proposed is based on the Richter model (Richter 1990). The algorithm generates the three-dimensional atmosphere by considering the transmittance and radiance functions of different height, included atmospheric conditions alteration in horizontally. Consequently, it is able to solve the problem of haze through the statistical algorithm, which helps to manipulate the optical depths in different altitude.

ATCOR3 is a new approach implemented in PCI Geomatica 10.3 image processing software and the technique based on the ATCOR2 model. But ATCOR2 model restricted for a flat terrain to calculate surface reflectance. Both of these models have the similarity, where the approaches apply the technique from dense dark vegetation approach (Liang et al. 1997) and the modified dense dark vegetation approach (Song et al. 2001).

A database contains radiative transfer code is acquired from the calculation of direct and diffuse solar flux, path radiance and atmospheric transmittance with the wide range of weather conditions, in terms of atmospheric conditions (Richter 1998). Beside, ATCOR3 requires the information, such as slope, orientation and surface elevation, to eliminate the topographic effects. The range cover the mountainous terrain should not greater than 3.5 km above the sea level. Furthermore, if ATCOR3 apply to the area involved rugged terrain, the algorithm need to calculate the specified atmospheric conditions based on Lambertian assumption. Overall, the method implemented here only restricted to high spatial resolution satellite sensors with small swath angle like Landsat and Systeme pour I'Observation de la Terre (SPOT).

The advantage of ATCOR3 is the module can generate a surface reflectance map using reference elevation data. In addition, retrieval of surface reflectance thematic map becomes more accurate, especially in high mountain terrain, because it considers the slope and aspect images. In this study, the reference DEM model known as ASTER Global Digital Elevation Map (GDEM) was used to retrieve the a surface reflectance values (Fig 2). And then the reflectance measured from the satellite [reflectance at the top of atmospheric, ρ(TOA)] was subtracted by the amount given by the surface reflectance to obtain the atmospheric reflectance.

Fig. 2. GDEM data used in this study.

SPOT satellite data was selected corresponding to the ground truth measurements of the pollution levels (Fig. 3). The satellite image was acquired on 30 January 2005. The corresponding PM2.5 measurements were collected simultaneously during the satellite overpass. And then the atmospheric reflectance was related to the PM10 using the regression algorithm analysis.

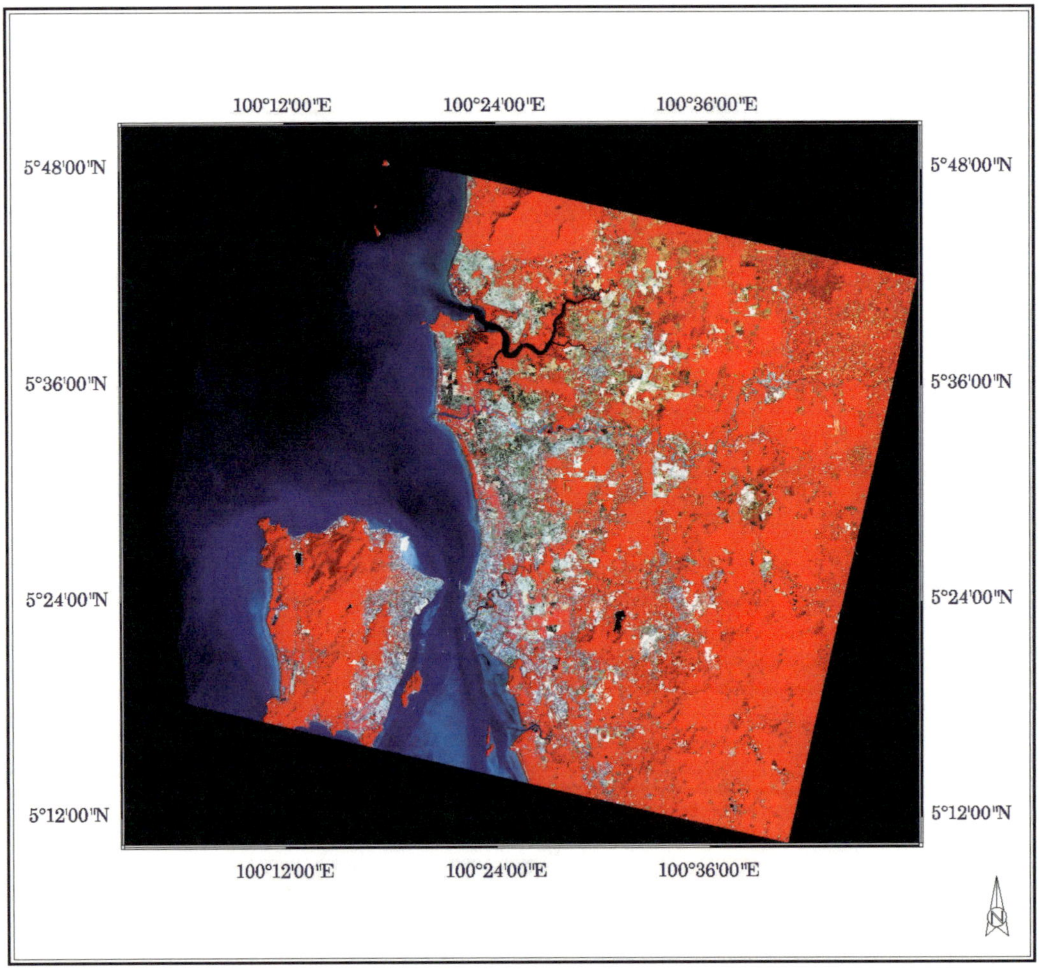

Fig. 3. Raw satellite image.

In this study, SPOT signals were used as independent variables in our calibration regression analysis. A good result was produced by the proposed model, which achieved the correlation coefficient of about 0.89. The PM10 map was generated using the proposed algorithm (Figure 4).

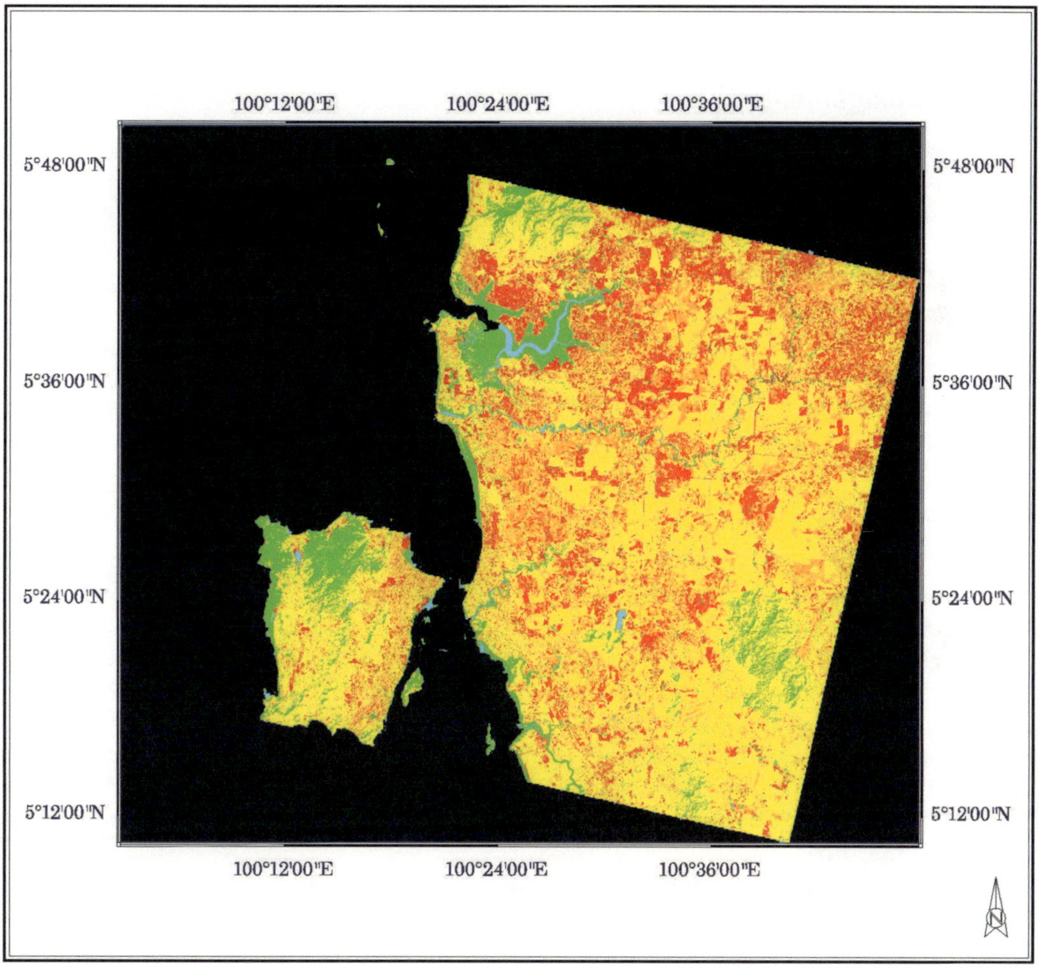

Fig. 4. Map of PM2.5 around Penang Island, Malaysia [Blue < 40 ug/m3, Green = (40-80) ug/m3, Yellow = (80-120) ug/m3, Orange = (120-160) ug/m3, Red = (>160) ug/m3) and Black = Water]

6. Conclusion

The result produce by this study indicated the used of SPOT satellite image for PM2.5 mapping over Penang Island. A good agreement was found in this study between PM2.5 measured by the DustTrak Aerosol Monitor 8520 and PM2.5 estimated using the newly proposed developed algorithm. The developed algorithm produced high R and low RMS values. This study indicates that satellite observation can be used for air quality mapping using SPOT data. The use of remotely sensed data produced better spatial resolution air quality map compared to the spacing between ground stations. Further study will be carried out to verify the results and a multi regression algorithm will be developed and used in the analysis.

7. Acknowledgements

This project was carried out using the USM short term grants and Science Fund. The authors gratefully acknowledge the financial support from the RU grant, Relationship Between Heavy Rain, Flash Floods And Strong Winds In Malaysia, Southeast Asia During Tropical Cyclones, account number: 1001/PFIZIK/811152. We would like to thank the technical staff who participated in this project. Thanks are extended to USM for support and encouragement. Special thanks are extended to Jet Propulsion Laboratory, California Institute of Technology, USA for providing free online ASTER DEM data used in this study through GDEM WIST url that follows. https://wist.echo.nasa.gov/wist-bin/api/ims.cgi?mode=MAINSRCH&JS=1. Thanks are extended to USM for support and encouragement.

8. References

Camagni. P. & Sandroni, S. (1983). Optical Remote sensing of air pollution, Joint Research Centre, Ispra, Italy, Elsevier Science Publishing Company Inc

Fukushima, H.; Toratani, M.; Yamamiya, S. & Mitomi, Y. (2000). Atmospheric correction algorithm for ADEOS/OCTS acean color data: performance comparison based on ship and buoy measurements. Adv. Space Res, Vol. 25, No. 5, 1015-1024

Gulliver, J. and Briggs, D. J. 2004. Personal exposure to particulate air pollution in transport microenvironments. Atmospheric Environment, 38:1-8.

Janzen DT, Fredeen AL, Wheate RD (2006) Radiometric correction techniques and accuracy assessment for Landsat TM data in remote forested regions. Can J Remote Sens 32(5):330–34

Kabbara N, Benkhelil J, Awad M, Barale V (2008) Monitoring water quality in the coastal area of Tripoli (Lebanon) using high- resolution satellite data. ISPRS J Photogramm Remote Sens 63(5):488–495

Kokhanovskya,b, A.A., Mayer, B. and Rozanov, V. V., 2005, A parameterization of the diffuse transmittance and reflectance for aerosol remote sensing problems, Atmospheric Research 73, 37 - 43

Liang S, Fallah-Adl H, Kalluri S, Jaja J, Kaufman YG, Townshend JRG (1997) An operational atmospheric correction algorithm for Landsat Thematic Mapper imagery over the land. Journal of Geophysical Research, 102:173-186King, M. D.; Kaufman, Y. J.; Tanre, D. & Nakajima, T. (1999). Remote sensing of tropospheric aerosold form space: past, present and future, Bulletin of the American Meteorological society, 2229-2259

Liu, C. H.; Chen, A. J. ^ Liu, G. R. (1996). An image-based retrieval algorithm of aerosol characteristics and surface reflectance for satellite images, International Journal Of Remote Sensing, 17 (17), 3477-3500

Makra, L. and Brimblecombe, P. 2004. Selections from the history of environmental pollution, with special attention to air pollution. Part 1. International Journal of Environment and Pollution (IJEP), 22(6):641-656

Makra, L., Horváth, Sz., Taylor, C.C., Zempléni, A., Motika, G. and Sümeghy, Z. 2001a. Modelling air pollution data in countryside and urban environment, Hungary. The 2nd International Symposium on Air Quality Management at Urban, Regional and Global Scales. Istanbul Technical University, Istanbul, Turkey, 25-28 September 2001. Proceedings.189-196. Eds: Topcu, S., Yardim, M.F. and Incecik, S

Makra, L., HorvՁth, Sz., Zempléni, A., CsiszՁr, V., Rózsa, K. and Motika, G. 2001b. Air quality trends in Southern Hungary. "3rd International Conference on Urban Air Quality and 5th Saturn Workshop. Measurement, Modelling and Management." Institute of Physics, Canopus Publishing Limited. Loutraki, Greece. Extended Abstracts CD-ROM. [2001. március 19-23.]

Popp, C.; Schläpfer, D.; Bojinski, S.; Schaepman, M. & Itten, K. I. (2004). Evaluation of Aerosol Mapping Methods using AVIRIS Imagery. R. Green (Editor), 13th Annual JPL Airborne Earth Science Workshop. JPL Publications, March 2004, Pasadena, CA, 10

Quaidrari, H. dan Vermote, E. F. (1999). Operational atmospheric correction of Landsat TM data, Remote Sensing Environment, 70: 4-15

Retalis, A.; Sifakis, N.; Grosso, N.; Paronis, D. & Sarigiannis, D. (2003). Aerosol optical thickness retrieval from AVHRR images over the Athens urban area, [Online] available:http://sat2.space.noa.gr/rsensing/documents/IGARSS2003_AVHRR_Re talisetal_web.pdf

Richter R (1990) A fast atmospheric correction algorithm applied to Landsat TM images. Int J Remote Sens 11(11):159–166

Schroeder TA, Cohen WB, Song C, Canty MJ, Yang Z (2006) Radiometric correction of multi-temporal Landsat data for characterization of early successional forest patterns in western Oregon. Remote Sens Environ 103:16–26

Song C, Woodcock CE, Seto KC, Lenney MP, Macomber SA (2001) Classification and change detection using Landsat TM data: When and how to correct atmospheric effects? Remote Sensing of Environment, 75, 230-244

Tan, K. C., Lim, H. S., MatJafri, M. Z. and Abdullah, K., 2010, Landsat data to evaluate urban expansion and determine land use/land cover changes in Penang Island, Malaysia, Springer, Environmental Earth Sciences, 60(7), p. p.1509–1521, ISSN: 1866-6280 (Print)18666299(Online),Availableonline:http://www.springerlink.com/content/d 00w220673408052/. Digital Object Identifier: 10.1007/s12665-009-0286-z

Vermote, E. & Roger, J. C. (1996). Advances in the use of NOAA AVHRR data for land application: Radiative transfer modeling for calibration and atmospheric correction, Kluwer Academic Publishers, Dordrecht/Boston/London, 49-72

Vermote, E.; Tanre, D.; Deuze, J. L.; Herman, M. & Morcrette, J. J. (1997). 6S user guide Version 2, Second Simulation of the satellite signal in the solar spectrum (6S),

[Online]available:http://www.geog.tamu.edu/klein/geog661/handouts/6s/6smanv2.0_P1.pdf
Vicente-Serrano SM, Perez-Cabello F, Lasanta T (2008) Assessment of radiometric correction techniques in analyzing vegetation variability and change using time series of Landsat images. Remote Sens Environ 112:3916-3934

7

Photoacoustic Measurements of Black Carbon Light Absorption/Scattering Coefficients and Visibility Degradation in Jordan During 2007/2008

Khadeejeh M. Hamasha
Physics Departement, University of Tabuk, Kingdom of Saudi Arabia

1. Introduction

Air is the name given to atmosphere used in breathing and photosynthesis. Air supplies us with oxygen which is essential for our bodies to live. Air consists of 79% nitrogen, 20% oxygen, 1% water vapor and inert gases. Air pollution is the introduction of chemicals, particulate matter, or biological materials that cause harm or discomfort to organisms into the atmosphere. Air pollutants are known as substances in the air that can cause harm to humans and the environment. These substances are not naturally found in the air at greater concentrations or in different locations from usual. Pollutants can be in the form of solid particles, liquid droplets, or gases. In addition, they may be arising from natural processes or human activities.

Pollutants can be classified as primary air pollutants or secondary air pollutants according to their sources. Usually, primary air pollutants are directly emitted from a process, such as ash from a volcanic eruption, sulfur dioxide released from factories or the carbon monoxide gas from a motor vehicle exhaust. Secondary pollutants are not emitted directly. But, they form in the air when primary pollutants interact or react. An example of a secondary pollutant is ground level ozone, which is one of the many secondary pollutants that make up photochemical smog. Some pollutants may be both primary and secondary: that is, they are both emitted directly and formed from other primary pollutants.

The primary air pollutants found in most urban areas are dispersed throughout the world's atmosphere in concentrations high enough to gradually serious health problems. This problems can occurs quickly when air pollutants are concentrated. The main sources of pollutants in urban areas are transportation and fuel composition in stationary sources, such as commercial, coal-burning power plant, cooling, and industrial heating.

One type of air pollution is the release of particles (aerosols) into the air from burning fuel for energy. Aerosols are defined as the relatively stable suspensions of solid or liquid particles in gas. There are many properties of particles that are important for their role in the atmospheric processes. These include number concentration, mass, size, chemical composition, and aerodynamic and chemical properties (Chang et al. 1982; Walker 1966). Of these, size is very important. It is related to the source of particles and their impact on health (Harber et al. 2003; Puntoni et al. 2004; Borm et al. 2005), visibility, and climate (Finlayson-Pitts and Pitts 2000).

2. Black carbon

Light absorbing carbon particles (organic carbon and black carbon) are the most abundant and efficient light absorbing component in the atmosphere in the visible spectrum. It typically depends inversely on wavelength (Horvath 1993; Horvath 1997). Organic carbon is strongly wavelength dependent, with increased absorption for UV and short wavelength visible radiation, but hardly at all at 870 nm. Black carbon is very likely to dominate at 870 nm (Lewis et al. 2008). When aerosols absorb light, the energy of the light is transferred to the particles as heat and eventually is given to the surrounding gas. Aerosol particles in the atmosphere have a great influence on fluxes of solar energy and the accompanied fluctuations in temperature caused by changes in the aerosol (Horvath 1993).

Black carbon, the main constituent of soot, is almost exclusively responsible for aerosol light absorption at long wavelength visible radiation and near infrared wavelengths. This type of pollution is sometimes referred to as black carbon pollution. Air pollution caused by black carbon particles has been a major problem since the beginning of the industrial revolution and the development of the internal combustion engine. Scientific publications dealing with the analysis of soot and smoke date back as early as 1896 (Arrhenius 1896). Mankind has become so dependent on the burning of fossil fuels (petroleum products, coal, and natural gas) that the sum total of all combustion-related emissions now constitutes a serious and widespread problem, not only to human health (Gillmour et al. 2004, Gardiner et al. 2001, Parent et al. 2000), but also to the entire global environment (IPCC 1996, Finlayson-Pitts and Pitts 2000).

Absorption of solar radiation by black carbon is expected to lead to heating of the atmosphere since the light energy is converted into thermal energy (Finlayson-Pitts and Pitts 2000). This is the opposite effect of scattering of light by particles into the upper atmosphere. This heating effect would be expected to be most important in polluted urban areas (Liu and Smith 1995, Horvath 1995). Black carbon aerosol light absorption reduces the amount of sunlight available at the surface to drive atmospheric circulation and boundary layer development.

Even the burning of wood and charcoal in fireplaces and barbeques can release significant quantities of soot into the air. Some of these pollutants can be created by indoor activities such as smoking and cooking. So pollution also needs to be considered inside homes, offices, and schools. According to the world health report 2002 indoor air pollution is responsible for 2.7% of the global burden of disease (WHO 2010). We spend about 80-90% of our time inside buildings, and so our exposure to harmful indoor pollutants can be serious (Harber et al. 2003; Puntoni et al. 2004; Borm et al. 2005). It is therefore important to consider both indoor and outdoor air pollution.

3. Jordan

Jordan is located between 29°10' N - 33°45' N and 34°55' E - 39°20' E. The discovery of oil in the Arabian Peninsula has resulted in fast growth and social and economical development in the Gulf States and their neighboring countries including Jordan, which provides skilled workers. The social and economic development in Jordan has been accompanied by an increase in the consumption of oil for different needs, including residential, commercial, industrial, transportation, and power generation. According to figures published by the Department of Statistics, Jordan imported about six million tons of crude oil in 2005 (Department of Statistics, 2010).

Combustion of oil and other fossil fuel is recognized as a major source of air pollution in urban areas. Several airborne substances can remain in the atmosphere for weeks, and travel over hundreds of kilometers, making air pollution a global problem. Common pollutants that are generated through oil combustion are carbon oxides (CO and CO_2), sulfur oxides (SOx), nitrogen oxides (NOx), particulate matter (PM), and volatile organic compounds (VOCs). Tropospheric ozone is a secondary pollutant that is generated in the troposphere through a photosynthesis reaction of NOx and VOCs in the presence of solar radiation. It is becoming a major threat to air quality in metropolitan areas.

Emissions from motor vehicles account for 50–90 percent of air pollution in urban centers (Cooper et al. 1996; Gillies et al 2008). There are just over 750,000 vehicles licensed in Jordan, of which 77.5% are registered in the capital, Amman (Department of Driving and Vehicles Licensing 2010). More than 31% of the vehicles in Jordan are diesel-powered. Vans and trucks represent 33% and 42.7% of the total diesel-powered vehicles, respectively. Most public transportation vehicles work inside cities, especially Amman and Zarqa. Particles emanated from motor vehicles contain sulfate, carbonaceous particles, and a large number of chemicals (Kassel 2003).

Other sources of air pollution in Jordan include power generation, which uses heavy oil and natural gas; cement production, which uses oil shale; cooking; home furnaces fueled by diesel, natural gas, or kerosene; in addition wood stoves. The unexpected jump in oil prices experienced during winter of 2007 has forced people with low income in the countryside and mountainous areas to switch to wood stoves because they use either olive husk or wood, which are available at low, or no, cost in their immediate surroundings.

The negative health impact of air pollution has been widely studied in humans and animals. Findings of several epidemiology studies pointed out that high levels of air pollution may result in several health problems, including eye irritation, skin irritation, asthma, lung cancer, cardiovascular issues, high blood pressure, lung tumors, and increasing mortality rate (Pope et al. 1995; Künzli et al. 2000; Pope et al. 2002; Takano et al. 2002; Sanjay Rajagopalan 2008). Over 300,000 cases of chronic bronchitis, 500,000 asthma attacks, and 16 million lost person-days of activity recorded in Europe were blamed on vehicle emissions (Künzli et al. 2000). Exposure to high levels of SO_2 causes impairment of the respiratory function and aggravates existing respiratory and cardiac illnesses (Andre 2001). Long-term exposure to NO_2 lowers resistance to respiratory infections and aggravates existing chronic respiratory diseases. In addition to its adverse impact on humans, air pollution has adverse impacts on animals, and vegetation, in addition to loss of crops.

In spite of the fast growth of urban areas and industrial activities in Jordan, air pollution has not received due attention. Air quality is not routinely monitored anywhere except at Alhashameiah (to the northeast of Zarqa), which experiences high levels of sulfur oxides and particulates. There have been a few studies that tackled air pollution in Jordan, but they have been limited to three stations only: Downtown and Shmeisani areas in Amman, as well as Al-Hashemyeh. Those studies have pointed out that local air quality is poor where concentrations of criteria pollutants (NOx, SOx, CO, PM_{10}, TSP, Lead, and hydrogen sulfide) exceed the National Air Quality Standards (Asi et al. 2001; Hamdi 2008). The Jordanian ministry of environment has recently launched a project to establish an air quality monitoring network throughout the country, but actual steps towards that goal have not been taken yet.

4. Measurements of black carbon levels using photoacoustic technique

Photoacoustic instrument (Arnott 1999) is used to measure the black carbon light absorption coefficients. Data were displayed as absorption coefficients in 1/Mm, and were later converted to black carbon mass concentration. The photoacoustic instrument (figure 1) utilizes a microphone to record sound issuing from heat transferred from light absorbing aerosol to the surrounding air. A power meter records the laser power. The ratio of microphone pressure and laser power is used to obtain the light absorption coefficient. Photoacoustic instruments have a very large dynamic range of measurement, and are not influenced by artifacts due to filter loading and scattering aerosol associated with filter-based sampling methods (Arnott et al. 2005).

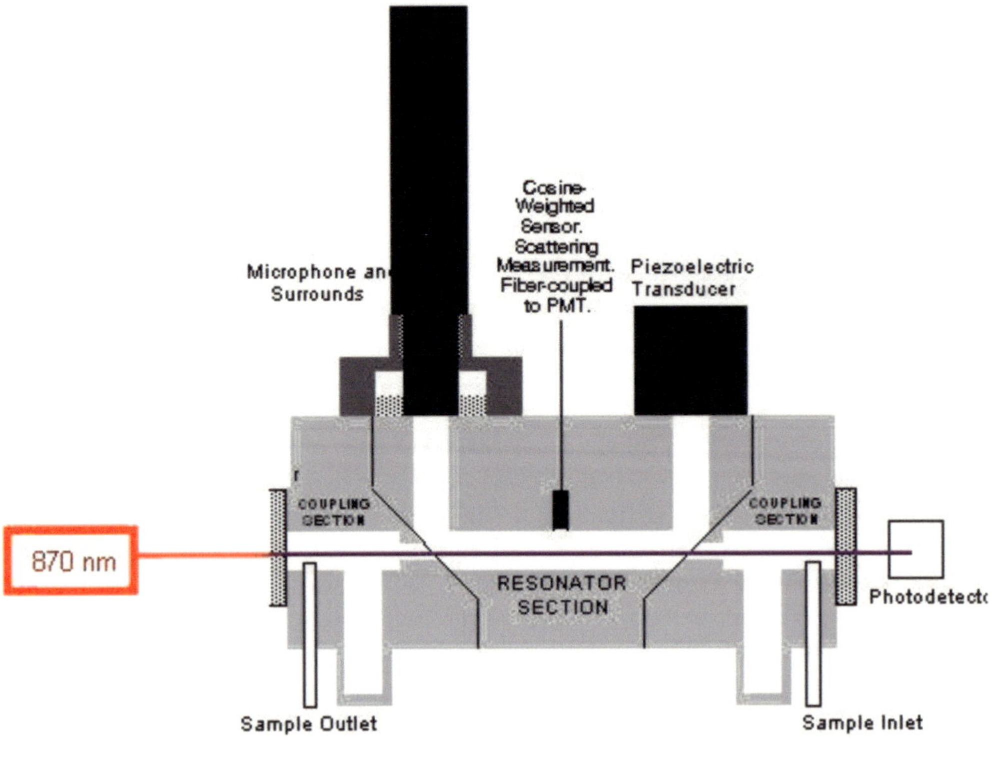

Fig. 1. A schematic view of the photoacoustic spectrometer instrument. (PMT is a photomultiplier)

Black carbon and organic carbon are the most efficient light-absorbing aerosol species in the visible spectral range. Organic carbon is strongly wavelength dependent, with increased absorption for UV and short wavelength visible radiation, but hardly at all at 870 nm. Black carbon is very likely to dominate at 870 nm (Rosen et al. 1978; Lindberg et al. 1993; Lewis et al. 2008). Thus the measurement of aerosol light absorption at wavelengths in the long visible wavelength is correlated to the measurement of black carbon. Light absorption by particles depends on the wavelength of the incident light. The relationship between the aerosol absorption coefficients, B_{abs} and the corresponding black carbon mass concentration (BC) is established by the aerosol specific mass absorption efficiency σ_{abs} via the relationship:

$$B_{abs} = BC\sigma_{abs} \tag{1}$$

The magnitude of σ_{abs} ranges from 2 to 20 m²/g (Liousse et al. 1993). Black carbon mass concentrations (BC) are calculated from B_{abs} using the light absorption efficiency for black carbon, α_a, such that (Arnott et al. 1999):

$$\mathrm{B_{abs}}\left(\mathrm{Mm^{-1}}\right) = \mathrm{BC}\left(\mu\mathrm{g}/\mathrm{m^3}\right) \times \alpha_\mathrm{a}\left(\mathrm{m^2}/\mathrm{gm}\right) \tag{2}$$

and,

$$\alpha_\mathrm{a} = 10\mathrm{m}^2/\mathrm{gm} \text{ for } \lambda = 532\mathrm{nm} \tag{3}$$

Since B_{abs} is proportional to $1/\lambda$ (Kirchstetter et al. 2004); then α_a is also proportional to $1/\lambda$. Therefore,

$$\begin{aligned} \alpha_a(870nm) &= \alpha_a(532nm)\left(\frac{870}{532}\right)^{-1} \\ &= 6.11m^2/g \end{aligned} \tag{4}$$

Substituting back in equation (2) yields

$$\mathrm{BC}(870\mathrm{nm}) = \mathrm{B_{abs}}(870\mathrm{nm})/6.11 \tag{5}$$

5. Black carbon levels in Jordan

Measurements of black carbon light absorption coefficients (B_{abs}) using photoacoustic instrument at the wavelength of 870 nm in different locations of Jordan show that B_{abs} is higher for the locations in the city centers than the locations in the industrial centers during summer 2007(Hamasha et al. 2010). Low black carbon concentrations in the vicinity of industrial zones are attributed to the efficiency of tall stacks in reducing ground level concentrations of emitted substances. However, tall stacks do not really make air cleaner; they only carry black carbon and other pollutants to distant locations as seen from the results at the location in Zarqa downtown. Measurements carried out at Zarqa downtown gave the highest levels of black carbon concentration during summer as well as winter (Hamasha et al. 2010); because of numerous air pollution sources concentrated in the city. Zarqa is a growing industrial city with a population of about half a million as 2008 estimate (Department of Statistics 2010). It hosts about 35% of the heavy industry in Jordan including the only oil refinery, an oil-based power plant, steel factories, a pipe factory, a wastewater treatment plant, to mention a few. A total of 2400 industrial activities are registered in the Zarqa Industrial Chamber.

B_{abs} in Zarqa city center is about 179 Mm^{-1}during summer day, 2007 And about 81Mm^{1}during winter day, 2008. While in Amman city center the measured values of B_{abs} were about 67Mm^{-1}during summer day, 2007 and about 23Mm^{-1}during winter day, 2008(Hamasha et al. 2010).

Measurements at Ibbeen city center on a winter day (28/2/2008) show that the city had relatively high levels of black carbon (about 72 Mm^{-1}) for such a small city that is not crowded with automobiles especially during winter. The city of Ibbeen is very cold in winter, and people usually use wood heaters. These heaters have chimneys outside that release significant amounts of black carbon particles as well as other pollutant gases.

Measurements of black carbon light absorption coefficients in six sites in Irbid city were done during summer 2007. The average value of B_{abs} of all the sites was about 40Mm^{-1}. While the largest value was about 61Mm^{-1} in the city center (Hamasha and arnott 2009).

6. Indoor air pollution by black carbon

Measurements of the black carbon light absorption coefficients (B_{abs}) using the photoacoustic instrument, at wavelength of 870 nm, were done inside different buildings at Yarmouk University/Jordan on summer 2007. The sources of black carbon inside buildings were the human activities and the incoming aerosol from outside that travel with air. Inside these buildings there were no kitchens, so no cooking source of black carbon. As the time of the measurements was summer, there was no source black carbon from heating systems. This measurements show that B_{abs} are low inside buildings with a max value of about 8Mm^{-1} and an average value of 6Mm^{-1} (Hamasha 2008). The building that has the highest level of black carbon is the closest building to very crowded main street. Crowded main street means a lot of automobiles and a lot of aerosol particles that could easily travel by air to the nearest building through the opened doors and windows.

Other indoor measurements of black carbon levels were conducting during the period, 20-26 January 2008 inside living rooms of different houses. During the period of measurements the temperatures were between 0^0C and 10^0C. Ventilation in these living rooms is few minutes during the day, while operation of heaters is about 15 hours. These measurements indicated that the daily indoor black carbon levels were high with average value of about 19 $\mu g/m^3$ (116Mm^{-1}) and max value of about 32 $\mu g/m^3$ (196Mm^{-1}) (Hamasha 2010a). The levels of the BC inside houses in winter were higher than that in summer. The reasons for that are: in summer doors and windows are opened most of the times which leads to a good ventilation, but in winter they are mostly closed to keep the warm inside. This means if there are pollutants species inside it stay inside. In addition, heaters in winter are another big source of pollutant species like black carbon caused by the incomplete combustion.

7. Impacts of serosols on the visibility in Irbid city

Diurnal aerosol visible light absorption and scattering coefficients at the wavelength of 870 nm were obtained using the Photoacoustic Instrument at two sites of Irbid city, urban site and suburban site. The diurnal absorption and scattering patterns showed a strong variability from day to day at both site. During most of the study days, the highest absorption peaks appeared in the early morning, while those of scattering appeared at later times. The earlier absorption peaks could be attributed to the elevated black carbon emissions during the heavy traffic hours whereas the later scattering peaks are attributed to secondary aerosol formed photochemically in the atmosphere. During the sampling period, the suburban site exhibited on the average a higher aerosol scattering and a lower aerosol absorption contribution to the total aerosol visible light extinction and a better visibility than the urban site. The average visibility attributed to aerosol at the urban site dominated by urban scale and regional scale was 44 km, while that of the suburban site was 115 km (Hamasha 2010b).

8. References

Andre, Nel, E., Diaz-Sanchez, David and Li, Ning, (2001). The role of particulate pollutants in pulmonary inflammation and asthma: evidence for the involvement of organic chemicals and oxidative stress. *Current Opinion in Pulmonary Medicine.* 7(1), 20-26.

Arnott, W. P., H. Moosmüller, C. F. Rogers, T. Jin, and R. Bruch. (1999). "Photoacoustic spectrometer for measuring light absorption by aerosols: Instrument description." Atmospheric Environment 33: 2845-2852.

Arnott, W P, Hamasha, K, Moosmüller, H, Sheridan, P J and Ogren, J A, "Towards aerosol light absorption measurements with a 7-wavelength Aethalometer: Evaluation with a photoacoustic instrument and a 3 wavelength nephelometer." Aerosol Science & Technology 39 (2005) 17-29.

Arrhenius, S., "On the Influence of Carbonic Acid in the Air upon the Temperature of the Ground," Philos. Mag., 41, 237-276 (1896).

Asi, R.; Anani, F.; Asswaeir, J. "Studying Air Quality in Alhashemeiah Area/Zarqa". A report prepared by the royal scientific association for the general institution for the protection of the environment, Amman, Jordan, 2001.

Borm,PJ., RP. Schins, and C. Alberecht. (2004)."Inhaled particles and lung cance, part B: Paradigms and Risk Assess. "Int J Cancer;110(1):3-14

Chang, S. G., R. Brodzinsky, L. A. Gundle, and T. Novakov. "Chemical and Catalytic Properties of elemental carbon", In Particulate Carbon: Atmospheric Life Cycle (G. T. Wolff, and R. L. Klimsch, Ends.), pp. 159- 181, Plenum, New York, 1982.

Cooper, C.D., and Alley, F.C., (1996). Air Pollution Control: A Design Approach. Sci. Total Environ. 146/147, 27–34. Boston, MA: PWS Publishers.

Department of Driving and Vehicles Licensing. Amman, Jordan, 2010.

Department of Statistics, Amman, Jordan.
http://www.dos.gov.jo/dos_home_a/main/index.htm, retrieved Dec,8, 2010.

Finlayson-Pitts, B. J. and J. James N. Pitts (2000). Chemistry of the Upper and Lower Atmosphere, Academic press.

Gardiner K., M. van Tongeren, and M. Harrington, "Respiratory Health Effects from Exposure to Carbon Black; Results of the Phase 2 and 3 Cross Sectional Studies in the European Carbon Black Manufacturing Industry," Occup. Environ Med. 2001;58(8)496-503.

Gillies, J.; Abu-Allabanb, M.; Gertler, A; Lowenthal, D: Jennison, B; Goodrich, A. (2008). Enhanced PM2.5 Source Apportionment Using Chemical Mass Balance Receptor Modeling and Scanning Electron Microscopy. JJEES, 1:(1) 1-9.

Gillmour, PS., A. Ziesenis, ER. Morrison, MA. Vickers, EM. Drost, I. Ford, E. Karg, C. Mossa, A. Schroeppel, GA. Ferron, J. Hayder, M. Greaves, W. MacNee, and K. Donaldson, "Pulmonary and Systematic Effects of Short-Term Inhilation Exposure to Ultrafine Carbon Black Particles," Toxicol Appl. Pharmacol. 2004 :195(1): 35-44

Hamasha, K. M., (2008), "Measurements of black carbon levels using photoacoustic technique inside different buildings at Yarmouk University/ Jordan", Jordan Journal of Physics, Vol. 1 No. 2, pp 1- 8.

Hamasha, K. M. and W. P., Arnott, (2009), "Photoacoustic measurements of carbon light absorption coefficients in Irbid city, Jordan, Environ. Monit. Assess, Doi 10.1007/s10661-009-1017-3

Hamasha, K. M., M. S. Almomani, M. Abu-Allaban and W.P.Arnott (2010) "Study of black carbon levels in city centers and industrial centers in Jordan", Jordan Jornal of Physics,volume3,No1, pp1-8.

Hamasha, K. M., (2010a), "Black carbon indoor air pollution from space heating in winter", Abhath al-Yarmouk Basic Sciences and Engineering, Vol. 19 No. 2, pp 47 – 53.

Hamasha, K. M., (2010b), "Visibility Degradation and light Scattering/Absorption Due to Aerosol Particles in Urban/Suburban Atmosphere of Irbid, Jordan", Jordan Journal of Physics, Vol. 3 No. 2

Hamdi, M. R., Bdour A.; Tarawneh, Z. (2008). Diesel Quality in Jordan: Impacts of Vehicular and Industrial Emissions on Urban Air Quality.

Harber, P., H. Muranko, S. Solis, A. Torossian, and B. Merz. (2003). "Effect of carbon black exposure on respiratory function and symptoms." J Occup Environ Med;45(2):144-155.

Horvath, H. (1993). "Atmospheric Light Absorption-A Review." Atmospheric Environment 27A: 293-317.

Horvath, H., "Size Segregated Light Absorption Coefficient of the Atmospheric Aerosol," Atmos. Environ., 29, 875-883 (1995).

Horvath, H. (1997). "Comparison of the light absorption coefficient and carbon measures for remote aerosols: An independent analysis of data from the improve network I and II: Discussion." Atmospheric Environment 13: 2885-2887.

IPCC, Intergovernmental Panel on Climate Change, Contribution of Working Group I to the Second Assessment Report (J.T. Houghton, L. G. Meira Filho, B. A. Callender, N. Harris, A. Kattenberg, and K. Maskell, Eds), Climate Change 1995: The Science of Climate Change, Cambridge Univ. Press, Cambridge, UK, 1996.

Kassel, R., (2003). Dump Dirty Diesel: The health and Air Quality Benefits of Cleaner Diesel Engines. Diesel Retrofit Workshop, Oct 21.

Kirchstetter, T.W., Novakov, T. And Hobbs, P.V. (2004), Evidence that the spectral dependence of light absorption by aerosols is affected by organic carbon. J. Geophysics. Res. 109(D21):D21208. doi:10.1029/2004JD004999.

Künzli, N., R. Kaiser, S. Medina, M. Studnicka, O. Chanel, P. Filliger, M. Herry,F. Horak, V. Puybonnieux-Texier, P. Quénel, J. Schneider, R. Seethaler, JC.Vergnaud, and H. Sommer, (2000). Public-health impact of outdoor and trafficrelated air pollution: a European assessment. The Lancet. 356(9232), 795-801.

Lewis, K., W.P. Arnott, H. Moosmüller, and E. Wold (2008) "Strong spectral variation of biomass smoke light absorption and single scattering albedo observed with a novel dual-wavelength photoacoustic instrument." Journal of Geophysical research, 113, D16203, doi:10.10292007JD009699.

Lindberg, J. D., Douglass, R. E., and Garvey, D. M. (1993). Carbon and the optical properties of the atmospheric dust. Applied Optics, 32, 6077-6081.

Liousse, C., Cachier, H., and Jennings, S. G. (1993). Optical and thermal measurements of black carbon aerosol content in different environments: Variation of the specific attenuation cross section, sigma (σ). Atmospheric Environment, 27A, 1203-1211.

Liu, L., and M. H. Smith. "Urban and Rural Aerosol Particle Optical Properties," Atmos. Environ. , 29, 3293-3301 (1995).

Parent ME, J. Siemiatycki, and L. Fritschi, " Workplace Exposures and Oesophagealcancer," Occup Environ Med 2000; 57:325-34

Pope, C.A., Thun, M.J., Namboodira, M., Dockery, D.W., Evans, J.S., Speizer, F.E., Health Jr., C.W., 1995. Particulate air pollution as a predictor of mortality in a prospective study of US adults. American Journal of Respiratory Critical Care Medicine 151, 669–674.

Pope, C.A., Burnett, R.T., Thun, M.J., Calle, E.E., Krewski, D., Ito, K., and Thurston, G.D. (2002): Lung Cancer, Cardiopulmonary Mortality, and Long-term Exposure to Fine Particulate Air Pollution. Journal of American Medical Association, Vol 287, No. 9, 1132-1141.

Puntoni,R., M. Ceppi, V.Gennaro, D. Ugolini, M. Puntoni, G. La Manna, C. Casella, and D. Merlo. (2004). "Occupational exposure to carbon black and risk cancer." Cancer Causes Control; 15(5):511-6

Rosen, H., Hansen, A. D. A., Gundel, and Novakov, T. (1978). Identification of the optically absorbing component in urban aerosols. Applied Optics, 17, 3859-3861.

Sanjay Rajagopalan; Ohio State University (2008, July 29). Exposure To Bad Air Raises Blood Pressure, Study Shows. ScienceDaily. Retrieved October 9, 2008, from http://www.sciencedaily.com /releases/2008/07/.htm

Takano H., Yanagisawa R, Ichinose T, Sadakane K, Yoshino S, Yoshikawa T, (2002.(Diesel exhaust particles enhance lung injury related to bacterial endotoxin through expression of proinflammatory cytokines, chemokines, and intercellular adhesion molecule-1. Am J Respir Crit Care Med. 165(9),1329–1335.

Walker, P. L., "Chemistry and physics of carbon". vol. 2, Marcel Dekker Inc., NewYork, USA (1966)

WHO, Indoor air pollution, http://www.who.int/indoorair/en/, URL, Dec 8th 2010.

8

$PM_{2.5}$ Source Apportionment Applying Material Balance and Receptor Models in the MAMC

V. Mugica[1], R. Vallesa[1], J. Aguilar[1], J. Figueroa[1] and F. Mugica[2]
[1]*Universidad Autónoma Metropolitana-Azcapotzalco,*
[2]*Universitat Politècnica de Catalunya,*
[1]*Mexico*
[2]*Spain*

1. Introduction

The expansion of urban areas and their surroundings suburbs has been increased in the last decades. Many of these cities, particularly in the developing world, experience an uncontrolled growth and face unprecedented severe air quality problems, due to the high demand of energy, industrial activity and transportation (Molina et al., 2010). Policy makers have the challenge to plan and govern, having as one of their priorities the reduction of air pollution with the aim to protect the health's population, providing at the same time infrastructure and services.

Air quality models or source models are important tools in the environmental assessment since they estimate receptor concentrations from source emissions and meteorological measurements. One of the problems when dispersion models application is considered is that they use estimates of pollutant emissions rates and often rely on meteorological measurements from distant airports and emission rate estimates which stand little resemblance to those applicable to the area under study. As a result of this lack of data, dispersion models cannot be applied in many places or their results have large uncertainties.

On the other hand, receptor models include a range of multivariate analysis methods that use ambient air measurements to infer the source types, locations, and contributions that affect ambient pollutant concentrations. Receptor models use the environmental concentration of the studied pollutants, as well as the composition of the chemical compounds emitted by the different sources to determine the source apportionment (Watson et. al., 2002a). These models are used also to evaluate the efficiency of specific control strategies associated with local programs to improve the air quality and also to estimate the emission inventory uncertainty, since they correlate the pollutants with their sources of emission. This article presents the importance to determine the main sources of $PM_{2.5}$ through the use of receptor models. As a case study, the Principal Component Analysis (PCA), the UNMIX and the Chemical Mass Balance (CMB) models were applied for the source reconciliation of $PM_{2.5}$ in the Metropolitan Area of Mexico City (MAMC). The results obtained by the three models are compared and discussed showing the advantages of the different models.

2. Airborne particles

Suspended particles in the atmosphere can be originated from natural sources, such as wind-driven erosion dust, sea spray, and volcanoes, or from anthropogenic activities such as combustion of fuels (by vehicles, food cooking, wood burning or industries). Airborne PM is composed of inorganic salts, organic material, crustal elements and trace metals and possess a range of morphological, physical, chemical and thermodynamic properties. Airborne particles can change in the atmosphere in size and/or composition through condensation of vapor species or by evaporation, by coagulating with other particles, by chemical reaction, or by activation in the presence of supersaturated water vapor to become cloud and fog droplets (Raes et al., 2000). When particles are emitted directly they are known as primary aerosols, but if particles are formed in the atmosphere as a consequence of physical or chemical interactions among gases, particles and/or water vapor they are called secondary aerosols. Many organic secondary aerosols are formed in the atmosphere by incomplete combustion or by photochemical reactions. The most common inorganic secondary aerosols are the ammonium nitrate and sulfate originated by the reactions among dissolved sulfuric and nitric acids (formed also in the atmosphere by the reaction between water and sulfur oxides and nitrogen oxides respectively, with ammonia gas).

An important characteristic of atmospheric particles is their size distribution, as it strongly affects particle behaviour, may determine their fate in atmospheric systems as well as their deposition in the human respiratory tract, and determines the equipment to be used for sampling. As atmospheric particles are not spherical and have a range of densities, the aerodynamic diameter (diameter of a spherical particle with an equal gravitational settling velocity but a material density diameter of 1 gcm^{-3}) is used to define their size (Mugica & Ortiz, 2006). With this in mind, PM_{10}, $PM_{2.5}$ and PM_1 refer to particles with aerodynamic diameter less or equal to 10 μm, 2.5 μm or 1 μm respectively. They are known also as respirable, fine and ultrafine particles, respectively.

Crustal species from mineral dust, such as Si, Fe, Al, Ca, K, and Mg, are often present in large quantities in the coarse fraction of PM (particles with aerodynamic diameter larger than 2.5 μm but smaller than 10μm). Usually organic aerosols can account for 50% or more of the fine PM, and inorganic secondary aerosols are an important fraction of fine particles.

2.1 Health adverse effects of PM

It has been well established that exposure to PM can cause cardiovascular and respiratory problems, and inclusive increase the premature mortality. For that reason the improvement of human health is the priority objective of air quality programs (McKinley, 2003). Fine and ultrafine particles are poorly captured by the lung macrophages and are able to introduce into the epithelia and the interstitial tissue. Then, the possibility of natural cleaning of lungs is diminished, with an increasing of lung toxicity (Schwartz et. al., 1996). It was observed also, than mortality rate is higher in polluted cities, associating the pollution by fine particles with lung cancer (Dockery et. al., 1993; Maynard & Maynard, 2002), as well as with cardiac and respiratory illness (Samet el al., 2000).Pope et al. (2002) reported tan an increase of 10 μgm^{-3} in the average concentrations of $PM_{2.5}$ implicates the increase of lung cancer and cardiorespiratory risk diseases in 8 and 6% respectively.

The precise chemical and physical properties and toxicological mechanisms by which PM causes adverse health effects are still uncertain. Significant differences exist in the chemical composition and size distribution of PM based on the wide range of sources, meteorological conditions, atmospheric chemistry, diurnal and seasonal factors. PM aerodynamic size is a relevant element when studying PM toxicity due to its variable ability to penetrate the respiratory system; fine particles can reach the deep regions of the lungs, whereas coarse PM may be deposited early within the nasal-pharyngeal passages of the airways. Fine PM potentially may owe the type and intensity of the toxic response to organic compounds, metals and other reactive chemical compounds, since several of those species can promote oxidative stress through the generation of reactive oxygen species (ROS) (Tao et al, 2003; De Vizcaya et al., 2006). ROS can also damage cellular proteins, lipid, membranes, and DNA and PM exposure is also linked to inflammation through the generation of ROS, particularly those PM derived from combustion of fossil fuels (Nel, 2005).

2.2 Adverse effects of PM in the environment

Fine particles and some pollutant gases scatter and absorb light reducing the visibility and generating a haze that has negative effects on the visibility. Visibility can be defined as the maximum distance at which the outline of the farthest target can be recognized against a horizon background (Horvath, 1981). Although absorbing particles remove light transmitted from the target and make it appear darker, they do not scatter much light into the sight path, and they generally have a lower effect on contrast reduction than light-scattering particles. The particles that are most efficient at scattering light are roughly the same size as the wavelength of visible light (about 0.5 μm) (Horvath, 1981).The correlation between fine and ultrafine particles with the decreasing of visibility has been measured in some studies showing that those PM are responsible of the light scattering. (Watson, 2002b).

Other effects of PM and pollutants have been found in materials, damage forests and crops, ecosystems, due to the abrasion, deposition, direct and indirect chemical attack and electrochemical corrosion (Davis & Cornwell, 1998). In addition, visible haze change the earth's radiation balance

3. Receptor models

Receptor models infer contributions from different source types using multivariate measurements taken at one or more receptor locations. Receptor models use ambient concentrations and the abundances of chemical components in source emissions to quantify source contributions. They are based on the same scientific principles as source models, but they are explanatory rather than predictive of source contributions. (Watson et al, 2002a).While source models need spatial and temporal resolution and accurate emissions rates, receptor models need only a seasonal or annual average, area wide inventory to identify potential source categories. Contributions are quantified from chemically distinct source-types rather than from individual emitters. Sources with similar chemical and physical properties cannot be distinguished from each other (e.g., it is quite difficult to differentiate the diesel exhaust emissions of heavy, cars, trucks, stationary generators and

engines or off-road equipment, thus they can be grouped in one diesel exhaust category). Nevertheless, with appropriate chemical analysis of organic and inorganic compounds of detailed profiles, more chemical markers from sources could be detected and the separation in sub-categories become possible.
Receptor models are based on the chemical mass balance equation and the main assumption is that composition of PM remains constant and chemical species do not react with each other. The source apportionment is accomplished by solving the mass balance equations expressing the measured ambient elemental concentrations as the sum of products between the source contributions and the elemental abundances in the source emissions, e.g. the source profiles. There are different receptor models which differ in the mathematical approaches that they have to solve the mass balance equations, as well as in the different degrees of knowledge about source profiles they need for source apportionment analysis. Receptor models are not statistics methods, and maybe the misunderstanding partially arises to the fact that much of the receptor modeling mathematics is also used to determine and test statistical associations in other scientific fields (Watson & Chow, 2004).
Among the receptor models, Multiple Linear Regression have been widely used from more than three decades due to they have the advantage to be implemented by many statistical packages; identification of markers is required. The application of Enrichment factor is one of the first methods used to identify presence or absence of anthropogenic sources or processes responsible of the different atmospheric chemical species. Sometimes the reference geological material could be different to the sampling site. Multivariate models based in eigenvector analysis but using different normalization and rotation schemes have also been applied the last two decades; the most important are: Principal component analysis (PCA), Empirical orthogonal functions (EOF) and Factor Analysis (FA).The Positive Matrix Factorization (PMF) model was developed by Paatero & Tapper (1993) as a new approach to factor analysis, where the principal components explaining the variance of the speciated data are extracted and then interpreted as possible sources. The CMB model has been widely used to determine source contribution estimates for PM_{10} and $PM_{2.5}$. This model calculates the source contributions by determining the best combination of source profiles needed to simulate the chemical composition of the ambient data. The model is able to estimate the source reconciliation for every day. Table 1 shows most of the common receptor models used in air quality studies to develop pollution control strategies.
Watson and Chow (2004) specify the following qualities which are desirable in any data base of source and receptor measurements: 1) a full range of chemical species in specified size fractions (for solid-phase pollutants); 2) specification of operating parameters (for source measurements), locations and sampling periods (for source and receptor measurements);3) documentation of sampling and analysis methods; 4) results of quality control activities and quality audits; 5) precision and accuracy estimates for each measurement; 6) data validation summaries and flags; and 7) availability in well-documented computerized formats.
Source and receptor models are complementary rather than competitive. Each has strengths and weaknesses that compensate for the other. Both types of models can and should be used in an air quality source assessment on outdoor and indoor air.

Receptor Model	Description
Enrichment Factors (EF)	The ratios of atmospheric concentrations of elements to a reference element are compared to the same ratios in geological or marine material. Differences are explained in terms of anthropogenic sources. It is more useful for identification of anthropogenic processes than for quantification.
Multiple linear regression (MLR)	Mass of chemical compounds is expressed as the linear sum of regression coefficients. The regression coefficients represent the inverse of the chemical abundance of the marker species in the source emissions. They can easy implemented in statistic packages, but limited to sources with marker species. The product of the regression coefficient and the marker concentration for a specific sample is the tracer solution to the mass balance that yields the source apportionment. Requires large data set.
Eigenvector multivariate models: Principal component analysis(PCA), Empirical orthogonal functions (EOF), Factor Analysis (FA)	Temporal correlations are calculated from a time series of chemical concentrations at one or more locations. These are eigenvector analysis multivariate models which can confirm and identify unrecognized source types. Eigenvectors of this correlation matrix are determined and a subset is rotated to maximize and minimize correlations of each factor with each measured species. The factors are interpreted as source profiles by comparison of factor loadings with source measurements. Source profiles from direct measurements are needed to interpret these eigenvectors. Easy implementation in statistic packages, but limited to sources with marker species. Requires large data set.
UNMIX Form of Factor Analysis	The UNMIX model "unmixes" the concentrations of chemical species measured in the ambient air to identify the contributing sources. Chemical profiles of the sources are not required, but instead are generated internally from the ambient data by UNMIX, using a mathematical formulation based on a form of factor analysis. UNMIX uses "edge detection" in a multidimensional space. The edges represent the samples that characterize the source. It can be run feasibly and easily on some statistical software. Requires large data set.
Positive Matrix Factorization [PMF]	The PMF technique is a form of factor analysis where the underlying co-variability of many variables is described by a smaller set of factors (PM sources) to which the original variables are related. The PMF assumption is that the concentration of specie in a site can be explained by the source matrix and contribution matrix. Both matrixes are obtained by an iterative minimization algorithm. A restriction of no-negativity ensures positive abundances and contributions. The main problem with PCA is that it does not provide a unique solution.
ChemicalMass Balance (CMB)	Ambient chemical concentrations are expressed as the sum of products of species abundances and source contributions and the equations are solved for the source contributions. Ambient concentrations and source profiles are supplied as input.The chemical characterization of the possible emission sources together with an estimation of the uncertainties for the species concentrations, are used as input for the CMB model. The main drawback of this model is that the accuracy of the source apportionment depends on the representativeness of the selected sources for the emission types in the area.

Table 1. Most used Receptor Models in Air Quality Studies

4. Sampling and chemical analysis

The Metropolitan Area of Mexico City (MAMC) is located in an elevated basin surrounded by mountains which do not favour the dispersion of air pollutants, especially during the cold season when frequent thermic inversions are present. The MAMC megacity has nearly 20 million inhabitants, more than 4 million of vehicles and around 35,000 industries. A total of 132 aerosol samples were collected from January 2002 to December 2003, every six days, at the Azcapotzalco Campus of the Metropolitan University, located in an industrial-residential area in the Northern. In addition, other three sites studied in previous campaigns (Chow et al, 2002) were sampled in March 2003 during ten days in order to determine the spatial variation. These sites were: 1) La Merced, located in the downtown with high commercial activity and high traffic activity; 2) Xalostoc, located at the Northeast is an industrial district surrounded for very important avenues with heavy traffic, and 3) Pedregal, is a residential neighborhood located at the Southwest.

Samples were collected onto Teflon and quartz 47 mm filters using PM_{10} and $PM_{2.5}$ Minivol samplers (Airmetrics, Eugene, OR). Teflon-membrane filters (Gelman Scientific, Ann Arbor, MI) with 2 mm pore size collected samples for mass and subsequent elemental analysis, whereas precalcinated Quartz fiber filters (Pallflex, Products Corp.,Putnam, CT) collected samples for water-soluble anions (Cl^-, NO_3^-, SO_4^{2-}) and cations (Na^+, K^+, NH_4^+), organic carbon and elemental carbon analyses. Filters were equilibrated for two weeks in a relative humidity (25–35%) and temperature (20±0.5°C) controlled environment before gravimetric analysis to minimize particle volatilization. Filters were weighed before and after sampling with a Mettler Toledo (MT-5) microbalance. The balance sensitivity is 0.001 mg. Subsequently, the filters were stored in a freezer until aerosol sampling and chemical analyses. Quartz filters were split into two using plastic scissors: the first part was for ion analysis and the second one for the quantification of organic and elemental carbon.

Soluble ions were extracted ultrasonically (Branson bath, USA) with Milli-Q deionized water during 20 min. Sulfate (SO_4^{2-}), water-soluble ammonium (NH_4^+), nitrate (NO_3^-), water-soluble sodium (Na^+), and potassium (K^+), were quantified by ion chromatography, with a Perkin Elmer-Alltech 550 instrument fitted with a conductivity detector), using specific anion and cation Alltech columns. Organic and elemental carbon was determined by an automated thermal-optical transmittance (TOT) carbon analyzer, Sunset Lab, USA, using method 5040 (NIOSH protocol) (Birch and Cary, 1996).

Inductively Coupled Plasma-Atomic Emission Spectrometry, ICP-AES, from Atom Advantage Thermo Jarrel Ash, was used to analyze the elemental components of the PM collected on the teflon filters. Filters were digested in a microwave oven (OI-Analytical, USA) using high-pressure Teflon digestion vessels with 2 ml of HF, 1 ml HCl and 2 ml HNO_3 (67%). The average filter blank value was used as a background subtraction for each sampled filter. 20 mg extractions of a well-characterized urban dust (SRM 1649a standard reference material NIST), field samples and filter blanks were handled and analyzed under the same procedure as filters with air samples. Quality audits of the sample flow rates were conducted each week of the study period. Data were submitted to three levels of data validation (Watson et al., 2002a.), so intercomparison and performance tests were carried out between CICATA-Altamira and UAM-Azcapotzalco. For the purposes of calculating weight fractions, elements were normalized for oxygenated species as described by Mc Donald (2000).

5. Mass of $PM_{2.5}$

Table 2 shows the basic statistic of the total mass of $PM_{2.5}$ in the four sampling sites. Traditionally (GDF, 2008), Xalostoc is the most polluted site due to the high industrial and vehicular activities. Winds use to blow from Northeast to Southwest, and although Pedregal is the less polluted place by PM, usually exceed the ozone standard.

Site	N	Mean	Max	Min
Azcapotzalco (N)	132 Two whole years 2002-2003	56.9±13.9	93.1	34.5
Merced (Center)	10 March 2003	58.1±19.3	74.2	39.6
Pedregal (Southwest)	10 March 2003	26.8±11.7	47.2	21.6
Xalostoc (Northeast)	10 March 2003	69.2±23.4	105.7	47.2

Table 2. Levels of $PM_{2.5}$ in the MAMC

For CMB model application is necessary to select fitting species, as well as the adequate sources profiles, thus, in this study the strategy was to use the Factor Analysis Models (PCA) and UNMIX to identify the main emission sources and marker elements, and subsequently apply the CMB model with speciated source profiles for a more robust source apportionment.

6. Factor analysis: principal component analysis

PCA model belongs to the category of factor analysis (FA) techniques, i.e. it is a multivariate method used to study the correlations among the measured elemental concentrations at the receptor. With this method, the principal components explaining the variance of the chemical species data, and then they interpreted as possible sources. Assuming a linear relationship between the total mass concentration and the contributions of each specie, PCA factors the data in several steps. First, the chemical composition data are transformed into a dimensionless standardized form

$$Zij = \frac{Cij - Cj}{\sigma j} \tag{1}$$

where i=1, …, n samples; j=1, …, m elements; Cij is the concentration of element j in sample i; and Cj and σj are the arithmetic mean concentration and the standard deviation for element j, respectively. The PCA model is expressed as:

$$Zij = \sum_{k=1}^{p} gik\ hkj \tag{2}$$

where k=1,p sources, and gik and hkj are the factor loadings and the factor scores, respectively. This equation is solved by eigenvector decomposition. Varimax rotation is often used to redistribute the variance and provide a more interpretable structure to the factors. PCA not provide a unique solution mainly because of its simple approach to factor analysis. Despite this drawback, known as rotational ambiguity, PCA has been applied as a tool for source apportionment in many air quality studies (Karar and Gupta, 2007).

With the chemical data obtained from the chemical analysis of samples, a data base was prepared for the PCA. The ambient data were normalized with media=0 and standard deviation = 1, to reduce the excessive influence of the species with mass. The statistic software SPSS v.12 for windows was used to obtain the number of factors, the mass matrix and the Varimax Rotation. The selection of chemical species was performed to get the better fittings. Maatlab 6.5 package was used to execute the matrix operations. Matlab estimated the not scaled contributions for further lineal regression to convert them in mass unities. Finally the mass balance matrix was cleared to determine the profiles. Model performance was evaluated with the mass percentage and the linear regression coefficient R^2.

PCA resulted to be very useful to determine the potentially contribution of source types, including those with small data set (as was de case of Merced, Pedregal and Xalostoc with only ten samples). The fitting species were: sulfate, ammonium, organic carbon, elemental carbon, aluminum, silicon, sulfur, calcium, and iron. Table 3 shows the factor loadings normalized with the VARIMAX rotation, which maximizes the variances of the squared normalized factor loadings across variables for each factor, thus making the interpretation easier. The final solution of PCA reported three values higher than 1, suggesting three main factors (sources) in the four sites: Vehicular, soil and secondary aerosols. These three sources accumulated more than the 90% of the system variance.

The markers related to the first factor associated with "soil" that explained 34% of variance were Al, Si, Ca, and Fe, which are crustal elements. The markers associated to the second factor "secondary aerosols" are SO_4^{2-} and NH_4^+ related with ammonium sulfate, a secondary aerosol which can be formed in the atmosphere. The third factor "vehicular", is mainly represented by organic and elemental carbon.

	Rotated Component Matrix*		
	Component		
	Soil	**Sec Aerosols**	**Vehicle**
SO4	0.005	0.994	0.042
NH4	-0.123	0.963	0.190
OC	0.412	0.197	0.830
EC	-0.004	0.067	0.964
AL	0.982	-0.094	0.065
SI	0.988	-0.048	0.101
SU	0.000	0.990	0.055
CA	0.984	0.008	0.089
FE	0.964	-0.012	0.173
% Total Variance	34.210	28.541	27.453
% AccumulatedVariance	34.210	62.750	90.204
Extraction Method: Principal Component Analysis.			
Rotation Method: Varimax with Kaiser Normalization.			
* Rotation converged in 4 iterations.			

Table 3. PCA final solution in Azcapotzalco site

Figure 1 shows graphically the apportionment of $PM_{2.5}$ considering the three sources mentioned above, obtained with PCA for the different sites. In all cases the most important contributor to $PM_{2.5}$ was the mobile sources with more than 45% of the total mass, followed by secondary aerosols. Pedregal had the lowest contribution of soil. It is important to highlight that the results from Merced, Pedregal and Xalostoc represent only the apportionment of PM measured in March 2003 that is part of the warm dry season in the MAMC, whereas the measurements in Azcapotzalco were carried out during two years, so these results are the average of measurements done in the dry and rainy seasons.

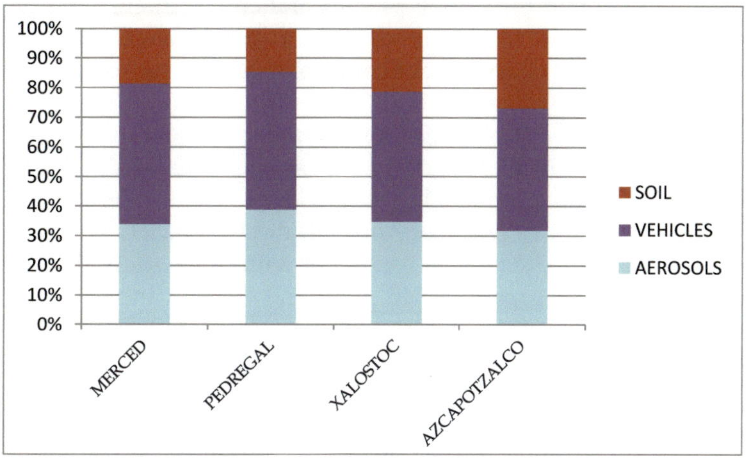

Fig. 1. Source apportionment results from PCA at the four sites

7. UNMIX model

The UNMIX model is a refined multivariate receptor model that uses a new transformation method based on the self-modeling curve resolution technique toderive meaningful factors. UNMIX incorporates user-specified non-negativity constraints and edge-finding algorithms to derive a physically reasonable apportionment of source contributions (Henry, 2001; Poirot et al., 2001). The edges are constant ratios among chemical components that are detected in multi-dimensional space. The edges detected by this model are translated into source profile abundances.This model does not require a previous knowledge about emission sources, although it is necessary a big number of measurements to estimate the different factors, as well as the magnitude of their contributions (Chen et al., 2002; Hellén et al. 2003). UNMIX try to solve the problem of the chemical species mixture with the assumption that the data of each sample has a lineal combination of an unknown number of sources which contributes with an unknown mass concentration to the total mass. Another assumption is that all values are positive (> 0).

UNMIX uses the singular value decomposition (SVD) method to estimate the source number by reducing the dimensionality of data space m to p (Henry, 2001). The UNMIX model can be expressed as

$$Cij = \sum_{i=1}^{p}\left(\sum_{k=1}^{p} Uik\ Dkl\right)Vlj + \varepsilon ij \tag{3}$$

Where *U*, *D*, and *V* are n×p, p×pdiagonal, and p×mmatrices, respectively; and εij is the error term consisting of all the variability in *Cij* not accounted for by the first *p* principal components.
Geometrical concepts of self-modeling curve resolution are used to ensure that the results obey (to within error) the nonnegative constraints on source compositions and contributions.The data are then projected to a plane perpendicular to the first axis of p-dimensional space. The edges represent the samples that characterize the source. Such edges in point sets are then used to calculate the vertices, which are used with the matrices decomposed by SVD to obtain the source profiles and contributions. The stand-alone EPA UNMIX version 5.0 was used in this study. For a given selection of species, UNMIX estimates the number of sources, the source compositions, and source contributions to each sample.
UNMIX has been applied to several studies for source apportionment of particulate matter (Chen et al., 2002; Song et al. 2006). One of the first applications was performed by Lewis et al. (2003) in a three years data set in Phoenix, Arizona. The model estimated the source profiles for five source categories (gasoline-vehicles, diesel-vehicles, secondary sulfates, soil and wood burning), and the results were consistent with other study that applied the PMF model. Maykut et al. (2003) compared CMB, PMF and UNMIX in Seattle to determine the $PM_{2.5}$ sources with the coincidence of three sources: wood burning, mobile sources and secondary aerosols. Larsen y Baker (2003) applied UNMIX and PMF models to determine the origin of polycyclic aromatic hydrocarbons in Baltimore.
When UNMIX model was applied to the MAMC samples, the same three sources obtained in the PCA were clearly identified. Table 4 shows the output of the model for Azcapotzalco site, where not only the total mass contributions are displayed, but also the contribution of the most abundant species to the total mass of $PM_{2.5}$.

Calculated Composition ($\mu g \cdot m^{-3}$) - **Azcapotzalco**

	Soil	Vehicle	Sec. Aerosol
Total	13.5358 ± 3.3193	21.0042 ± 4.4152	18.8298 ± 3.2149
SO4	0.0014 ± 0.0386	0.0754 ± 0.0186	0.3383 ± 0.0367
NH4	-0.0202 ± 0.0227	0.0555 ± 0.0093	0.1457 ± 0.0136
OC	0.2382 ± 0.0345	0.3856 ± 0.0279	0.2075 ± 0.0271
EC	0.0704 ± 0.0229	0.2492 ± 0.0118	0.0861 ± 0.0130
Si	0.1110 ± 0.0229	0.0087 ± 0.0049	0.0072 ± 0.0049
S	0.0006 ± 0.0149	0.0314 ± 0.0071	0.1296 ± 0.0140
Ca	0.0495 ± 0.0106	0.0014 ± 0.0022	0.0033 ± 0.0020
Fe	0.0354 ± 0.0065	0.0062 ± 0.0018	0.0044 ± 0.0019
Br	0.0001 ± 0.0001	0.0007 ± 0.0001	0.0003 ± 0.0001

Table 4. Output of UNMIX model for Azcapotzalco site.

Figure 2 shows the contribution of the three mentioned sources to the total mass of $PM_{2.5}$ at the three sites. It is possible to appreciate some difference of the apportionment yield by PCA. UNMIX apportioned a higher quantity due to mobile sources than PCA.

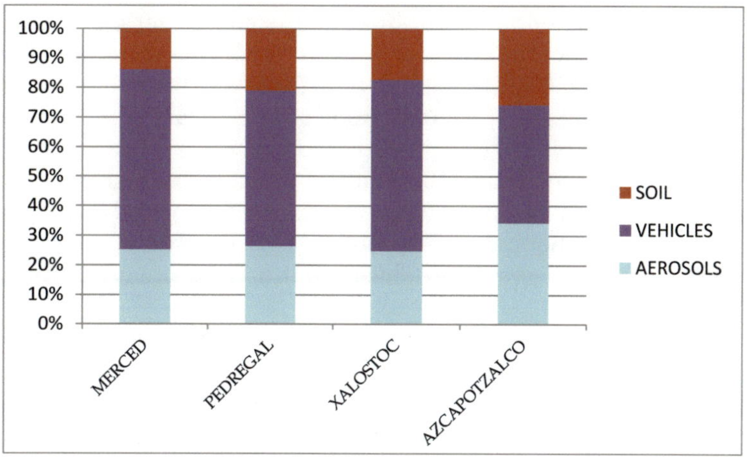

Fig. 2. Source apportionment results from UNMIX at the four sites

8. Chemical Mass Balance receptor model (CMB)

The CMB model is similar to a tracer model, in which a specific compound, that is associated with a particular type of source, is used to identify and quantify the contributions of each source. The model uses the complete model of chemical emissions of a category of specific source to determine its contribution. For the application of the CMB model is necessary to have the databases of the ambient and the source emission profiles. The first one is obtained by collecting samples of ambient air at different locations with the purpose of obtaining information of the population that is investigated. When taking the samples it is expected that they are representative and reflect the properties of the site. On the other hand, source profiles are obtained directly inside the source or as near as possible. The quality of the data will depend on the number of taken samples, used devices, the place and time of the sampling. Equation 4 is the fundamental base of the receptor model, this expresses the relationship between the concentrations of the chemical species measured in the receptor with those emitted in the source.

$$Ci = \sum_{j=1}^{p} Fij \cdot Sj \tag{4}$$

Where

Ci = Ambient concentration of the species "i" measured in the receptor site

p = Number of sources that contribute $j = 1, 2, \ldots j$

Fij = Fraction of the emissions of the species "i" starting from the source "j"

Sj = Impact to the receptor (calculated contribution) of the source "j"

These equations are solved for the source contributions. Several different solution methods have been applied, but the effective variance least squares estimation method is most commonly used because it incorporates precision estimates for all of the input data into the solution and propagates these errors to the model outputs

The CMB model provided values for several performance measures to evaluate the solution. These measured values included chi-square, the weighted sum of the squared differences between calculated and measured fitting species concentrations divided by the effective

variance and degrees of freedom (ideally chi-square would be zero, but values up to 4 are acceptable). R^2 is the fraction of the variance in the receptor concentrations. R^2 ranges from 0 to 1, when R^2 is less than 0.8 the source contribution estimated did not explain the observations clearly with the fitting source profiles. The calculated mass should be in the range of 100 ± 20 (Watson et al., 1991).
The chemical mass balance model, CMB, which is based upon regression analysis of PM chemical composition, is the fundamental receptor model to find the most appropriate combination of source apportionment. This model has been used in other countries (Chow and Watson, 2002) with the aim to establish control measurements for the main PM contributors.
In this study, each of the daily ambient concentrations of $PM_{2.5}$ and elemental components were submitted as input to the CMB model (Henry, 1997). The source profiles for fugitive dust (Vega et al., 2001), food cooking (Mugica et al., 2001) and combustion source profiles developed for Mexico City (Mugica et al., 2008) were used also as input. The most common inorganic components were included as fitting species in the CMB model as well as organic and elemental carbon (OC and EC). In order to account for secondary aerosol contributions to $PM_{2.5}$, ammonium sulfate, and ammonium nitrate profiles were introduced in the analysis. Each result was evaluated by using the regression statistical parameters available for each CMB output.
CMB model could identify six different sources: soil, gasoline vehicles exhaust, diesel vehicles exhaust, food cooking, ammonium sulfate and ammonium nitrate. This means that CMB could separate two different types of vehicles (e.g. those which use gasoline and those that use diesel), as well as the two types of inorganic secondary aerosols. Table 5 displays the average of the statistical parameters of the model in the $PM_{2.5}$ source reconciliation in the four sites. In general, the parameters of R^2, Chi^2 and percentage of mass were in the acceptable interval. The values of R^2 fluctuated between 0.92 and 0.96. Likewise, the values of Chi^2 were smaller than 4. The percentages of mass calculated when applying the model varied from 88.1 to 104.5, with an average of 93.5%.

Site	R^2	CHi^2	%Mass	Meas. Conc. [μgm-3]	Calc. Conc. [μgm-3]
Azcapotzalco	0.95	0.95	95.7	56.92	54.17
Merced	0.96	2.34	94.3	51.25	48.04
Pedregal	0.96	3.49	94.6	26.32	25.74
Xalostoc	0.97	2.86	91.6	68.32	70.74

Table 5. Average statistical parameters of the CMB model applied to $PM_{2.5}$

The estimated contributions in μgm^{-3} by CMB model vary considerably from one day to another in every site, although in all the cases the major emission sources were the vehicles (sum of diesel plus gasoline exhaust) with contributions between 50 and 66%, followed by aerosols (ammonium sulfate plus ammonium nitrate) and soil (Figure 3).
Figure 4 shows the source contribution of the six sources separated by CMB model in some selected samples of the Azcapotzalco site. In this graphic the separation between gasoline exhaust (with around 28% of the total of $PM_{2.5}$) and diesel exhaust (with 26%) is visible. The new source due to food cooking was also identified with contributions up to 10%, and it was possible to detect that ammonium sulfate concentration is more than four times greater than ammonium nitrate.

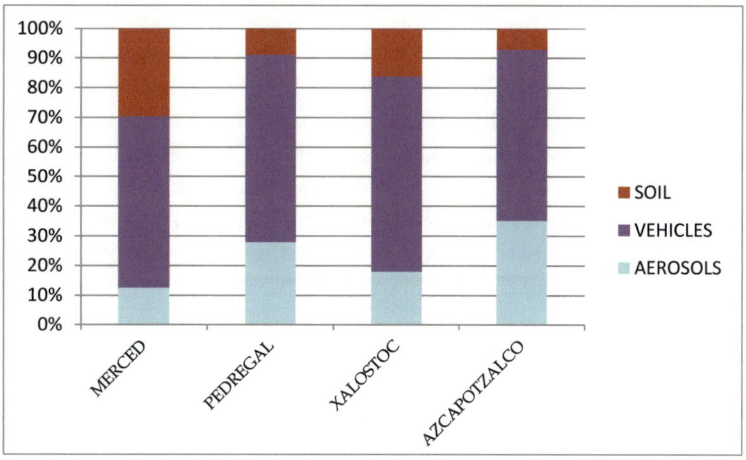

Fig. 3. Source apportionment from CMB at the four sites.

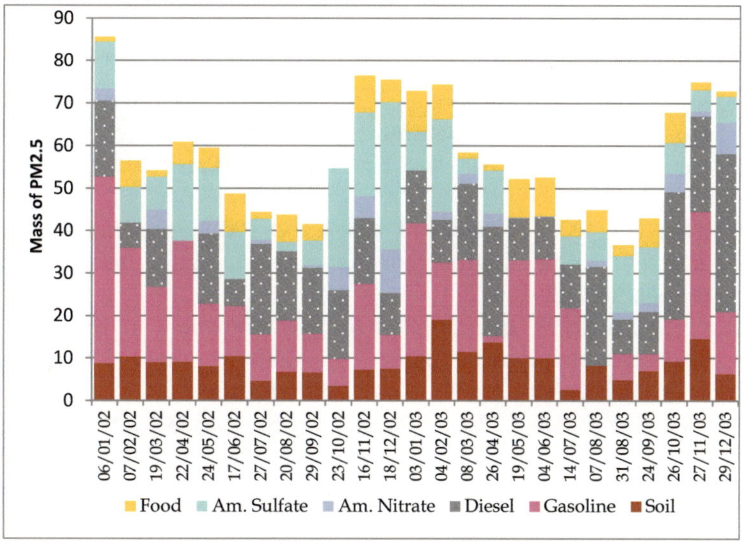

Fig. 4. Source apportionment of PM$_{2.5}$ (μgm^{-3}) in Azcapotzalco

Mann-Whitney U test was used to determine differences among the results obtained for the three models. The findings showed that the contributions of soil, vehicles and secondary aerosols estimated by the three models are statistically equivalent, with ($p > 0.05$). CMB fully apportions receptor concentrations to chemically distinct source-types depending upon the source profile database, while UNMIX and PMF internally generate source profiles from the ambient data.

9. Conclusion

In this paper, the principles of different receptor models were revised and the performances of CMB, PMF and PCA were evaluated in their application to PM$_{2.5}$ samples from different sites of the MAMC. The use of several types of models helps to identify and quantify model inaccuracies and focus further investigation on the areas of greatest uncertainty. PCA and UNMIX apportioned one single source of mobile sources, but the CMB model was able to distinguish between the two main sources of mobile sources (gasoline and diesel exhaust) in the four sites. In addition CMB could separate the two different types of secondary aerosols. Thus, in this study was demonstrated the capability of CMB model to better apportion on PM mass. Nevertheless the use of PCA and UNMIX was fundamental to identify the main sources as well as the marker elements which were further used during the CMB application as fitting species. The use of three models improve the source reconciliation and allows a better knowledge of the suspended $PM_{2.5}$ in the MAMC.

10. Acknowledgements

The authors wish to express their thanks for the chemical analysis to the Applied Chemistry laboratories at the Metropolitan University-Azcapotzalco, and CICATA/IPN. V. Mugica and J. Aguilar gratefully acknowledge the SNI for the distinction of her membership and the stipend received.

11. References

Chen, L.W.A., Doddridge, B.G.; Dickerson, R.R.; Chow, J.C.; Henry, R.C. 2002. Origins of Fine Aerosol Mass in the Baltimore–Washington Corridor: Implications From Observation, Factor Analysis, and Ensemble Air Parcel Back Trajectories; Atmos. Environ. 36, 4541-4554.

Chow, J.C., Watson, J.G., 2002. Review of $PM_{2.5}$ and PM_{10} apportionment for fossil fuel combustion and other sources by the chemical mass balance receptor model. Energy& Fuels 16, 222–260.

Davis ML, Cornwell DA. 1998. Introduction to environmental engineering. McGrawHill, Singapore. e in atmospheric aerosols. Atmos. Environment. 38: 1387-1388.

De Vizcaya-Ruiz A., Gutiérrez-Castillo M.E., Uribe-Ramirez M., Cebrián M.E., Mugica-Alvarez V., Sepúlveda J., Rosas I., Salinas E., Garcia-Cuéllar C.M., Martínez F., Alfaro-Moreno E., Torres-Flores V., Osornio-Vargas A., Sioutas C., Fine P.M., Singh M., Geller M.D., Kuhn T., Eiguren-Fernandez A., Miguel A., Schiestl R., Reliene R., Froines J. 2006. Characterization and in vitro biological effects of Concentrated particulate matter from Mexico City. Atmospheric Environment. 40, 2: 583-592.

Dockery DW, Pope CA III, Xu X, Spengler JD, Ware JH, Fay ME, Ferris Jr BG, Speizer FE. 1993. An association between air pollution and mortality in six US cities. The New England Journal of Medice 329: 1753-1759.

Hellén H, Hakola H, Laurila T. 2003. Determination of source contribution of NMHC in Helsinki (60ºN, 25ºE) using chemical mass balance and the UNMIX Multivariate receptor models. Atmospheric Environment. 37: 1413-1424.

Henry, R.C., Willis, R.D., 1997. Chemical mass balance receptor model version 8 (CMB8) user´s manual. Prepared for US Environmental Protection Agency, Research Triangle Park, NC, by Desert Research Institute, Reno, NV.

Henry, R. C. UNMIX Version 2.4 Manual; U.S. Environmental Protection Agency: Research Triangle Park, NC. 2001.

Karar, K., Gupta, A.K., 2007. Source apportionment of PM10 at residential and industrial sites of an urban region of Kolkata, India. Atmospheric Research 84, 30–41.

Larsen RK III, Baker JE. 2003. Source apportionment of polycyclic aromatic hydrocarbons in the urban atmosphere: a comparison of three methods. Environ. Sci. Technol. 37: 1873-1881.

Maynard AD, Maynard RL. 2002. A derived association between ambient aerosol surface area and excess mortality using historic time series data. Atmospheric Environment 36: 5561-5567.

McKinley G., Zuk M, Hojer M, Avalos M, González I, Hernández M, Iniestra R, Laguna I, Martínez MA, Osnaya P, Reynales LM, Valdés R, Martínez J. 2003. The Local Benefits of Global Air Pollution Control in Mexico City: Final Report of the Second Phase of the IntegratedEnvironmentalStrategies Program in Mexico. IntitutoNacional de Ecología - InstitutoNacional de SaludPública, México.

Maykut NN, Lentas J, Kim E, Larson TV. 2003. Source apportionment of $PM_{2.5}$ at an urban IMPROVE site in Seattle, Washington. Environ. Sci. Technol. 37: 5135-5142.

Mc Donald J., Zielinska B., Fujita E., Sagebiel J., Chow J. and Watson J. (2000). Fine particle and gaseous emission rates from residential wood combustion. Environ. Sci. Technol. 34, 2080-2091.

Mugica V., Vega E., Chow J., Reyes E., Sanchez G., Arriaga J., Egami R., Watson J. 2001. Speciated non-methane organic compounds emissions from food cooking in Mexico. Atmospheric Environment 35, 1729-1734.

Mugica V. & Ortiz E. 2005. Elemental composition of airborneparticles Analytical techniques and application in decision-making for air quality management in Applications of Analytical Chemistry in Environmental Research, 219-261. ISBN: 81-308-0057-8 M. Palomar (Ed). Research Signpost. 37/661 (2). India.

Mugica V., Mugica F., Torres M., Figueroa J. 2008. $PM_{2.5}$ Emission Elemental Composition in the Metropolitan Area of Mexico City. From diverse Combustion Sources in the Metropolitan Area of Mexico City. The Scientific World.8: 275-286.

Nel A. 2005. Atmosphere. Air pollution-relatedillness: effects of particles. Science. 308 (5723): 804-6.

Paatero, P.&Tapper, U., 1993. Analysis of different modes of factor analysisas least squares fit problems. Chemometrics and Intelligent LaboratorySystems 18, 183–194.

Poirot, R.L.; Wishinski, P.R.; Hopke, P.K.; Polissar, A.V. 2001. Comparative Application of Multiple ReceptorMethods to IdentifyAerosol Sources in Northern Vermont; Environ. Sci. Technol. 35, 4622-4636.

Pope III C.A., Burnett R.T., Thun M.J., Calle E.E., Krewski D., Ito K., Thurston G.D. 2002 Lung cancer, cardiopulmonarymortality, and long-termexposure to fine particulate air pollution. JAMA. 287: 1132-41.

Raes F., Van Dingenen R, Vignati E., Wilson J, Putaud JP, Seinfeld JH, Adams P. 2000. Formation and cycling of aerosols in the global troposphere. Atmos. Environ. 34: 4215-4240.

Song Y., Xie S., Zhang Y., Zeng L. Salmon L., Zheng M. 2006. Source apportionment of $PM_{2.5}$ in Beijing using principal component analysis/absolute principal component scores and UNMIX.The Science of the Total Environment. 15: 372(1):278-86.

Samet JM, Dominici F, Curriero FC, Coursac I, Séller SL. 2000. Fine particle air pollution and mortality in 20 US cities, 1987-1994. The New England Journal of Medicine 343: 1742-1749.

Schwartz J, Dockery DW, Neas LM. 1996. Is daily mortality associated specifically with fine particles? J. Air & Waste Manage. Assoc. 46: 927-939.

Tao F, Gonzalez-Flecha B, Kobzik L. 2003. Reactiveoxygenspecies in pulmonary inflammation by ambient particulates. Free Radic.Biol. Med. 35:327-40.

Vega E., Mugica V., Reyes E., Sánchez G., Chow J., Watson J. 2001.Chemical Composition of Fugitive Dust Emitters in Mexico City. Atmos. Environ., 35, 23, pp 4033-4039.

Watson, J., Chow, J., Pace, T., 1991. Chemical mass balance. In: Hopke, P.K. (Ed.),ReceptorModeling for Air Quality Management. Elsevier Press, New York, NY,pp. 83–116.

Watson JG, Zhu T, Chow JC, Engelbrecht J, Fujita EM, Wilson WE. 2002a. Receptor modeling application framework for particle source apportionment. Chemosphere. 49:1093-1136.

Watson J. 2002b. Visibility: Science and regulation. J. Air Waste Manag Assoc. 52 : 628-713.

Watson J. & Chow J. 2004. Receptor Models for Air Quality Management.EM. October:15-24.

Part 3

Air Pollution in Office and Public Transport Vehicles

Emission and Formation of Fine Particles from Hardcopy Devices: the Cause of Indoor Air Pollution

David D. Massey[1]* and Ajay Taneja[1,2]
[1]*School of Chemical Sciences, Department of Chemistry, St John's College,*
[2]*Department of Chemistry, Dr. B.R. Ambedkar University,*
India

1. Introduction

The last few decades have seen major changes in the home and work environments. The economies of the Indian and other industrialized nations have shifted from manufacturing towards services that engage information technologies. Advances in information technology have increased the quantity and transformed the nature of equipment used in proximity to office worker due to which electronic media used for entertainment, telecommunications and data processing have become widespread in daily life. Typical examples are television sets, video recorders, hi-fi systems, and computers with their peripherals such as monitors and printers, scanners and copiers. Tabletop printers serve individual users in their workspace or home, or clusters of users in an office suite. Scanning machines and photocopiers are prevalent in office environments (Newburger, 2001). In addition, the use of notebook computers spanning both work and non work environments is on the rise. These devices are predominantly made of polymeric components and materials which can contain not only additives, such as flame retardants and plasticizers (Wensing et al., 2005), but also chemical residues from production processing aids (Wensing et al., 2002).

There is growing concern about the levels of potentially harmful pollutants that may be emitted from office equipment and for which either toxicological effects or potentially significant exposures have been described in the literature. Office equipment has been found to be a source of ozone, particulate matter, volatile organic compounds (VOCs) and semi volatile organic compounds (SVOCs). VOC, SVOC and particles can also be emitted by the paper processed during printing and copying (Wolkoff et al., 1993). Many studies have investigated the health effects of photocopier toner dust and concluded that siderosilicosis and sarcoidosis-like pulmonary diseases are associated with human exposure to photocopier toner dust (Armbruster et al., 1996). Black and Worthan (1999) have described the VOC/ TVOC, particle and ozone emissions of laser printers, dryprocess photocopiers and personal computers. Wolkoff (1999) study dealt with photocopiers and indoor air pollution. Later on Lee et al. (2001) characterized VOC, ozone and PM_{10} emissions from office equipment. Today discussion focuses in particular on particle release from hardcopy devices, printers and photocopiers and its impact on the health of office workers (Roller, 2006). Recent advances in measurement techniques have enabled researchers to measure the

ultrafine particles of nanoscale range and have provided evidence that the smaller particles typically emitted from sources such as internal combustion engines may have more severe impact on the human respiratory system than the bigger particles (Newburger, 2001).
Ozone and particulate matter have been associated with occupational symptoms such as eye, nose or throat irritation, headache and fatigue (Wolkoff et al., 2006). The results of He et al., (2007) suggested that there is potential harm to human beings because of breathed in toner particles. A recent study by Gatti, 2008 using in-vitro and in-vivo experiments with 5 types of nanoparticles found chemical evidence of particulate matter in human pathological tissues from patients who had suffered diseases of unknown origin. It was pointed out in this study that inhaled and ingested nanoparticles can penetrate through the alveolar as well as the digestive walls to enter the blood system and subsequently be transported to any organ in the body. Only about 20% of nanoparticles are removed once deposited in alveolar regions in animal subjects after 24 hour exposure, in contrast to about 80% removal for particles above 500 nm (Oberdörster et al., 2005). In related work, Chalupa et al., (2004) found about 74% deposition of carbon ultrafine particles in asthmatic human subjects for a 2 hour exposure.
With possible adverse health effects, the question of the chemical characterization of the ultra-fine particles released by such devices is of special importance (BfR, 2008). However, due to the low mass of the UFPs and high volatility nature, they evaporate as soon as they are released; it has not yet been possible to ascertain their chemical composition (Wensing et al, 2008). Previous publications by Bake and Moriske, (2006) and Wensing et al., (2006) have shown that hardcopy devices often emit UFPs while larger particles (e.g. toner dust) could only be detected in low concentrations. Moreover, adverse health effects from exposure to nanoparticles have been found to be more closely related to particle number concentration than to particle mass concentration (Oberdorster,2000). These health effects may differ substantially depending on the size, morphology, composition (both bulk and surface), and concentration of airborne particles (Oberdorster, 2005).
Many studies of photocopier-related emission have been carried out in test chambers (Lee et al., 2001). Field studies on the impact of photocopiers and printers on indoor air quality are relatively limited. Though specific printer-emitted VOCs and PM has been studied (Kagi et al., 2007), no report has addressed the PM concentration in photocopier and printer centers. Since the size of individual particles influences the degree to which they can be inhaled and the effects that they can cause (Lee et al., 2001), the characteristics of such-emitted particles are needed to evaluate whether exposure control and reduction efforts are necessary. Furthermore, the main difference between the photocopier centers in India and other developed countries is that such centers in India are generally small and serve as both businesses and residences. Thus, the pollutants emitted during such processes would affect the indoor air quality and potentially have adverse health effects on the employees as well as the residents of the workplace. The objective of this study is to investigate size distributed particle number and mass concentration in 250 to 1000 nano meters range in some representative commercial photocopier and printer centers in the Northern Central India. The sources of these particles are also discussed in the indoor air.

2. Materials and methods

Air sampling was conducted at 2 photocopier centers A and B (Fig.1) in the Agra city in the month of June 2009. Measurements were made for eight days, four days each at each sampling center. In India, most photocopiers and printer centers are located in multi-storey street houses. The area of each center is approximately 30 to 36 m^3. No forced ventilation systems used during the measurement. However, the door to the experimental rooms were opened and closed often by the users and the customers whenever they entered the room to use the equipment. In a typical street house, the ground floor is the work area and the upper floors are living areas. Typical interior materials used in photocopier centers include ceramic tile floor, painted concrete ceiling, painted concrete walls and sliding aluminum-framed glass doors. Usually only some metal desks and chairs, and no other furniture are present in the confined space of a photocopier center. Basic information of each center, including business hours, room dimensions, environmental conditions, types of ventilation and entrance, number of photocopiers, printers and number of copies made were collected. Table 1 lists there characteristics of the centers.

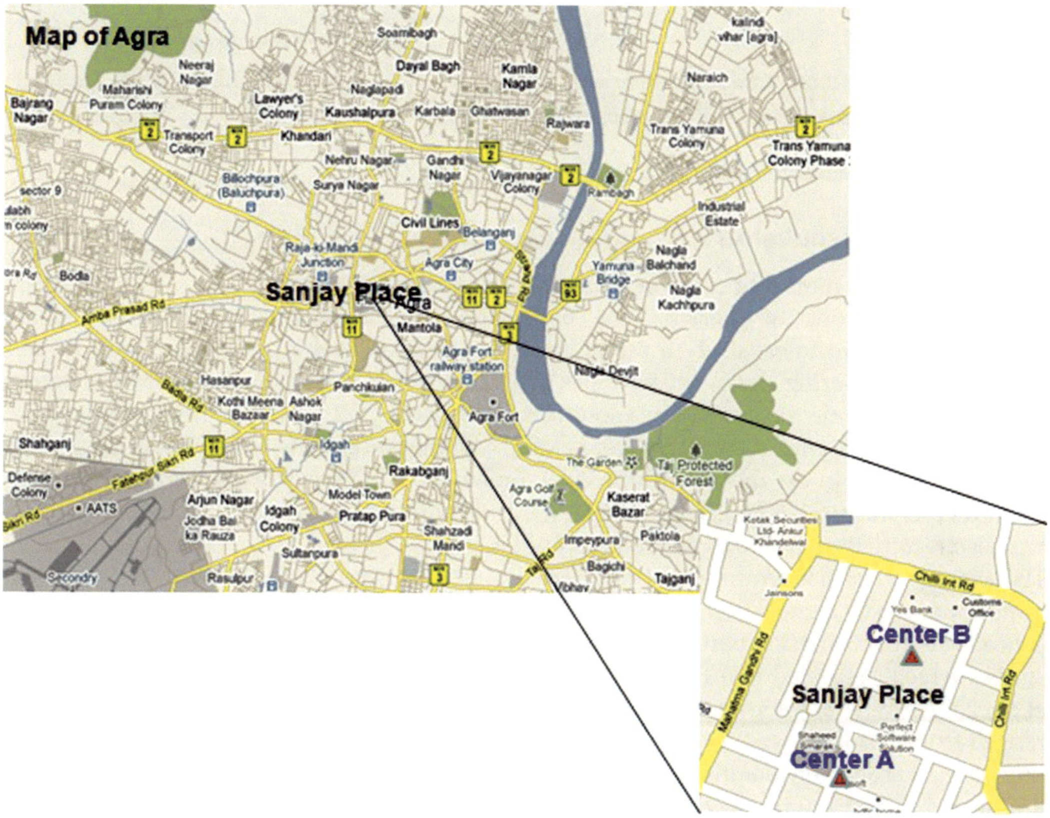

Fig. 1. Map of Agra Showing the Sampling Centers

Grimm 31-Channel Portable Aerosol Spectrometer model No.1.109 was selected for monitoring the particle mass and size distribution in the range of 250 - 1000 nm, at a flow rate of 1.2 L/min ± 5% constant with controller for continuous measurement during the sampling period. The instrument was set to collect data at 10 minute intervals and it store the data in data memory logger card from which data can be downloaded to computer and can be analysed. Particles are collected close by the analyzer from a dedicated 5 cm long vertical sampling head (no sampling tubes and therefore no particle loss). The instrument works on dual technology i.e. the principle of scattering of light at 90° to give the real-time measurements and total particles can be collected on 47- mm PTFE filter paper for chemical analysis. Its real time measuring range is from 0.25 μm to 32 μm or 250 nm to 32,000 nm in 31 channel sizes, each unit is with NIST (National Institute of Standards and Technology) certified, monodisperse latex on the size of channels calibrated [www. GRIMM-aerosols.com].To improve the time resolution, the range was limited to 0.25 to 1 μm or 250 to 1000 nm in 12 channels. The sampling tube of the analyzer was positioned as close as possible to head height in the center of the sampling room. The GRIMM particle measuring system is equipped with GRIMM 1174 Software for data acquisition.

Physical characteristics of the photocopy centers investigated							
Center	Room volume m^3	Number of Hardcopiers		Number of copies made Per day	Number of Measurements day	Type of entrance	Ventilation
		Photocopiers	Printers				
A	36	1	3	12500 pages	4	push-and-pull door	a
B	30	2	2	7000 pages	4	wide-opened door	b

a = air-conditioned, b = natural ventilation open door

Table 1. Physical Characteristics of the Photocopier Centers

3. Result and discussion

3.1 Experimental methods used in emission characterization

Sampling was done to measure the ultra fine particle size distribution and mass concentration at an interval of 10 minutes, in two periods at the photocopier and printer centers A and B during the business hours and background hours. The back ground values were obtained inside the centers by monitoring the particles 2 hours before opening of the centre and 2 hours after the centers were closed. All measurements were conducted during eight days period in the month of June 2009. Emissions from hardcopies were measured at room temperature (21–30°C), with an intermediate humidity conditions (45–60% RH). CO_2 was also recorded up to 650 PPM while sampling.

The average mass concentration of the particles ranged from 1.86 μgm^{-3} to 11.71 μgm^{-3} at center A and 4.89 μgm^{-3} to 46.46 μgm^{-3} at center B as shown in table 2 A. Increase in the concentration of ultra fine particles in this study seems to be in consistent with the results of studies which suggested that PM emitted by hardcopiers are aerosolized toner powder (Lee et al., 2007). Table 2 A shows the particle mass concentrations measured in back ground air ranged from 0.87 μgm^{-3} to 9.10 μgm^{-3} at center A and 0.87 μgm^{-3} to 9.13 μgm^{-3} at center B and during the hardcopier making they ranged from 2.43 μgm^{-3} to 13.71 μgm^{-3} at center A and 8.33 μgm^{-3} to 80.16 μgm^{-3} at center B which were much higher at both the sites from the background values. Increase in the particulate concentration at the center B was observed more in comparison to A.

Table 2 B, shows the particle number concentration obtained in photocopier centers A and B. The table reveals that the number of particle increased when hardcopier devices began to make copies. During the operational mode of these hard copying machines the number concentration of the particulate varied in center A from 68223/Lit to 569896/Lit with an average of 258485/Lit, where as in the center B it varied from 82612/Lit to 2580941/Lit in the range of 250-1000 nm with an average of 1504133/Lit respectively. A high number concentration of ultra fine particles was found with a peak value of 569896/Lit particle at center A and 1504133/Lit at center B particularly during business hours. The number concentration of particles in 250–1000 nm was significantly higher than mass concentration of the same range at both the centers. It was 3-7 times higher during operational hours than background values obtained before and after the machine was operational at both the centers (table 3). As the machines were nonoperational before the opening of the centers in morning and after the closing of the centers at night, there were significant decrease in the particulate numbers and mass concentrations from the working hours of the machine during the day. While during the working hours, the number of particulate and mass concentration kept on increasing during the first hour of hardcopying; however they decreased after a period of time as shown in Fig. 2.

The number and mass concentrations of the particles in the sampling centers were found to increase significantly from background values during the sampling of 4 hrs in a definite trend as shown in Fig. 2. This increase with respect to background values in centers A and B was 27% for first hour which decreased to 23% in the second hour, 20% in the third and 19% in the fourth hour with respect to background values. Thus settling time for the particles emitted from the photocopier was analyzed from the background values measurement. The photocopier and printers were not in use during the night hours that is before opening and after closing of the centers. The total particle count in a cubic centimeter and mass concentration in μgm^{-3} of air was estimated for the working hours as shown in Fig 2. Activity resumed from the morning by photo printing of the machines. Hence, the settling time of the particles could be estimated using the data from the background values. It can be seen that the total particle count and mass concentration dropped to low levels over two hours of working and then remained constant during the further working hours.

Centers also have other individual sources than the hardcopiers itself for particles generation. Other chemical constituents, as well as mechanical processes, can also influence the emission behavior during operation (Wensing et al., 2006). Characterizing emissions from hardcopier equipment are also difficult due to the diversity of available equipment, the rapid evolution and turnover of product lines and the variability in environmental and operating conditions. Lee (2001) have pointed out earlier in his laboratory study about 75% of photocopier toner is transferred to the photoconductive drum and that which does not adhere to the drum becomes available for emission to indoor air. The toner particles are about 10 μm. It needs further consideration but is indicating (Kagi et al., 2007) in the study that fine particles were not directly generated from toner particles but by the secondary formation of the VOCs and the water mists emitted during the operation of the printers.

Finally, the path by which the UFPs leave the printer is also an important aspect describing emission behavior. As an example, the maximum total concentration of particles ($d<1\mu m$) and the sampling points are displayed by using a printer in Fig.3 (Wensing et al., 2008). The results show that most particles leave the printer near the paper tray and at the back. Release through the fan above the toner waste bottle is considerably lower. Consequently, a retrofitted filter system (designing of air flow system in such a way that the majority of the released UFP leave the casing through a definite opening) may be a possible way to reduce the overall UFP emissions from the appliances. However, the results of this experiment are limited to the printer examined because every type of laser printer—even from the same manufacturer—can have different ventilation and air flow paths. The air flow direction will not be the same for every printer. In some cases air is blown into the printer to cool the internal components (such as printed circuit boards). In such a case a fitted filter would only come in contact with the emitted particles via the backflow.

4. Formation of particles

The particle size distribution obtained in this study indicated the formation of fine particles during photocopying and printing. Many studies have suggested different mechanism of formation of fine and ultra fine particles (UFP) (Lee et al., 2007).

a. Physical process of nucleation and condensation. The first possible formation mechanism of UFP is the nucleation/condensation of low vapor pressure substances, which were vaporized at high temperature and condensed at low temperature to form particles. Some substances from the heated toner or paper were vaporized during the fusing stage, in which the fuser temperature reached around 200°C, and their concentrations exceeded their saturation vapor concentration (Jang and Kamens, 2001). Therefore, particles may form when the saturated vapor condenses at a lower temperature.
b. Oxidation of VOCs. The second possible mechanism of UFP formation during photocopying is the oxidation of indoor VOCs. The byproducts of corona charging during photocopying, such as ozone, NO_x and OH-radicals, are both strong oxidants for the oxidations of emitted VOCs. Many studies have demonstrated that photo-oxidation products of aromatic hydrocarbons can undergo various reactions to produce secondary organic aerosols (SOA) in the presence of O_3, OH radicals, and NO_x (Edney et al., 2001; Jang and Kamens, 2001). The microenvironment inside the photocopier is very similar to a photochemical smog chamber that contains a light source and higher concentrations of reaction agents. Therefore, SOA formation inside photocopiers might be an important source of indoor UFP and FP during photocopying. Furthermore, many studies have confirmed that ozone may react with unsaturated VOCs (such as terpenes and styrene), causing secondary emission of UFP and FP in an indoor environment (Wolkoff and Nielsen, 2001; Fan et al., 2005). Even though UV irradiation is not present in indoor environment (except the spaces inside the photocopiers), SOA may form when ozone reacts with those unsaturated VOCs presented in photocopy center.
c. Ion-induced nucleation. Ions, which are generated by corona devices during photocopying, may play a role in the formation of UFP and FP by ion-induced nucleation of organic vapors. Many works have confirmed the effect of ionizing radiation on aerosol formation (Ramamurthi et al., 1993). Ion-induced nucleation is the gasto-particle process causing supersaturated vapors to condense on ions. During ion-induced nucleation processes, the higher particle growth rates are observed because electrostatic forces would enhance the stability of electrically charged clusters (Yu and Turco, 2001). Ichitsubo et al., (1996) reported an experimental study of UFP generated from organic vapors by corona ionizers. Among the organic compounds tested (aromatics, alcohols, ketones and others), only aromatic compounds undergo gas-to-particle conversion process and yield unstable clusters, which may grow into detectable particles (42 nm) during corona discharge. Based on the results of the above studies, UFP could be formed rapidly during photocopying by the ion-induced nucleation of emitted aromatic hydrocarbons.

Range	Center A (0hr)	Center B (0hr)	Average	Center A (I hr)	Center B (I hr)	Average	Center A (II hr)	Center B (II hr)	Average	Center A (III hr)	Center B (III hr)	Average	Center A (IV hr)	Center B (IV hr)	Average	Maximum	Minimum	Total Average
250 nm	0.87	0.87	0.87	2.08	5.27	3.67	1.95	8.33	5.14	2.02	5.06	3.54	2.43	4.94	3.68	8.33	0.87	3.36
250-280 nm	1.28	1.68	1.48	3.97	10.34	7.15	3.71	14.89	9.30	3.79	10.03	6.91	4.46	9.02	6.74	14.89	1.28	6.29
280-300 nm	2.23	2.47	2.35	5.88	17.45	11.66	5.44	20.47	12.95	5.38	16.41	10.90	6.05	12.68	9.36	20.47	2.23	9.45
300-350 nm	3.56	3.77	3.66	8.86	31.33	20.10	8.15	27.70	17.92	7.78	27.74	17.76	8.08	17.74	12.91	31.33	3.56	14.58
350-400 nm	5.74	5.15	5.45	11.19	47.45	29.32	10.17	35.28	22.73	9.67	40.76	25.22	9.61	23.29	16.45	47.45	5.15	20.07
400-450 nm	6.37	6.07	6.22	12.07	57.39	34.73	10.90	40.05	25.47	10.41	47.84	29.12	10.23	26.44	18.34	57.39	6.07	23.09
450-500 nm	6.84	6.55	6.70	12.40	62.59	37.50	11.16	42.58	26.87	10.70	50.94	30.82	10.49	27.93	19.21	62.59	6.55	24.58
500-580 nm	7.82	7.33	7.58	12.86	70.40	41.63	11.50	46.53	29.02	11.14	55.36	33.25	10.88	30.07	20.47	70.40	7.33	26.81
580-650 nm	7.17	7.97	7.57	13.80	74.57	44.18	11.79	48.67	30.23	11.42	57.34	34.38	11.15	31.01	21.08	74.57	7.17	27.95
650-700 nm	8.75	8.25	8.50	13.31	76.26	44.79	11.85	49.51	30.68	11.55	58.05	34.80	11.30	31.33	21.31	76.26	8.25	28.49
700-800 nm	8.08	8.78	8.43	13.54	78.61	46.08	12.02	50.64	31.33	11.76	58.97	35.37	11.51	31.75	21.63	78.61	8.08	29.06
800-1000 nm	9.10	9.13	9.12	13.71	80.16	46.94	12.17	51.38	31.78	11.90	59.61	35.76	11.67	32.02	21.85	80.16	9.10	29.60

0 hr = Background Value. 1-4 hrs = Business hours

A. Mass concentration in µgm-3

Range	Center A (0hr)	Center B (0hr)	Average	Center A (I hr)	Center B (I hr)	Average	Center A (II hr)	Center B (II hr)	Average	Center A (III hr)	Center B (III hr)	Average	Center A (IV hr)	Center B (IV hr)	Average	Maximum	Minimum	Total Average
250-280 nm	68223	82612	75418	385430	385430	385430	157540	345548	251544	89317	365189	227253	94463	305666	200065	385429	68223	229933
280-300 nm	78235	174748	126491	840121	840121	840121	145044	743344	444194	151031	739818	445425	148129	646567	397349	840121	78235	454528
300-350 nm	92458	316955	204706	1526854	1526854	1526854	193377	1349793	771585	197763	1276894	737329	187516	1172732	680124	1526854	92458	791548
350-400 nm	149115	435597	292356	2071146	2071146	2071146	369045	1841427	1105236	219930	1693623	956777	222187	1611708	916948	2071146	149115	1079317
400-450 nm	153605	490071	321838	2315659	2315659	2315659	381400	2064461	1222930	227794	1876354	1052074	229257	1813262	1021260	2315659	153605	1198573
450-500 nm	154846	513021	333934	2418167	2418167	2418167	385129	2158172	1271651	230282	1951297	1090790	231429	1898177	1064803	2418167	154846	1248088
500-580 nm	155936	535012	345475	2516456	2516456	2516456	388509	2248001	1318255	232572	2021432	1127002	233445	1979546	1106495	2516456	155936	1295325
580-650 nm	156293	543730	350012	2553531	2553531	2553531	389989	2282667	1336328	233695	2047185	1140440	234440	2011802	1123121	2553531	156293	1313370
650-700 nm	156381	546679	351530	2565524	2565524	2565524	390526	2294118	1342322	234144	2055961	1145053	234836	2022713	1128775	2565524	156381	1319346
700-800 nm	156441	549760	353101	2577481	2577481	2577481	391133	2305796	1348465	234692	2064963	1149828	235331	2034111	1134721	2577481	156441	1325433
800-1000 nm	156450	550750	353601	2580941	2580941	2580941	391402	2309359	1350381	234951	2067613	1151282	235570	2037777	1136674	2580941	156450	1327283

0 hr = Background Value. 1-4 hrs = Business hours

B. Particulate count/liter

Table 2. Particulate Mass and Number at Centers A and B

Range	0hr	I hr	II hr	III hr	IV hr
250 nm	0.9	4.2	4.2	4.1	4.2
250-280 nm	1.5	4.8	4.7	4.7	4.5
280-300 nm	2.4	5.0	4.9	4.6	4.0
300-350 nm	3.7	5.5	4.9	4.8	3.5
350-400 nm	5.5	5.4	4.2	4.6	3.0
400-450 nm	6.2	5.6	4.1	4.7	2.9
450-500 nm	6.7	5.6	4.0	4.6	2.9
500-580 nm	7.6	5.5	3.8	4.4	2.7
580-650 nm	7.6	5.8	4.0	4.5	2.8
650-700 nm	8.5	5.3	3.6	4.1	2.5
700-800 nm	8.4	5.5	3.7	4.2	2.6
800- 1000 nm	9.1	5.1	3.5	3.9	2.4

Mass difference

Range	0hr	I hr	II hr	III hr	IV hr
250-280 nm	75418	5.1	3.3	3.0	2.7
280-300 nm	126491	6.6	3.5	3.5	3.1
300-350 nm	204706	7.5	3.8	3.6	3.3
350-400 nm	292356	7.1	3.8	3.3	3.1
400-450 nm	321838	7.2	3.8	3.3	3.2
450-500 nm	333934	7.2	3.8	3.3	3.2
500-580 nm	345475	7.3	3.8	3.3	3.2
580-650 nm	350012	7.3	3.8	3.3	3.2
650-700 nm	351530	7.3	3.8	3.3	3.2
700-800 nm	353101	7.3	3.8	3.3	3.2
800- 1000 nm	353601	7.3	3.8	3.3	3.2

Count difference

Table 3. Mass and Number Difference at the Two Centers

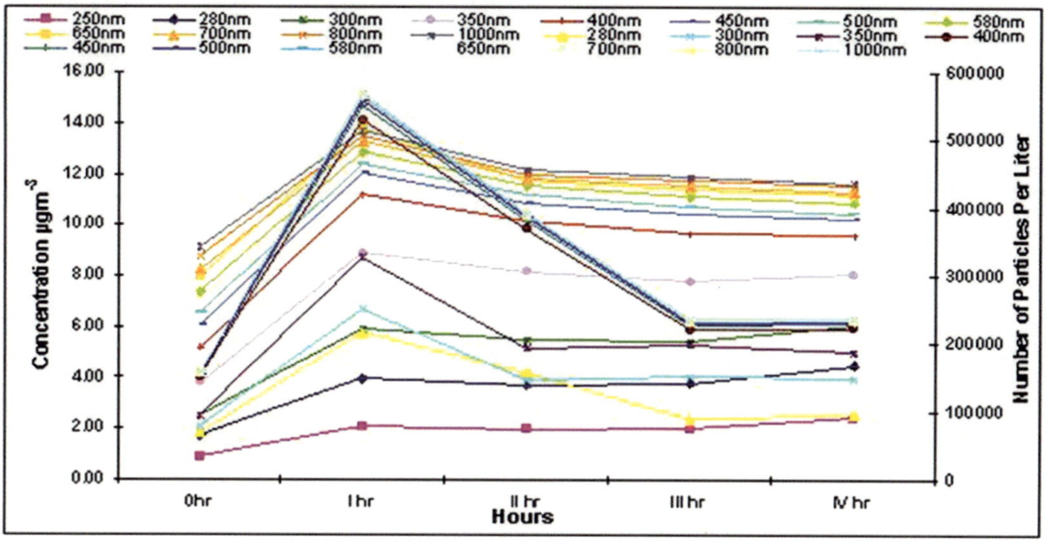

Center A

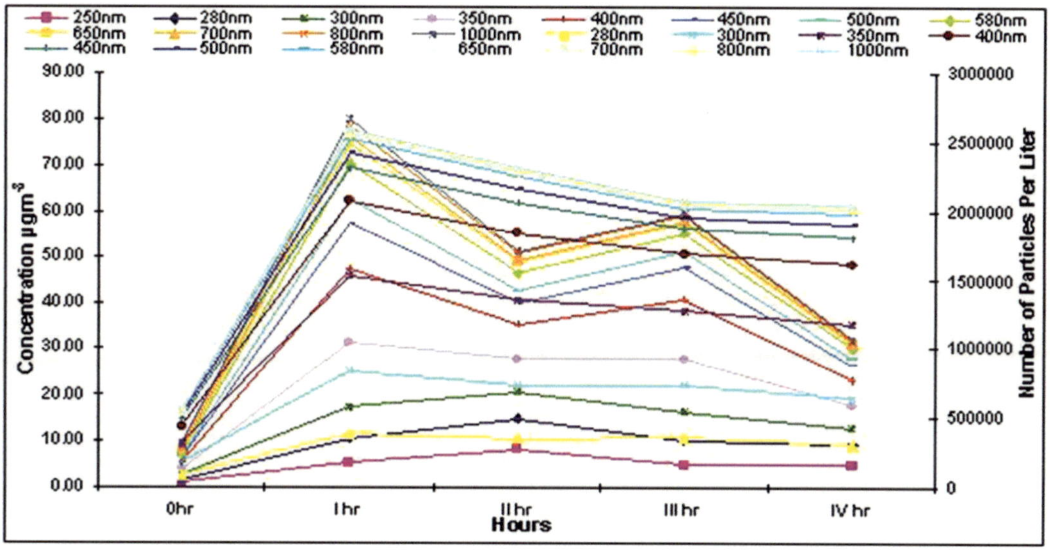

Center B

Fig. 2. Trends in number and mass concentration of particles in photocopier center A and B

To date, the information regarding the formations of UFP and FP during photocopying is still limited. The mechanism of UFP and FP formation is far from being well understood and a single process is not likely to explain all the phenomena's. Although the formation mechanism remains unclear, Fig. 3 summarizes the possible mechanisms for the formation of UFP during photocopying, including condensation, oxidation and ion-induced nucleation. Corona devices, which can generate ozone, NOx, radicals and ions during photocopying, may be the key element of UFP formation and particle removal in photocopy centers.

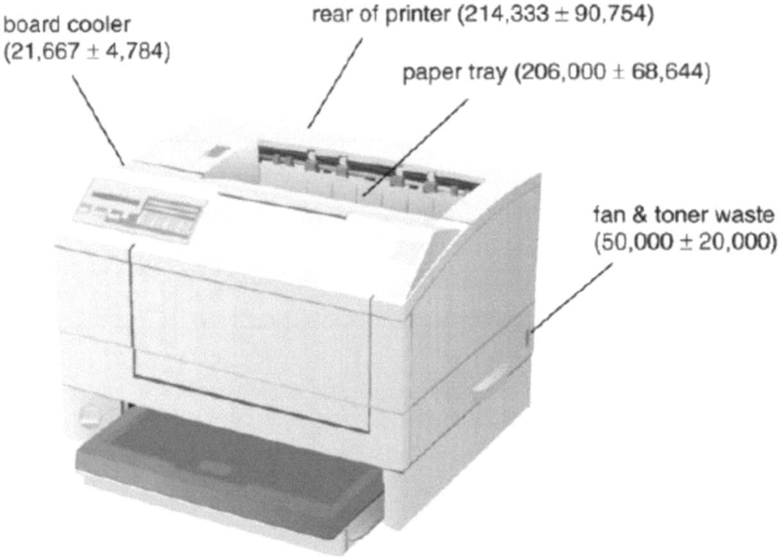

Fig. 3. Example of paths of UFP release from a laser printer Taken from (Wensing et al., 2008)

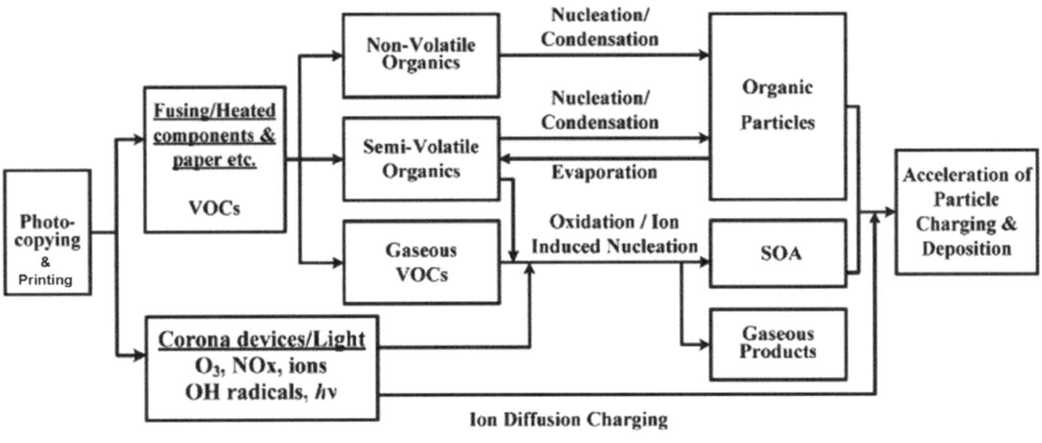

Fig. 4. Conceptual model of indoor air chemistry and particle formation and removal during photocopying and Printing Taken from (Lee et al., 2007)

Although the formation mechanism remains unclear, Fig.4 summarizes the possible mechanisms for the formation of UFP and FP during photocopying, including condensation, oxidation and ion-induced nucleation (Lee et al., 2007). Corona devices, which can generate ozone, NOx, radicals and ions during photocopying, may be the key element of UFP and FP formation and particle removal in photocopier centers.

5. Conclusion

The unexpected phenomenon namely declined in particle mass and number concentration as operation proceeded for few hours is likely attributable to the surface deposition of charged particles, which are charged primarily by the diffusion charging of corona devices equipped inside the hardcopies devices. Particle charging is a function of the ion concentration. Based on the monitored results in centers, particle number and mass concentrations increased immediately as the operations proceeded. During the first hour of operation, ions emitted from corona devices might not be high enough to charge particles indoors; therefore, the increasing trends of particles were consistent. However, after the first hour of operation, the ion concentrations in indoor environment might reach to a point that can accelerate the speed of diffusion charging and increase the deposition rates of charged particles to nearby surfaces. After this point, the particle removal rates were higher than the particle formation rates and therefore the particle number concentrations decreased, although hard copying process was consistently being conducted under the same ventilation conditions. This decrease was less in center A than in comparison to center B because center A was fully air-conditioned. So the doors and windows were kept close where as center B was naturally ventilated.

The results of the these real room measurements are not sufficient to permit classification of possible health related issues with printer and photocopier generated aerosols for this purpose both a more detailed chemical characterization of the particles and a model for exposure assessment would be required. The fact that hardcopy devices are not the only source of fine particulate in indoor environment also needs to be accounted for. In Agra photocopy centers usually open at 10 am and close at 10 pm. If the background particulate value is taken as the particle mass concentration in close hours then the 24 h average PM concentration can be calculated for each photocopy centre by assuming 12 hrs for business and 12 for close hours respectively. Additionally most photocopy centers in Agra open 6 days a week and 52 weeks per year. Therefore based on the results of this study, the PM in the range of 250 nm to 1000 nm should be concerned in view of annual human exposure. Personal exposures may be significantly larger than those estimated through average pollutant indoor concentrations, due to proximity of users to the sources over extended periods of time. The magnitude of emissions, the link from emissions to personal exposure, the toxicological significance of the chemicals emitted, and the costs and impacts of alternate materials should all be considered in order to evaluate potential importance of human exposures and health risks. The policy for precautionary reasons for example developing ecolables for low emitting products can be a possible solution to it.

Finally, it is important to put this work in the context of exposure, health implications, energy costs, and technology options. Considering the diversity of equipment, the rapid evolution and turnover of product lines, changes in manufacturing processes and variability in operating conditions, the values summarized in this study represent initial estimates of emissions and their implications. This study also highlights the importance the need for evaluating long term effects of exposure to toner particles since these are yet to be fully understood. Further studies are recommended to measure the direct adverse effects of these particles to human health.

6. Acknowledgements

The authors like to thank the CSIR (COUNCIL OF INDUSTRIAL RESEARCH) project no: 231065/2K10/1, Dr. F.M.Prasad, Principal of St. John's College Agra and Dr. Ashok Kumar, Head, Department of Chemistry, St. John's College Agra for providing us the facilities.

7. References

Newburger E. C. (2001): "Home computers and Internet use in the United States: August 2000", Special Studies, US Census Bureau, Washington, DC

Wensing M., Uhde E., Salthammer T. (2005): ''Ultra-fine particles release from hardcopy devices: Sources, real-room measurements and efficiency of filter accessories", Science of The Total Environment 339: 19–40

Wensing M., Kummer T., Riemann A., W Schwampe W. (2002): "Emissions from electronic devices: examination of computer monitors and laser printers in a $1m^3$ emission test chamber", The 9th International Conference on Indoor Air and Climate, 2, Monterey, p. 554–9

Wolkoff P., Wilkins C.K., Clausen P.A., Larsen K. (1993): "Comparison of volatile organic compounds from processed paper and toners from office copiers and printers: methods, emission rates, and modeled concentrations", Indoor Air, 3: 113–123

Armbruster C., Dekan G., Hovorka A. (1996): "Granulomatous pneumonitis and mediastinal lymphadenopathy due to photocopier toner dust", Lancet 348: 690

Black M.S., Worthan A.W. (1999): "Emissions from office equipment", The 8th International Conference on Indoor Air and Climate. 2, Edinburgh. p. 454–9

Wolkoff P. (1999): "Photocopiers and indoor air pollution", Atmospheric Environment, 33: 2129–30

Lee S.C., Lam S., Fai H.K. (2001): "Characterization of VOCs, ozone, and PM_{10} emissions from office equipment in an environmental chamber", Building and Environment 369, (7): 837–42

Roller. (2006): "Quantitative risk assessment for the exposure to toner emissions from copiers", Gefahrst Reinhalt Luft, 66: 211–6

Wolkoff P., Wilkins C.K., Clausen P.A., Nielsen G.D. (2006): "Organic compounds in office environments—sensory irritation, odor, measurements and the role of reactive chemistry", Indoor Air, 16, 7–19

He C., Morawska L., Taplin L. (2007): "Particle Emission Characteristics of Office Printers, Environment Science and Technology, 41 (17)

Gatti A.M. (2008): "Nanopathology: a new vision of the interaction environment-human", Available on line from address: http://ec.europa.eu/research/quality-oflife/ka4/pdf/report_nanopathology_en.pdf

Oberdorster G., Oberdorster E., Oberdorster (2005): "Nanotoxicology: An emerging discipline evolving from studies of ultrafine particles Environ" J. Health Perspect, 113: 823–839

Chalupa D.C., Marrow P.E., Oberdorster G., Utell M.J., Frampton M.W. (2004): "Ultrafine particle deposition in subjects with asthma", Environmental Health Perspectives. 112: 879-882

BfR–Federal Institute for Risk Assessment, (2008), Gesundheitliche Bewert Organic compounds in office environments—sensory irritation, odor, measurements and the role of reactive chemistry, ung Nr. 014/2008 vom 31, März

Bake D., Moriske H.J. (2006): "Investigations about emissions of fine and ultrafine particles by using laser printers. Umweltmed Forsch Prax 11: 301–8

Wensing M., Pinz G., Bednarek M., Schripp T., Uhde E., Salthammer T. (2006): "Particle measurement of hardcopy devices", Healthy Buildings, 2: Lisbon. pp. 461–464

Oberdorster G. (2000): "Toxicology of ultrafine particles: In vivo studies", Philos.Trans. R. Soc. Lond, A 358: 2719–2740

Oberdorster G., Oberdorster E., Oberdorster (2005): "Nanotoxicology: An emerging discipline evolving from studies of ultrafine particles Environ" J. Health Perspect, 113: 823–839

Kagi N., Fujii S., Horiba Y., Namiki N., Ohtani Y., Emi H., Tamura H., kim Y.S. (2007): "Indoor air quality for chemical and ultrafine particle contaminants from printers", Building and Environment, 42: 1949-1954

www.GRIMM-aerosol.com

Lee C.W., D J Hsu D.J. (2007): "Measurements of fine and ultrafine particle formation in photocopy centers in Taiwan", Atmospheric Environment, 4: 6598-6609

Wensing M., Schripp T., Uhde E., Salthammer T. (2008): "Ultra-fine particles release from hardcopy devices: Sources, real-room measurements and efficiency of filter accessories", Science of The Total Environment, 407: 418-427

Jang M., Kamens R.M. (2001): "Characterization of secondary aerosol from the photooxidation of toluene in the presence of NO_x and 1-propene", Environmental Science and Technology, 35: 3626-3639

Edney E.O., Driscoll D.J, Weathers W.S., Kleindienst T.E., Conver T.S., Mclver C.D., W Li W. (2001): "Formation of polyketones in irradiated toluene/propylene/NO_x/air mixtures", Aerosol Science and Technology, 35: 998-1008

Wolkoff P., Nielsen G.D. (2001): "Organic compounds in indoor air—their relevance for perceived indoor air quality?", Atmospheric Environment, 35: 4407-4417, 2001

Fan Z.H., Weschler C.J., Han I.K., Zhang J.F. (2005): "Conformation of hydroperoxides and ultra-fine particles during the reactions of ozone with a complex VOC mixture under simulated indoor conditions", Atmospheric Environment, 39: 5171-5182

Ramamurthi M., Strydom R., Hopke P.K., Holub R.F. (1993): "Nanometer and ultrafine aerosols from radon radiolysis", Journal of Aerosol Science, 24: 393–407

Yu F., Turco R.P. (2001): "From molecular clusters to nanoparticles: role of ambient ionization in tropospheric aerosol formation", Journal of Geophysical Research, 106: 4797-4814, 2001

Ichitsubo H., Alonso M., Ishii M., Endo Y., Kousaka Y., Sato K. (1996): "Behavior of ultrafine particles generated from organic vapors by corona ionizers", Particle and Particle System Characterization, 13: 41-46, 1996

10

In-Office Dispersion and Exposure to Contaminants Originating from an Unfolded Letter

Alfred D. Eisner[1], Russell W. Wiener[2] and Jacky Rosati[2]
[1]*Alion Science and Technology,*
[2]*National Homeland Security Research Center,*
U.S. Environmental Protection Agency,
USA

1. Introduction

Incidents of bioterrorism that have occurred over the past decade have demonstrated a need to understand the transmission and exposure risks of daily activities to potential biological agents (NATO, 2005; de Armond, 2002; Block, 2001). Based on experience since September 11, 2001, the mail has become a significant means of bioagent dispersion. This chapter seeks to further advance our understanding of fluid and aerosol dynamic processes of exposures resulting from dust lying on the surface of a letter or a table being resuspended by air flow, (Richmond-Bryant, et al., 2006).

Transmission of aerosols from an unfolded letter, (Duncan et al., 2009), is dependent on the motion of the air in the environment in which the letter resides (Dull et al., 2002). The primary source of fluid motion in most buildings is the heating, ventilation, and air-conditioning (HVAC) system. Several reports suggest that numerous pathogens may survive such airborne transport (e.g., Nardell et al., 1986; Mangili and Gendreau, 2005). Others show how contaminants can be dispersed into the indoor environment (e.g., reviews by Wallace, 1996, and Nazaroff, 2004; Price, et al., 2009; Reshetin & Regens, 2003; Reshetin & Regens, 2004). These reviews and many papers cited therein show that indoor particle transport is subject to complex interactions of dispersion, deposition, and resuspension. Understanding these processes is predicated on understanding the interaction between turbulent airflow and particles. Rooms often have complex geometries that result in extremely complex turbulence because of flow phenomena such as flow separation, recirculation, and buoyancy (Posner et al., 2003; Rim and Novoselac, 2009).

Contamination and exposure resulting from a localized source such as a contaminated letter has received some recent attention. (Agranovski et al., 2005; Ho et al., 1993; Ho et al., 2005; Kornikakis et al., 2001; Kornikakis eta l., 2009; Kornikakis et al., 2010; Lien et al., 2010).

In many offices, outlets from the HVAC system are positioned in the ceiling and often generate a substantial downward blowing of air, (Nardell, et al., 1986). Ceiling fans can have a similar effect. This airflow will almost certainly incorporate flow separation and recirculation zones. Advancing the understanding of dispersion of particulate contaminants under such complex conditions can provide useful input for decontamination efforts directed toward contaminated individuals or objects. To this end, the study described in this paper investigated dispersion and surface contamination resulting from contaminated material being reentrained from flat letter lying on a table top under a vent.

2. Methods

2.1 Experimental setting

This investigation was designed to explore a hypothetical situation in which a person seated at a table is exposed to reentrained dust from the surface of a letter that is lying flat on a desk under an HVAC vent. In our experimental simulation, an individual, represented by a manikin, was seated at a table adjacent to an office wall and positioned under an HVAC vent (Fig. 1). A surrogate letter was made of Rosco cine foil ™, matte black, and gauge 0.002 in. It's thickness, stiffness and roughness were similar to a standard paper sheet. Using this surrogate was necessary to prevent PIV cameras over saturation and to obtain images of particles moving very close to the surface. From now on we will refer to this surrogate letter as letter. It was placed flatly on the table in front of the manikin and sprinkled with test dust. The dust was Arizona dust with particle size ranging from 1-5 microns.

2.2 Experimental systems

Several fundamental experimental systems were used in this research: a thermal articulated manikin (TAM), an environmental walk-in chamber (EWC) used as an office space simulator, and a particle imaging velocimetry (PIV) system. Each of these systems is described briefly below.

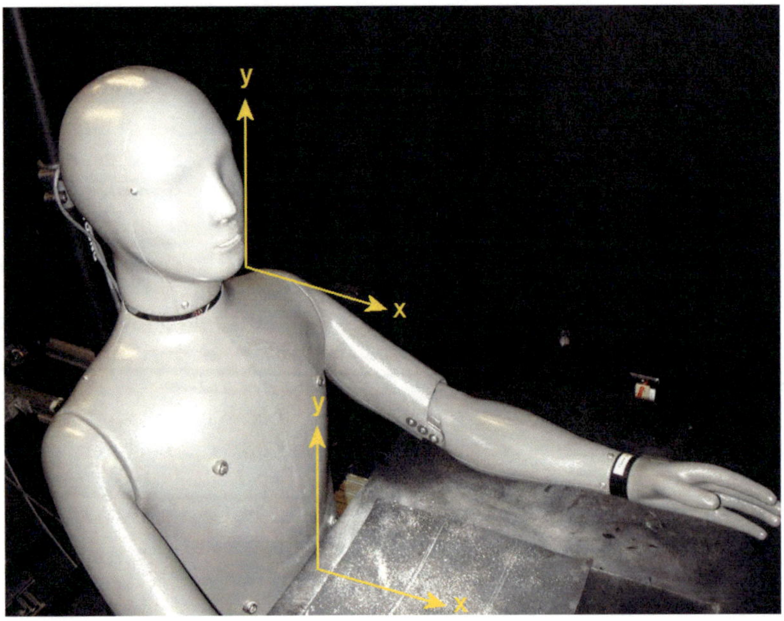

Fig. 1. Manikin confronted by a cloud of contaminating dust blown from an unfolded letter by air from a ceiling vent. Piled-up dust of mostly agglomerated and shifted large particles can be seen on the letter section close to the manikin's chest. Two x,y coordinate systems reflect positions of the PIV test areas, namely, a table area and a head area.

2.2.1 Articulated manikin

An adult-size TAM (Model Newton, Measurement Technology Northwest, Seattle, WA, USA) with 18 heating zones was used in this study. The dimensions of the manikin were sized to match a 50th percentile U.S./European male. The TAM, designed as a repeatable instrument to evaluate various thermal conditions, has isothermal surfaces over each individual zone. All thermal zones are fitted with heaters to simulate metabolic heat output rates and a distributed temperature sensor to accurately measure the average temperature over each zone. For the purpose of this study all zones were set at 37 °C.

2.2.2 Environmental walk-in chamber

The EWC (297 by 216 by 221 cm) was made of industrial steel and was located inside a large laboratory facility with temperature and humidity kept at normal laboratory levels. The EWC was fitted with two ceiling openings (20 cm in diameter) located centrally 50 cm from the front and back walls. The openings were used as the HVAC system's air inlet and outlet and were connected to the recirculating air moving unit positioned on the roof of the EWC. The air mover speed could be controlled by a variac, and the blower fan could be turned on or off as needed. Aluminum corrugated duct work several meters long was connected to the blower to allow for quick heat dissipation by the blower fan, thus ensuring the temperature and humidity conditions inside the EWC were essentially those in the large laboratory space. A table measuring 122 by 70 by 91 cm and a TAM were positioned inside the EWC.

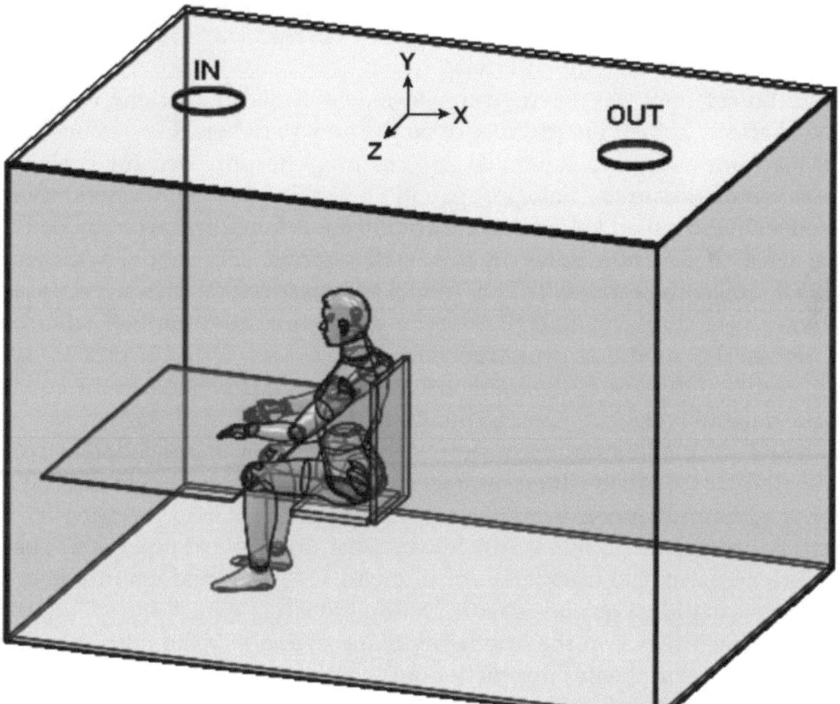

Fig. 2. Schematic view of TAM seated inside EWC. The table and chair are represented schematically by flat rectangles. Two openings in the ceiling represent the HVAC inlet (IN) (above the table) and outlet (OUT).

2.2.3 Particle imaging velocimetry (PIV)

A two-pulse laser technique such as PIV (TSI Inc., Shoreview, MN, USA) is commonly used to investigate particle-laden fluid flows. In these laboratory configurations, two laser pulses are fired in rapid sequence, typically 10 microseconds to 10 milliseconds apart. Usually two synchronized laser heads are used. In this study, a standard PIV configuration was used in which two laser beams following a common path form sheets that illuminate a plane of air, thus illuminating the location of particles in the flow. Two PIV digital cameras capable of recording two frames in one image were used. PIV was activated remotely to collect images of the dispersed dust particles. The images were then analyzed for particle displacement, allowing study of the flow. The images were analyzed using Insight 3G software provided by the PIV manufacturer (TSI, Inc.). This application can execute statistical analysis and generate 2-D and 3-D graphics in conjunction with applications such as TechPlot (Polysoft, Salt Lake City, UT, USA) and Matlab (The MathWorks, Inc., Natick, MA, USA).

2.3 Numerical methods

Computational fluid dynamics (CFD) numerical methods, (Darrell, et al., 2007), were used to simulate and analyze airflow patterns and thermal fields inside the chamber and around the manikin (Lu et al., 1997; Patankar, 1980). The CFD method is predicated on solving the Navier-Stokes equations, which are formulations of mass, momentum, and energy conservation laws for fluid flows. The equations are supplemented by fluid state equations defining the nature of the fluid and by empirical dependencies of fluid density, viscosity, and thermal conductivity on temperature.

To predict turbulent flow, the Favre-averaged Navier-Stokes equations were used, where time-averaged effects of the flow turbulence on the flow parameters were considered. In this procedure, the information on Reynolds stresses must be provided for the equations. To close this system of equations, transport equations for the turbulent kinetic energy and its dissipation rate, the so-called k-ε model, are employed. A laminar/turbulent boundary layer model was used to describe flows in near-wall regions. The model was based on the modified wall functions approach. This model is employed to characterize laminar and turbulent flows near the walls and to describe transitions from laminar to turbulent flow and vice versa. The modified wall function uses a Van Driest's profile instead of a logarithmic profile. If the size of the mesh cell near the wall is more than the boundary layer thickness, the integral boundary layer technology is used.

The CFD model calculates two-phase flows as a motion of spherical solid particles in a steady-state flow field. Their drag coefficient is calculated with Henderson's formula, derived for continuum laminar, transient, and turbulent flows over the particles and taking into account the temperature difference between the fluid and the particle. The gravity is also taken into account. The interaction of particles with the model surfaces is taken into account by specifying ideal or non-ideal reflection (which is typical for solid particles). The ideal reflection denotes that, in the impinging plane defined by the particle velocity vector and the surface normal at the impingement point, the particle velocity component tangent to the surface is conserved, whereas the particle velocity component normal to the surface changes its sign. A non-ideal reflection is specified by the two particle velocity restitution (reflection) coefficients.

Briefly, the CFD program solves the governing equations with the finite volume (FV) method on a spatially rectangular computational mesh designed in the Cartesian coordinate system with the planes orthogonal to its axes and refined locally at the solid/fluid interface and, if necessary in specified fluid regions, at the solid/solid surfaces and in the fluid region during calculation. Values of all the physical variables are stored at the mesh cell centers. In the FV method, the governing equations are discretized in a conservative form. The spatial derivatives are approximated with implicit difference operators of second-order accuracy. The time derivatives are approximated with an implicit first-order Euler scheme. The viscosity of the numerical scheme is negligible with respect to the fluid viscosity. All issues related to solution convergence, such as mashing or boundary flow convergence, are taken care of automatically or by user defined criteria.

A numerical (virtual) EWC (NEWC), as shown in Fig. 2, was used to model the airflow and aerosol dispersion inside the simulated office, (Rhie & Chow, 1983; Vlahostergios, et al., 2009). The dimensions of the NEWC were identical to the actual EWC. The NEWC is a fully functional meshed 3-D numerical model of the EWC and the articulated manikin seated at the table. The manikin's position and orientation could be changed and the chamber furnishings rearranged as desired. The NEWC was fitted with two ceiling vents that could be used to define air in-flow and out-flow as desired based on volume or pressure. For simulations, the wall temperature and the manikin's body temperature were 20 °C and 37 °C, respectively, based on actual experimental conditions.

2.4 Experiments

2.4.1 Table zone tests

In our experiments, the letter was folded as a trifold letter and then unfolded and placed on the desk. It was positioned at two locations in the EWC: (1) close to the vent with the center of the letter at 30 cm from the manikin's chest and (2) close to the manikin's chest with the center of the page at 20 cm from the manikin's chest. The letter was coated (contaminated) with dust. The dust coating was achieved by loading a small amount of fine test dust (Powder Technology, Inc., Burnsville, MN, USA) in the 5 μm or less size range onto a No. 270 sieve and vigorously shaking the sieve above the foil. Experiments were conducted with the EWC closed and no laboratory personnel present to reduce any uncontrolled disturbance to air motion.

The vertical test area measured 25 by 25 cm and was located directly above the letter in the vertical plane bisecting the manikin's chest, as indicated by the x,y coordinate system shown just above the table in front of the manikin's chest in Fig. 1. The coordinate system origin was located on the table surface 5 cm from the manikin's chest. Thus, the x-axis coincided with the table surface and extended from the manikin toward the wall of the test room, while the y-axis extended vertically upwards. (Note that because of various limitations, such as accessibility inside the EWC, PIV camera positioning and viewing orientation, and a separate CFD-defined calculational domain, several coordinate systems appear in images and figures in this paper.)

The HVAC system was activated simultaneously with the PIV system to capture the event of the dust being reentrained from the foil. The PIV system could collect 20 double images in real time (saved in ROM) at a frequency of up to 10 images per second. Thus, to increase the possibility of detecting particle liftoff from the letter, we kept the PIV frequency at 2–3 images per second. These experiments showed that dust particles were indeed blown from the letter and reached the breathing zone of the manikin, as discussed below.

2.4.2 Breathing zone tests

After demonstrating in the table zone tests that particles could be lifted from the contaminated letter, experiments were conducted to determine if these particles reached the manikin's breathing zone. For the purpose of these experiments, the PIV test section was positioned in front of the manikin's head. This positioning is reflected by the x,y coordinate system adjacent to the manikin's face (see Fig. 1). Experimental procedures were similar to those in the previous experiments.

3. Experimental results and analyses

3.1 Airflow pattern in table zone area

Several experiments were conducted with theatrical smoke particles fed into the air duct system to determine the airflow pattern above the table. When the blower was activated, the air velocity from the vent quickly reached approximately 1 m s^{-1}. PIV images of the entire test area were then analyzed. Representative velocity vector fields, measured within a second of each other, are shown in Fig. 3a and 3b.

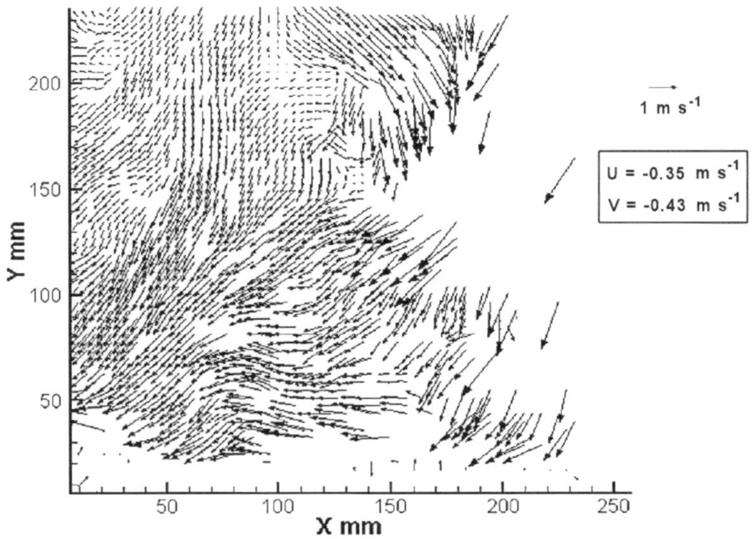

Fig. 3a. Airflow vector field in PIV test area just above the letter. Manikin's torso is to the left of the y-axis. For the investigation area shown, the average U (horizontal) velocity component was -0.35 m s^{-1}, and the average V (vertical) velocity component was -0.43 m s^{-1}.

Areas void of vectors, especially in Fig. 3a, most likely resulted from the lack of particles at the instant the image was taken. The smoke generator was delivering particles directly into the venting duct and images were captured before well-mixed conditions were achieved.
A comparison of Fig. 3a and 3b shows that the velocities in Fig. 3b are generally higher than in Fig. 3a, because these images were captured as the blower was speeding up. The higher velocity resulted from activation of the air mover and its rapid acceleration to the steady maximum rate. Part of the airflow is diverted by the table toward the manikin's chest, especially within the layer 10 cm from the table surface. Although the average velocity components during those seconds when the images were captured were on the order of 0.5 m s^{-1}, the increased vector lengths in Fig. 3a and 3b show that the velocity of the air flowing parallel to the table surface in that layer was higher and exceeded 1 m s^{-1}.

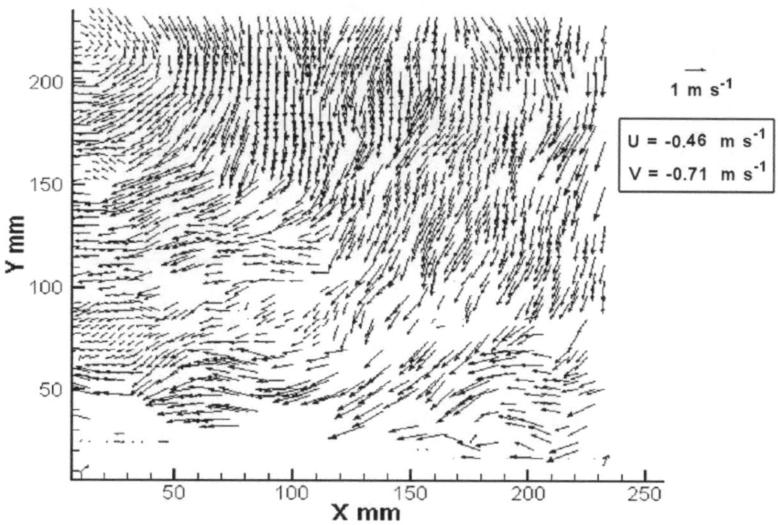

Fig. 3b. Airflow vector field in PIV test area just above the letter. The manikin's torso is to the left of the y-axis. For the investigation area shown, the average U (horizontal) velocity component was -0.46 m s^{-1}, and the average V (vertical) velocity component was -0.71 m s^{-1}.

3.2 Contaminated letter tests

In these experiments, the air mover and the PIV system were activated simultaneously to capture images of the dust being blown from the letter. Particle motion away from the edge of the letter is visible in Fig. 4a in the form of a particle cloud. This area was analyzed to produce the particle velocity vector field shown in Fig. 4b. Although particle motion toward the manikin's chest was a dominating characteristic of the transport, some particle motion was affected by the flow separation from the letter edge.

Suspecting that higher average horizontal air velocities may exist along the table surface farther from the vent's central axis stagnation area, the letter was positioned closer to the manikin, at 20 cm from the chest. The event is shown in Fig. 5a and its corresponding velocity field in Fig. 5b.

As expected, this experiment resulted in higher average horizontal particle velocities than in the previous case. Positioning the letter somewhat farther away from the vent resulted in particles being subjected to a less chaotic and more developed boundary airflow pattern. Such velocities can certainly be effective in delivering the dust to the manikin's chest.

Although these data produced clear, quantifiable evidence that particles on the contaminated foil can become airborne, some particles became deposited on the table surface, thus contaminating the table surface as shown in Fig. 6. The initial powder coating of the letter was very fine. The air jet affected the particles in a unique way: namely, particles traversed the surface and in the process agglomerated into much larger particles that are easily visible on the page surface and the table surface. Many particles followed that airflow below the table edge and contaminated lower parts of the manikin's torso in the process.

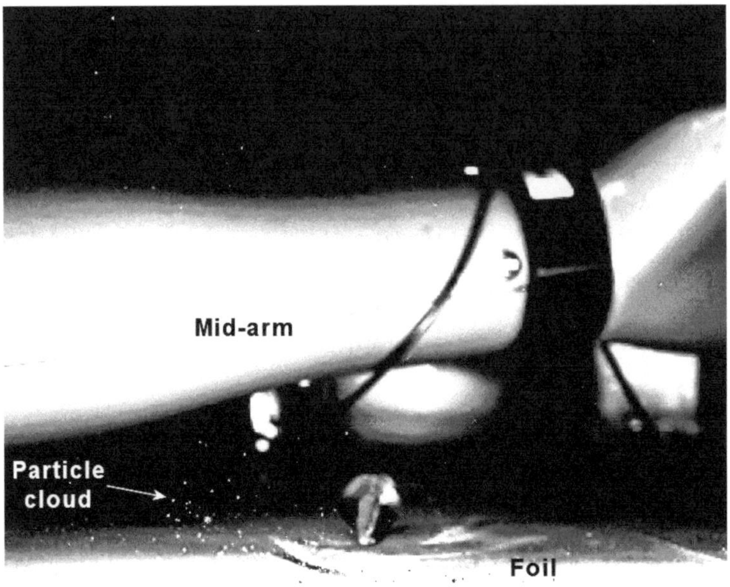

Fig. 4a. Cloud of particles (under mid-arm) moving away from the letter positioned 30 cm from the chest by air emerging from the ceiling vent.

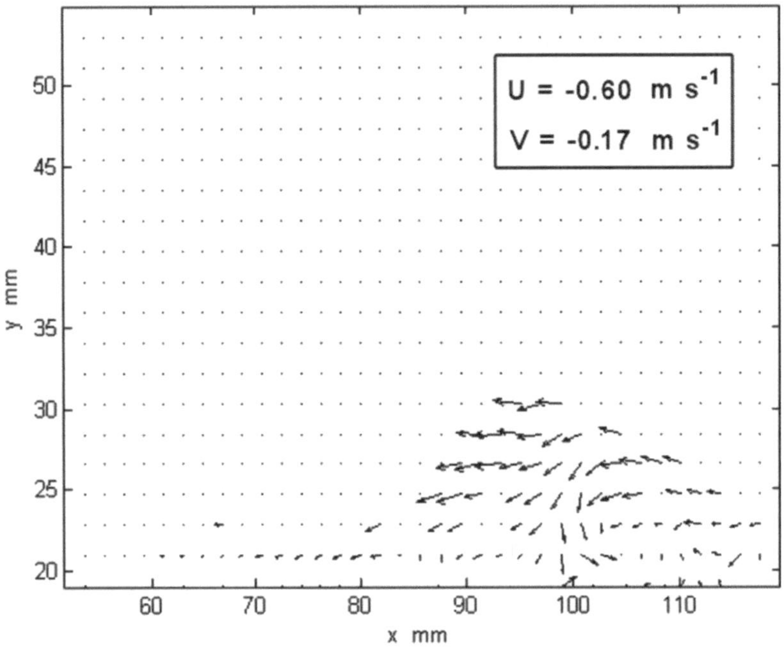

Fig. 4b. Air velocity field obtained using particles lifted from the letter positioned 30 cm from the chest. The average U (horizontal) velocity component was -0.60 m s^{-1}, and the average V (vertical) velocity component was -0.17 m s^{-1}. The letter surface is essentially at y = 20 mm. (Values smaller than y = 20 mm pertained to the part of the image irrelevant to particle transport.)

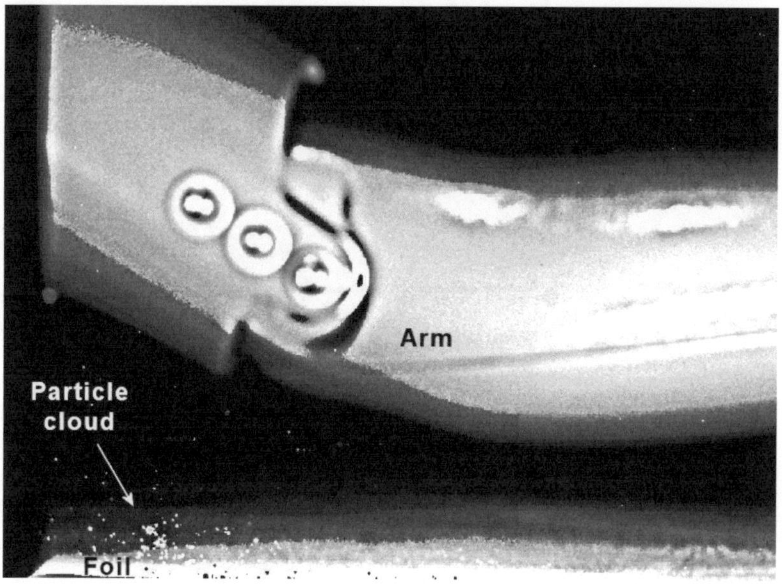

Fig. 5a. Contaminant particles (lower left corner of image) being moved from letter positioned 20 cm from the chest by air emerging from the vent.

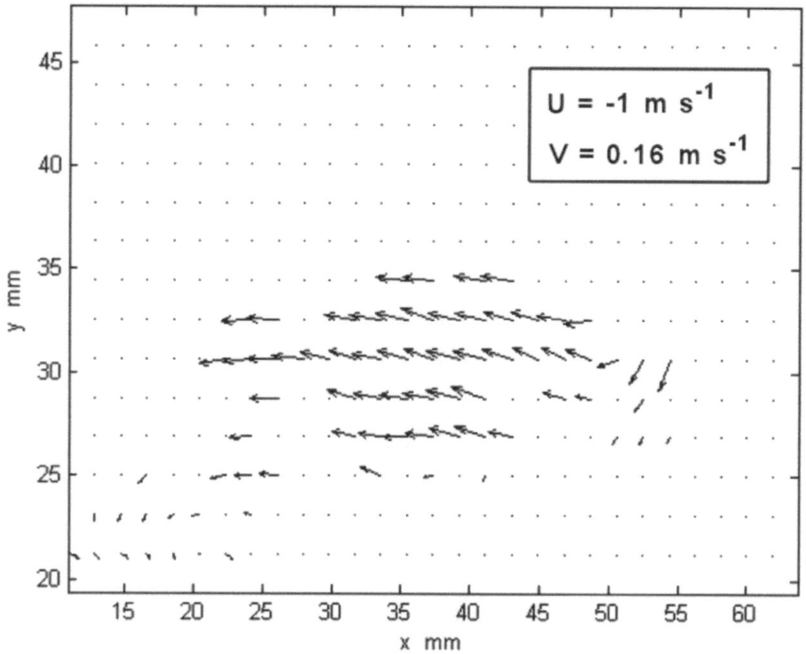

Fig. 5b. Velocity field within 10 mm of the letter surface obtained using particles lifted from the letter. The average U (horizontal) velocity component was approximately -1 m s^{-1}, and the average V (vertical) velocity component was 0.16 m s^{-1}. The letter surface is essentially at y = 25 mm. (Values smaller than y = 25 mm pertained to the part of the image irrelevant to particle transport.)

Fig. 6. Dust pattern created by air from the vent as dust is blown from the letter. The manikin is positioned to the left from the table edge visible in this figure.

3.3 Breathing zone tests

To determine whether particles reentrained from the letter reached the manikin's breathing zone, PIV images were analyzed to obtain a velocity vector field like the one in Fig. 7. A few seconds later a residual smoke aerosol entered the airflow and allowed detailed observation of the airflow in front of the manikin's face. The flow pattern is shown in Fig. 8. Strong deflection by the chin and other facial features is noticeable. In addition, the orientation of the flow vectors also suggests the possibility that a recirculation zone is created in front of

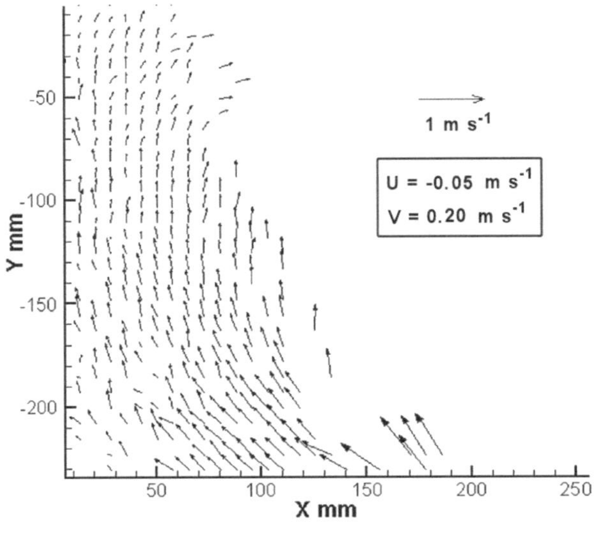

Fig. 7. Airflow vector field in the PIV test area in front of the face, created entirely by particles lifted from the contaminated letter. The manikin's head is positioned to the left of the y-axis. The average U (horizontal) velocity component was -0.05 m s^{-1}, and the average V (vertical) velocity component was 0.2 m s^{-1}. Y = 0 corresponds to the top of the manikin's head, a convenient point of reference. X = 0 is adjacent to the manikin's face.

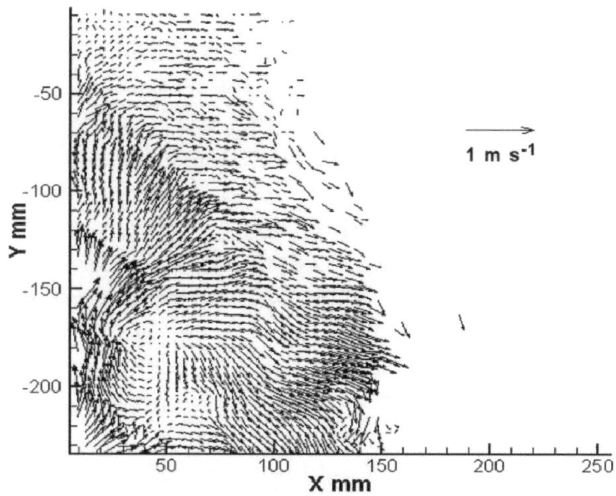

Fig. 8. Detailed airflow pattern in front of the manikin's face, as depicted by the smoke particles carried by the vent air. Y = 0 corresponds to the top of the manikin's head, a convenient point of reference. X = 0 is adjacent to the manikin's face

and slightly below the head zone. Such recirculation can lead to a longer exposure time. The very few vectors after approximately x = 130 is likely due to lack of light sheet in that area. Here the light sheet (shaped triangularly) originated from above the manikin's head to illuminate the area as close to the face as possible.

4. Numerical results and analyses

A numerical (virtual) EWC (NEWC) with the TAM seated at the table was created, as shown in the two cross-sectional views (side and front) in Fig. 9. CFD simulations were conducted using the NEWC. Boundary conditions were set based on actual experimental conditions as described in the experiments above. The simulations assumed that the wall's temperature was 20 °C and the manikin's body temperature was 37 °C. The inlet air velocity was assumed to be 1 m s^{-1}, which matched closely the inlet velocity during the experiments when PIV images were taken.

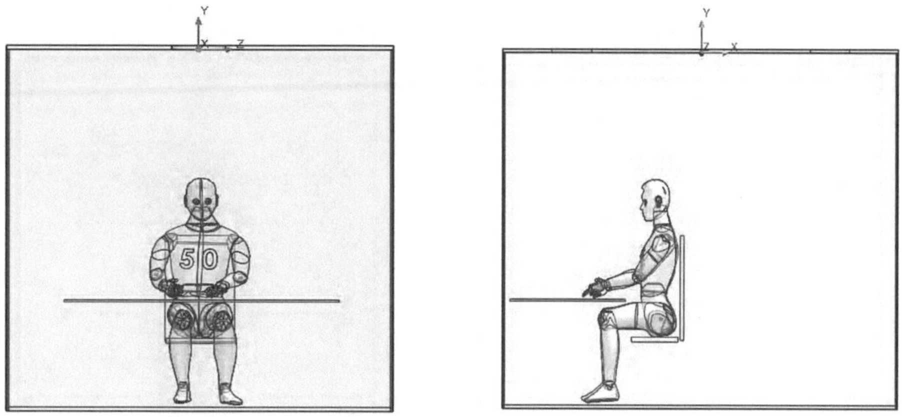

Fig. 9. Front and side views of the NEWC

4.1 Airflow patterns inside NEWC with seated TAM

Fig. 10 shows the air velocity profile (velocity vectors and color-coded velocity range zones) in the vertical middle cross section of the NEWC. Deflection of the air by the table toward the manikin and the opposing wall is clearly seen here as indicated by the vectors. A stagnation area around the inlet axis and accelerated airflow zones along the table surface are also quite visible.

A more detailed velocity field is presented in Fig. 11, with color-coded zones representing the temperature field in the NEWC. To avoid clutter, the velocity field is demonstrated using only 30 trajectories. This figure shows several remarkable stationary eddies that develop near the manikin. The vortex slightly below the head zone and the flow pattern above the table surface were also captured in the PIV experiments. An enlarged section of Fig. 11 is shown in Fig. 12 to elucidate the detailed structure of the flow near the manikin's body.

In the flow near the manikin, body heat contributes to convective airflow along the body, resulting in airflow in the upward direction. If this airflow is contaminated by particles, as from the contaminated foil under study, the flow pattern will certainly result in contamination of the torso. (In fact, Fig. 1 shows that the manikin's torso has become covered with the dust originating from the contaminated foil.) In addition, the stationary eddies that form in front of and behind the head may result in enhanced exposure to the contaminant. Turning the head away from the table may not be helpful in terms of avoiding exposure. Other eddies, such as the ones near the wall and under the table in Fig. 11, may require a longer time to clear the contaminating material from the air in the room.

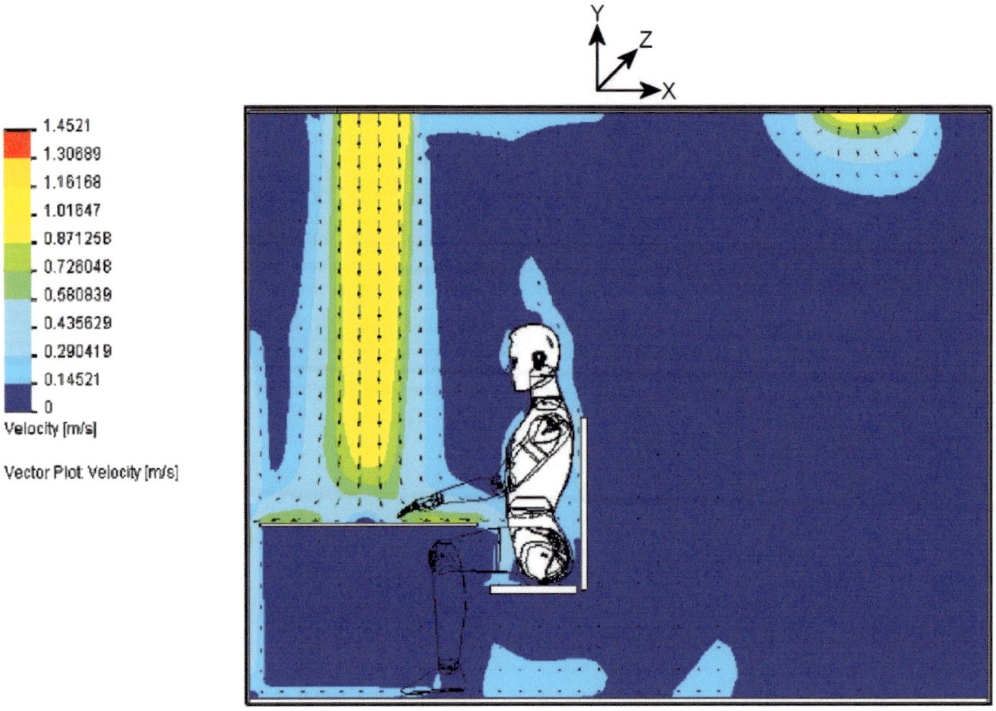

Fig. 10. Air velocity profile in vertical middle cross section of NEWC

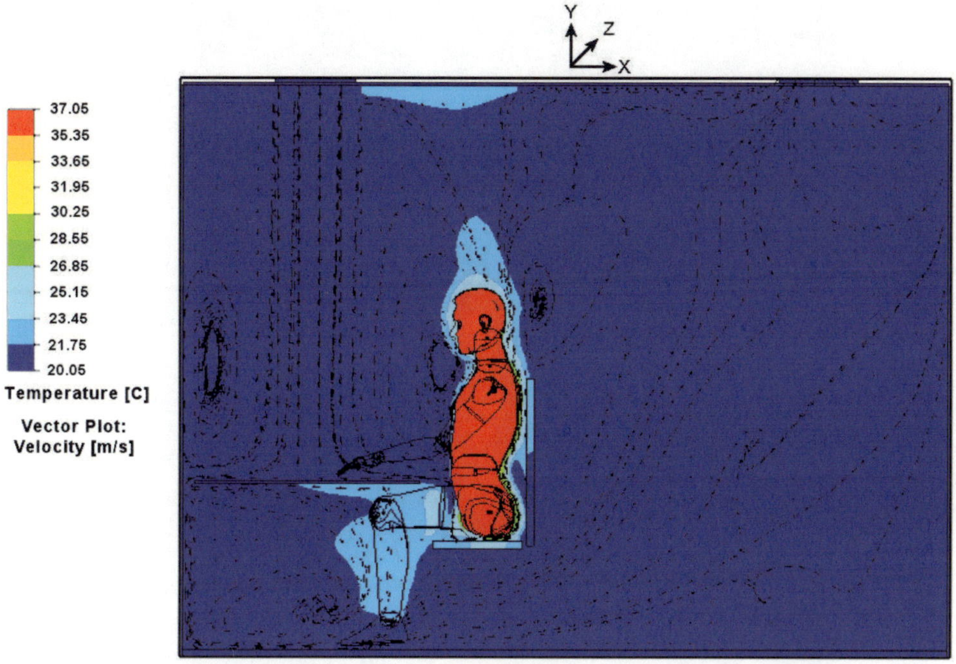

Fig. 11. 2-D air velocity field, represented by 30 trajectories, in vertical middle cross section of NEWC

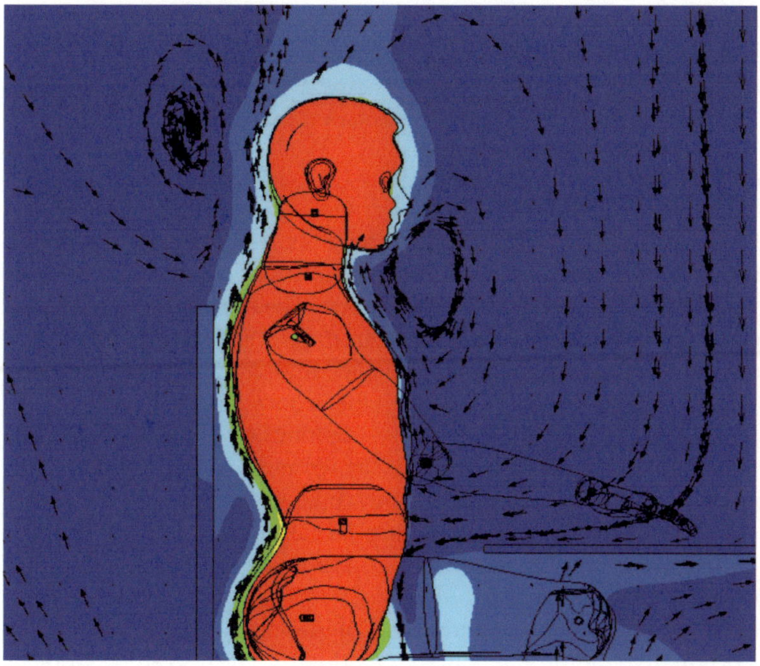

Fig. 12. 2-D flow patterns near manikin's body. Picture is flipped horizontally relative to previous one in order to match the PIV views shown in Fig. 7 and 8.

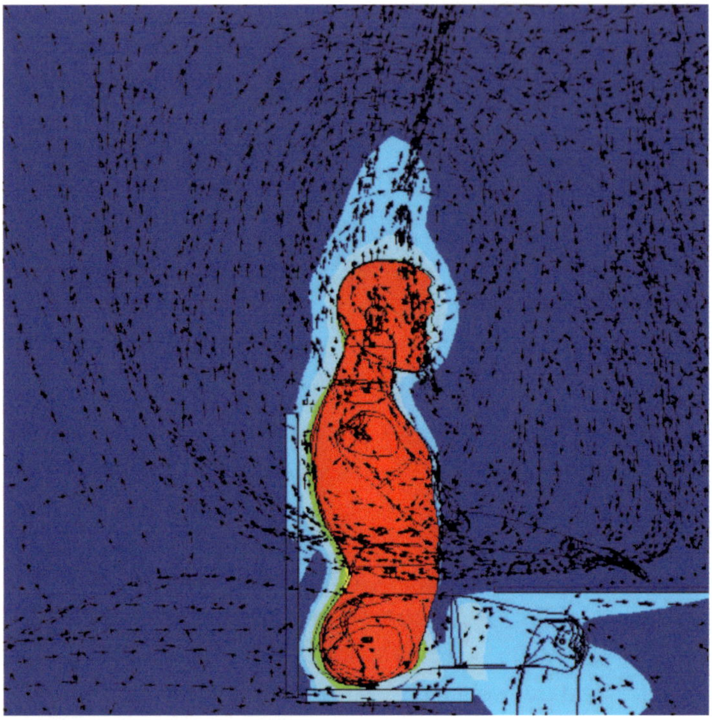

Fig. 13. 3-D flow trajectories inside the NEWC.

In Fig. 13, 3-D flow trajectories from the entire chamber volume are shown as projected onto the central plane. From this figure, one can surmise that the upper torso is essentially engulfed in a complex recirculating vortex. If this air was contaminated, the flow pattern suggests entrapment of the torso in the contaminated personal cloud.

5. Conclusion

The potential exposure of a person to biocontaminants residing on the surface of a letter was examined through experimental and computational evaluation of particle motion and fluid flow between a flat surface and the breathing zone. The scenario examined used an airflow pathway where air is released and withdrawn from ceiling vents on either side of a manikin sitting at a desk in a simulated office enclosure and a black anodized aluminum sheet sprinkled with fine test dust simulating the contaminated letter. Particle imaging velocimetry (PIV) showed that dust from the contaminated sheet could readily reach the torso and the breathing zone of a person sitting next to the letter. Further, extensive contamination of the table and person's body below the table is also likely. Numerical simulations suggest that several recirculating eddies may form in the proximity of the person's head. Such eddies may entrap contaminant particles and thus prolong air contamination and enhance exposure.

The experimental data analyses and numerical modeling described above demonstrate that dust particles originally coating the contaminated letter are dislodged from the letter when it is inadvertently positioned under a ceiling vent. Boundary airflow in the vicinity of the letter causes particle entrainment into the air. Subsequent airflow toward the person's chest and further vertical airflow is induced by body heat and is sufficient to deliver particles from the contaminated letter to the breathing zone in a couple of seconds. The presence of a recirculating vortex behind the manikin's head suggests that the air behind the manikin rapidly becomes contaminated as well. The flow patterns obtained in numerical simulations suggest that the entire torso will become engulfed in a complex recirculating cloud of particles that leads to its overall contamination.

Based on the experimental and numerical analyses conducted above, it is clear that contaminated dust residing on the surface of a letter would very likely be entrained and transported to the breathing zone of a subject. Further, due to the complex fluid motion generated by the HVAC system, this material is likely to be widely dispersed throughout the office.

6. Disclaimer

The U.S. Environmental Protection Agency through its Office of Research and Development funded and managed the research described here under Contract EP-D-05-065 with Alion Science and Technology. The views expressed in this paper are those of the authors and do not necessarily reflect the views or policies of the U.S. Environmental Protection Agency. Mention of trade names or commercial products does not constitute endorsement or recommendation for use.

7. References

Agranovski, E., Pyankov, V. & Altman, S. (2005). Bioaerosol Contamination of Ambient Air as the Result of Opening Envelopes Containing Microbial Materials. *Aerosol Science and Technology*, Vol. 39, No. 11, pp. 1048 – 1055.

de Armond, P. (2002). The anthrax letters. Albion Monitor, http://www.monitor.net/monitor/0208a/anthrax.html.

Block, S., (2001). The growing threat of biological weapons. *American Scientist*, Vol. *89*, pp. 28-37.

Ferziger, H. (1990). Approaches to turbulent flow computation: applications to flow over obstacles. *J Wind Eng Indus Aerodyn,* Vol. 35, pp. 1–19.

Darrell, W., Pepper, & Xiuling, W., (2007). A meshless model for rapid prediction of indoor contaminant dispersion ICCES, Vol. 13, No. 1, pp.15-21.

Dull, P., Wilson, K., Kournikakis, B., Whitney, E., Boulet, C., & Ho, J., (2002), Bacillus anthracis aerosolization associated with a contaminated mail-sorting machine. *Emerging Infectious Diseases*, Vol. *8*, pp. 1044-1047.

Duncan, S., Kournikakis, B., & Ho, J. (2009). Pulmonary deposition of aerosolized Bacillus atropheus (BG) in an awake, unrestrained swine model due to exposure from a simulated anthrax letter incident. *Inhalation Toxicology*, Vol. 21, No. 2, pp. 141–152.

Ho, J. & Fisher, G. (1993). Detection of BW agents: flow cytometry measurement of *Bacillus subtilis* (BG) spore fluorescence. Defense Research Establishment Suffield Technical Memorandum, DRES SM-1421.

Ho, J. & Duncan, S., (2005). Estimating aerosol hazards from an anthrax letter. *Journal of Aerosol Science*, Vol. *36, pp.* 701-719.

Lien, F., Ji, H., Yee, E., & Kournikakis, B., (2010). Prediction of aerosol hazards arising from the opening of an anthrax-tainted letter in an open office environment using computational fluid dynamics. *Journal of Engineering Science and Technology.* Vol. 5, No. 3, pp. 302 - 331.

Lu, W., Howarth, T, & Jeary, P. (1997). Prediction of airflow and temperature field in a room with convective heat source. *Build Environ,* Vol. 32, No. 6, pp. 541–550.

Kournikakis, B., Armour, J., Boulet, C.A., Spence, M. & Parsons, B. (2001). Risk assessment of anthrax letters. *Defense Research Establishment Suffield Technical Report,* DRES TR 2001-048.

Kournikakis, B., Walker, M., Ho, J., & Duncan, S. (2009). Statistical analysis of bacterial spore aerosols created by opening a spore containing "Anthrax Letter" in an office. *Journal of Aerosol Science,* Vol. 40, No. 6, pp. 514–522.

Kournikakis, B., Ho, J., & Duncan, S. (2010). Anthrax letters: personal exposure, building contamination and effectiveness of immediate mitigation measures. *Journal of Occupational and Environmental Hygiene,* Vol. 7, No. 2, pp. 71-79.

Mangili A. & Gendreau, A. (2005). Transmission of infectious diseases during commercial air travel. *Lancet,* Vol. 365, pp. 989–994.

Nardell, A., McInnis, B., Thomas, B. & Weidhaas S. (1986). Exogenous reinfection with tuberculosis in a shelter for the homeless. *New Engl J Med,* No. 315, pp. 1570–1571.

NATO Programme for Security through Science. (2005). Risk Assessment and Risk Communication Strategies in Bioterrorism Preparedness. In Green MS, Zenilman, J., Cohen, D., Wiser, I. & Balicer, D., editors. NATO Security through Science Series - A: Chemistry and Biology. The Netherlands: Springer. ISBN 978-1-4020-5807-3 (PB), ISBN 978-1-4020-5806-6 (HB).

Nazaroff, W. (2004). Indoor particle dynamics. *Indoor Air,* Vol. 14, pp. 175–183.

Patankar, S. (1980). *Numerical heat transfer and fluid flow.* Hemisphere Publishing Corporation.

Posner, D., Buchanan, R. &Dunn-Rankin D. (2003). Measurement and prediction of indoor air flow in a model room. *Energ Build,* Vol. 35, No. 5, pp. 515–526.

Price, P., Sohn, M., Lacommare, K. & Mcwilliams, J. (2009). Framework for evaluating anthrax risk in buildings. *Environmental Science and Technology,* Vol. 43, No. 6, pp. 1783–1787.

Reshetin, V., & Regens, J., (2003). Simulation modeling of anthrax spore dispersion in a bioterrorism incident. *Risk Analysis,* Vol. 23, pp. 1135-1145.

Reshetin, V., & Regens, D. (2004). Evaluation of malignant anthrax spore dispersion in high-rise buildings. *Journal of Engineering Physics and Thermophysics,* Vol. 77, No. 6, pp.1155–1166.

Rhie, C. & Chow, W., (1983). A numerical study of the turbulent flow past an isolated airfoil with trailing edge separation. *AIAA Journal,* Vol. 21, No. 11, pp. 1525-1532.

Richmond-Bryant, J., Eisner, A., Brixey, L., & Wiener, R. (2006). Short-term dispersion of indoor aerosols: can it be assumed the room is well mixed? *Building and Environment,* Vol. 41, No. 2, pp. 156-162.

Rim, D. & Novoselac, A. (2009). Transport of particulate and gaseous pollutants in the vicinity of human body. *Build Environ,* Vol. 44, pp. 1840–1849.

Vlahostergios, Z., Yakinthos, K. & Goulas, A. (2009). Separation-induced boundary layer transition: modeling with a non-linear eddy-viscosity model coupled with the laminar kinetic energy equation. *Int J Heat Fluid Fl,* Vol. 30, No. 4, pp. 617–636.

Wallace, L. (1996). Indoor particles: a review. *J Air Waste Manage Assoc,* Vol. 46, pp. 98–126.

Part 4

Effect of Air Pollutants on Historical Objects

11

Air Pollution and Cultural Heritage: Searching for "The Relation Between Cause and Effect"

Eleni Metaxa
School of Chemical Engineering, National Technical University of Athens, Greece

1. Introduction

Pollution of the natural environment is largely unintended and unwanted consequences of human activities in manufacturing, transportation, agriculture and waste disposal. High levels of pollution are largely a consequence of industrialization, urbanization and the rapid increase of human population in modern times. Pollutants are commonly classified according to the part of the environment primarily effected by them, either by air, water or land. Sub-grouping depends on characteristics of the pollutants themselves: chemical, physical, thermal and others. Many pollutants affect more than one resource. The substances that pollute the atmosphere are either gases, finely divided soils, or finely dispersed liquids aerosols. Five major classes of pollutants are discharged into the air: carbon monoxide, sulphur oxides, hydrocarbons, nitrogen oxides and particulates (dust, ash). The principle source of air pollution is the burning of fossil fuels, e.g., coal, oil and derivatives of the latter, such as gasoline, in internal combustion engines or for heating or industrial purposes.

The term *heritage* was used for first time from experts in the early seventies, to declare all the human creation with artistic features, which have been delivered to us as hereditary asset, namely as heritage. At the end of the same decade, the term *heritage* acquired collective sense and it was used to talk about *European Heritage* or later about *Universal Heritage;* in any case to indicate monuments, objects and places.

If in a sense *culture* is the evolution of human life in space and time, the "monuments-remnants" of the human creation of all the times form the prints, the signs, the evidences, the strides of the human-beings progress within the time: "past narrates its history…". Thus, monuments form an undivided entirety with time and place, with man, his surroundings and his history. These unique and unprecedented fingerprints of human civilization form the natural and cultural heritage of a place, of a country, of a people, the peculiar features of a nation which characterize its identity.

Cultural heritage is continuously undergoing numerical strains: anthropogenic and natural ones, from which the former can be anticipated or/and prevented, whereas the latter not. The result of these strains is the *deterioration* of all the materials. In fact, there is no material which is not to be downgraded. The Second Law of Thermodynamics inevitably intervenes and finally results in the deterioration of all the materials. For this reason, materials' deterioration is independent, in practice, on their surroundings and it is taking place in any environment, even *without* the direct contact of the materials with the constituents of a corrosive environment. Of course, the environment impacts *quantitatively* the deterioration or corrosion phenomenon taking place, by means of the impact on the *rate* of the deterioration process(es) and the kind of the produced substances.

Air pollution as an anthropogenic reason for materials' deterioration forms a problem of a great importance, because it has catastrophic consequences, universally, in health, in the environment and in the cultural heritage monuments and artifacts. The most famous kind of atmospheric pollution is the photochemical cloud, whose components are complicated chemical reactions in atmosphere, which have as principal reactants the hydrocarbons, nitrogen oxides, sulphur oxides, ozone and ultraviolet radiation.

The *conservation* of works of art and antiquities is intended to: (a) the preservation of cultural heritage, (b) the deceleration of their deterioration processes and (c) the restoration, in some cases, of their form in order to be comprehensible from the public. All of these purposes can be achieved with: (i) control of the environment, (ii) saving static interferences on the monument (i.e., structural conservation), which restore the static sufficiency of the monument, so it does not collapse; (iii) saving interferences on the surface of the monument (i.e., surface conservation), since all decay actions start from the surface of the monument. As the Nobel prize-winners Wolfgang Pauli and Enrico Fermi have felicitously worded: "if God made solids, surfaces were work of Devil"! Indeed, solid surfaces are not uniform, namely homogeneous, but they present *heterogeneity*, which in general arises from the existence of "imperfections" of various origins. These imperfections are distributed randomly on the surface of the solid material influencing its potential.

An absolutely serious scientific approach of the problem of confrontation of historic buildings and monuments decay because of air pollution presupposes the finding of the relation between "cause and effect", namely of "how and why air-pollutants interact with each other and with the solid surfaces". Then someone could interfere and inhibit a corruptive action on them, by restricting even minimizing the conditions are being responsible for. A scientific answer in the previous question presupposes knowledge of the mechanism of materials surfaces deterioration due to polluted surrounding atmosphere.

2. A scientific approach to the problem of cultural heritage deterioration due to air pollution

In order to study the action of air pollutants on cultural heritage monuments is important not only to obtain results by pure chemical analysis of monuments but also to clarify the mechanism of this action. This mechanism may consist of various steps in series, which are usually rate processes, with the deposition as the first step, or sometimes equilibrium states, such as the distribution of air pollutant(s) between the solid surface and the nearby atmospheric environment through adsorption-desorption phenomena. Thus, a simulation of various physicochemical actions of air pollutant(s) on the solid surface must be done followed by the experimental determination of various physicochemical parameters pertaining to the adsorption-desorption phenomena and possible surface heterogeneous reactions constants as well. A schematic representation of the possible physicochemical actions taking place between air pollutant(s) and monuments surface could be the following one:

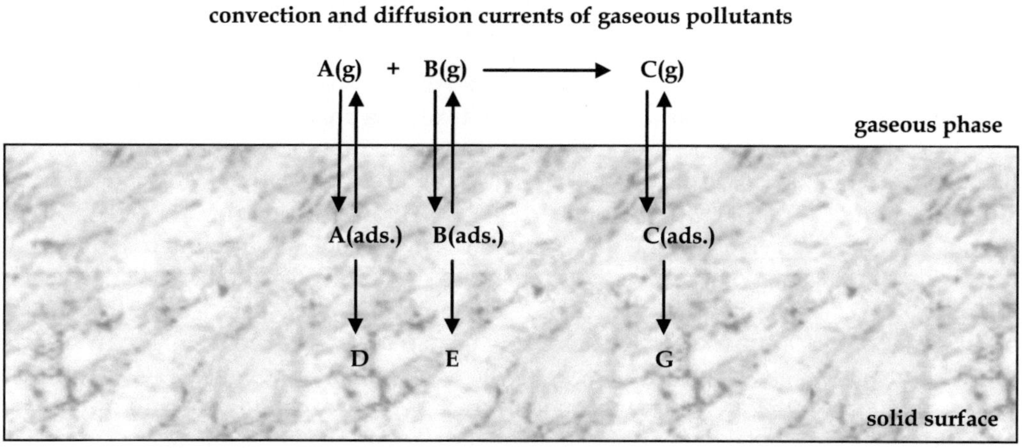

Fig. 1. A model describing the action of air pollutants with the surface of the monument

The model in Fig. 1 is based on the general concept of an open system, consisting of the exposed solid surface, above the which convection currents and diffusion currents as well are causing the transport of the gaseous pollutants A(g) and B(g) parallel and perpendicular to the solid surface, while a possible simultaneous interaction between them may produce another gaseous pollutant C(g), which may be also adsorbed onto the solid surface or/and desorbed back to the gaseous phase, or to be undergone a surface heterogeneous reaction, e.g. dissociation or isomerization. As soon as gaseous pollutants A(g) and B(g) are nearing the solid surface, adsorption phenomena are taking place, followed either by a surface chemical reaction between the adsorbed species producing D and E, or a desorption of them back to the gaseous phase. Therefore, the rate processes describing the above phenomena are the following ones: (i) diffusion of the pollutants from the gaseous to the solid surface, (ii) adsorption of them onto the solid surface, (iii) a possible surface heterogeneous reaction and (iv) desorption of the pollutants back to the gaseous phase.

Therefore an estimation of the crucial relations between environmental factors and materials' deterioration *cannot only* based on simple measurements of various physicochemical quantities which are validating the materials' decay, *but also* "time-resolved measurements" are necessary to be done, since only the latter can give information about the actual mechanism of materials' decay. The latter has, in fact, a "local" character, in the sense that it depends on the active sites of the solid surface which are available for adsorption at any particular time *t*. The achievement of this purpose could be done by using a dynamic experimental methodology, which could supply us with "real-time" measurements concerning the whole physicochemical phenomena taking place. To this direction, the novel method of the Reversed-Flow Inverse Gas Chromatography (RF-IGC) has already been successfully applied for various interacting systems gas–solid material or/and gas1/gas2–solid (e.g., gas=HCs, NO_x, SO_x, O_3, etc. and solid=a marble sample, a ceramic, a pigment, etc.). The results of these applications of RF-IGC in the investigation of the deterioration mechanism of cultural heritage caused by air-pollutants have already been published in high impact factor International Scientific Journals and reported in Scientific Symposiums both in Greece and abroad as well.

3. A brief overview of various methods and techniques used for studying environmental impacts on cultural heritage

Cultural heritage is comprised of a great variety of materials including buildings, monuments, pigments and art objects. Thus analytical data are essential for determining the state of conservation of the object, as well as the causes and mechanisms of its deterioration. The analytical methods used in this field of research are identical with those used at the cutting edge of modern science. Techniques developed for advanced physics and chemistry can apply to both of ancient and modern materials, since problems encountered in both the advanced technology and cultural heritage areas are similar. However, there is one essential difference between the analysis of ancient and modern materials, since an art or ancient object cannot be replaced and the consumption or damaging of even a small part of it for analytical purposes must be undertaken only where vital data cannot otherwise be obtained. Thus, a significant number of different modern instrumental methods for cultural heritage characterization are available and they have already been used for the investigation of the weathering effects of air pollution on them, supplying us with information on morphology, chemical composition and structure of the materials present in the monument, archaeological artifact, or art object. Depending on the information required and the procedure involved, the analysis can be considered destructive or nondestructive and it can be carried out on the bulk or the object surface. In addition, the obtained data can be panoramic or sequential and the measurements can be directly performed on the work itself or on a sample, depending on the instrumental technique used. In any case, however, one should aim at the maximization of information and the minimization of the consumed volume of the cultural object.

Materials characterization generally includes determination of chemical composition, of crystalline and molecular structure and of morphology of the object under investigation (A. Doménech-Carbó et al., 2009).

The major instrumental methods used for characterizing the *chemical composition* of the object either in layers or/and in its bulk include: (i) spectroscopic (e.g., XRF, AAS, ICP-AES, Mössbauer spectroscopy) or/and spectrometric techniques (e.g., ICP-MS, LA-ICP-MS), which have been widely used in the identification and determination of major, minor and trace-elements composing either inorganic or organic type cultural objects. The provided information and the application of each specific technique depends on the range of electromagnetic radiation and the phenomenon involved in its interaction with the materials present in the analyzed object (A. Doménech-Carbó et al., 2009; Jenkins, 2000; Putzig et al., 1994); (ii) activation methods (e.g., NAA, PAA), which are based on the interaction of the object material with (fast) neutrons or protons and provide information about the major, minor and trace element composition of the art and archaeological object, which, in turn, can be used to establish their provenance and temporal origin (A. Doménech-Carbó et al., 2009).

Concerning the characterization of the *crystalline and molecular structure* of cultural goods, the analytical techniques most frequently used are grouped into diffraction methods (XRD), spectroscopic (e.g., UV-VIS, FTIR, DRIFT, ATR, FTIR-PAS, Raman, NMR, EPR) and spectrometric methods (e.g., MS, DTMS, DPMS, MALDI), chromatographic methods and thermoanalytical methods (e.g., TG, DTA, DSC) (A. Doménech-Carbó et al., 2009; Jenkins, 2000; Putzig et al., 1994).

The majority of instrumental methods which yield morphological, topological and textural information of objects are mostly microscopy techniques (e.g., light microscopy (LM), electron microscopy (SEM, ESEM, TEM) and atomic force microscopy (AFM)) (A. Doménech-Carbó et al., 2009). By using light microscopy (either the low-magnification or the high-magnification technique), characteristics of materials such as the percentage of aggregates, pores, temper or specific minerals, pore or grain size and grain shape as well can be determined, allowing for a better analysis and interpretation of composition, technology, provenance, deterioration and conservation. In addition, the use of electrons instead of light in these instruments permits the characterization of the finest topography of the object surface and additional analytical information can be obtained. The AFM maps the topography of a substrate by monitoring the interaction force between the sample and a sharp tip attached to the end of a cantilever, so that the morphology of the surface of the studied solid sample can be reproduced at nanometer resolution (A. Doménech-Carbó et al., 2009). In addition, whenever a more elaborate surface analysis is pursued, methods based on the interaction of the incident energy provided by a microbeam of photons, electrons, or particles with the atoms or molecules located in the surface of the object sample are used. In such studies, the concept of "surface" should not considered in a strict sense, since the investigation concerns a depth in the range of a few μm on the solid surface. Such surface analysis techniques most frequently used in the characterization of cultural objects include high-resolution spatially resolved microspectroscopes, such as micro-FTIR (μFTIR), micro-Raman (μRaman), laser-induced breakdown spectroscopy (LIBS), micro-XRF (μXRF), XPS, PIXE, etc (A. Doménech-Carbó et al., 2009; Giakoumaki et al., 2007; Jenkins, 2000; Putzig et al., 1994).

It is worthy of noting that the time-resolved versions of the previous spectroscopic methods (TRS), although it is not new, has opened up a wide range of nascent application areas, including test and measurement in materials characterization. Though the basic technique differs little from the traditional spectroscopic methods, it allows us to measure the temporal dynamics and the kinetics of photophysical processes. The advantage of TRS over traditional spectroscopy is that it enables scientists to make more exact measurements of a sample's properties (Bhargava and Levin, 2003; Isnard, 2006; Miliani et al., 2010; Osticioli et al., 2009; Putzig et al., 1994; Quellette, 2004).

In what follows some representative examples of various analytical techniques commonly used in this field are reported for a better understanding of the particular contribution of each method used.

- FTIR-studies in materials decay: Infrared radiation is usually defined as that electromagnetic radiation whose frequency is between ~ 14300 and 20 cm^{-1} (namely, ~ 0.7 and 500 μm). Within this region of the electromagnetic spectrum, chemical compounds absorb IR-radiation providing there is a dipole moment change during a normal molecular vibration, molecular rotation, molecular rotation/vibration, or a lattice mode or from combination, difference and overtones of the normal molecular vibrations. The frequencies and intensities of the IR-bands exhibited by a chemical compound uniquely characterize the material and its IR-spectrum can be used to identify and quantify the particular substance in an unknown sample. Thus, FTIR and μFTIR-spectroscopy is useful for the study of degradation forms of cultural heritage, as it permits to identify the degradation phases and to establish the structural relationship between them and the substrate. A representative example of application of this method concerns the results obtained on marble from a Roman sarcophagus, located in the medieval cloister of St. Cosimato Convent in Rome (Italy) and on oolitic limestone from the façade of St. Giuseppe Church in Syracuse (Sicily). The IR-spectra of these samples showed the presence of degradation products composed of calcium sulphate hydrate, commonly called gypsum ($CaSO_4 \cdot 2H_2O$) and calcium oxalate, as well as the presence of organic matter probably due to conservation materials. The qualitative distribution maps of degradation products, obtained by means of micro-FTIR (µFTIR) operating in ATR-mode, revealed that the degradation process is present deep inside the stones also if it is not visible macroscopically (La Russa et al., 2009).

- SEM-studies in deterioration of glass: Deterioration of glass includes both chemical and structural changes. The initial stage of attack is a process that involves ion-exchange between alkali ions, which are present in the silicate structure of the glass, such as Na, K, and hydrogen from the environment. This leads to the formation of a leached or so-called "gel layer" in which alkaline elements are depleted. In case of atmospheric attack, the leached ions will interact with components from the ambient air such as carbon dioxide and sulphur dioxide which will lead to a crust formation including products such as a calcite ($CaCO_3$) and gypsum ($CaSO_4 \cdot 2H_2O$) (Adriaens, 2005).
- A combination of stereo-microscope, XRD and ICP-OES techniques was used (Elgohary, 2008) for the investigation of stone degradation due to air-pollution in Amman citadel of Liwān. The whole investigation and specific measurements showed that the damage produced on the surfaces of various calcareous stone samples of this region, either being physical or chemical, such as crustation, crystallization, dirties accumulations and other deteriorating forms, was essentially the result of the synergistic action of rain water and the various gaseous pollutants at prevailed in the region under study.

4. Gas chromatographic instrumentation for studying the impacts of air pollution on cultural heritage

4.1 A brief overview of gas chromatographic techniques

Chromatography is a separation method that combines separation and analysis. It is well-known that chromatographic separations are based on physicochemical processes such as diffusion, adsorption and chemical equilibrium of the studied solutes distributed among the mobile and the stationary phase. Gas chromatography (GC) is a technique that is used not only to separate substances from each-other, but also to study physicochemical properties. Some of these properties measured are concerned with the moving gaseous phase, giving emphasis on the determination of the properties of the solutes; for instance, diffusion coefficients of solutes into the carrier gas. Gas chromatographic analysis suffers from the so-called broadening factors, the majority of which is related to non-fulfilment of the assumptions under which the central chromatographic equation embraced by Van Deemter is derived; namely, the non-negligible axial diffusion of the solute gas in the chromatographic column, the non-linearity of the distribution (e.g. adsorption) isotherm and the non-instantaneous equilibration of the solute distribution among the mobile and the stationary phase. However, through these broadening factors gas chromatography is capable of making physicochemical measurements, which lead to very precise and accurate results, by using relatively cheap instrumentation and very simple experimental arrangements. Among the most widely used gas chromatographic methods for physicochemical measurements are the traditional techniques of elution development, frontal analysis and displacement development under constant gas flow-rate (Cazes, 2009).

The majority of gas chromatographic physicochemical measurements has been done by the inverse gas chromatography (IGC) technique, which uses the same experimental procedures employed in direct gas chromatography, but it focuses its interest on the stationary phase and its behavior towards known probe solutes; for instance, the catalytic properties of the solid stationary phase for reactions between gases. As in direct GC, the results used in IGC to derive information about the physicochemical properties of the stationary phase are based on net retention volumes, broadening of elution peaks and further on the analysis of the statistical moments of the peaks. The usual inverse gas chromatography (IGC), having the stationary phase of the system as the main object of investigation, is an integration method and not a time-resolved chromatography, since it totally ignores the heterogeneity of the adsorbing solid surface, it does not take into account the non-linearity of isotherms, the non-negligible axial diffusion in the chromatographic column and the kinetics of mass transfer across the gas/solid boundary (Cazes, 2009; Katsanos & Karaiskakis, 2004; Thielmann, 2004).

All the afore-mentioned chromatographic systems are not usually in true equilibrium during the retention period, so that extrapolation to infinite dilution and zero carrier-gas flow-rate is required to approximate true equilibrium parameters. Moreover, they have not a time-resolved character of the experimental procedure, since they provide measurements for physicochemical properties statistically weighed over time and enclosed by the chromatographic elution peaks; some of these properties are indeed independent of time, but there are other properties strongly dependent on the time variable. A new version of IGC is a flow perturbation method, the so-called Reversed-Flow Inverse Gas Chromatography (RF-IGC), which has been introduced in 1980 by N. A. Katsanos et al., and since then it is extensively used as a tool to study various physicochemical processes (Katsanos, 1988; Katsanos & Karaiskakis, 2004). It is a differential method depending neither on retention times and net retention volumes, nor on broadening factors and statistical moments of the elution bands. In addition, the results of RF-IGC do not need extrapolation to infinite dilution and zero carrier gas flow rate to approximate true physicochemical parameters. All the determinations achieved by RF-IGC are based on rate measurements over an extended period of time, thus constituting a time-resolved chromatography (Katsanos & Karaiskakis, 2004).

4.2 The novel method of RF-IGC: physical description and experimental setup

The Reversed-Flow Inverse Gas Chromatography (RF-IGC) method: (i) abandons the main role of carrier-gas in classical gas chromatography and substitutes it with gaseous diffusion currents inside a new diffusion column perpendicular to the conventional chromatographic current (sampling column), the latter being a little far from the solid bed in which all the desired physicochemical phenomena take place in the absence of gas running; (ii) by means of a four or six port valve the direction of carrier-gas flow is reversed from time to time for short time intervals, thus creating extra narrow chromatographic peaks which are deposited onto the conventional chromatographic signal. All the above described are schematically presented in Figs. 2 and 3. By introducing these modifications, the carrier gas flow does not intervene with the measurement of the desired physicochemical quantities, which describe step by step the entire physicochemical phenomena taking place inside the diffusion column where no carrier gas flows but only a static pressure of it exists.

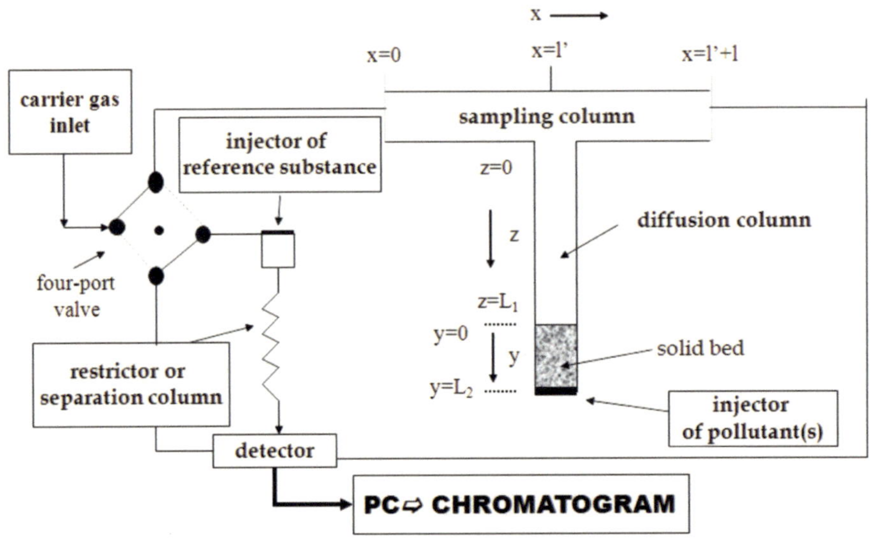

Fig. 2. Experimental setup of RF-IGC

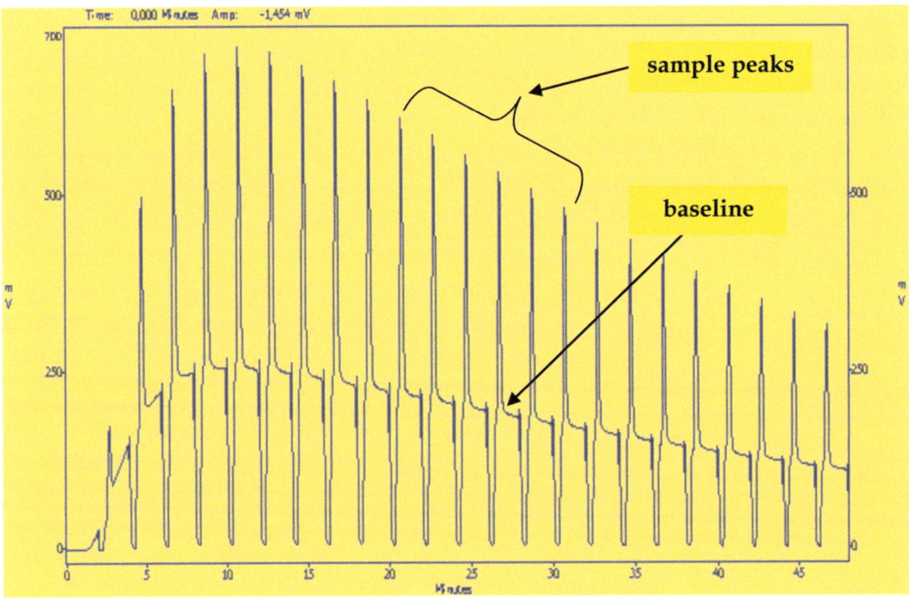

Fig. 3. A typical chromatogram obtained by RF-IGC

The extra chromatographic peaks (Fig. 3) obtained by repeatedly reversing the carrier gas direction for short time intervals are termed sample peaks, because they constitute samples of the phenomena taken from the region of their occurrence at various times, like small samples taken from a reaction occurring in a usual chemical flask containing the reactants. They have different heights depending on the time at which each flow reversal was made. Since this happens at various chosen times, it constitutes a time-resolved experiment like those in chemical kinetics. The experimental details by means of which the reversals are effected are shown in Fig. 2. From the series of the sample peaks obtained under various conditions, several physicochemical quantities have been determined and published (Agelakopoulou et al., 2009; Arvanitopoulou et al., 1994; Bakaoukas et al., 2005; Floropoulou et al., 2009; Katsanos et al., 1998,2003,2004; Metaxa et al., 2009a,2009b,2009c; Roubani-Kalantzopoulou, 2004,2009; Roubani-Kalantzopoulou et al., 1996; Sotiropoulou et al., 1995). The sample peaks are predicted theoretically by the so-called chromatographic sampling equation (1), which describes the concentration-time curve of the sample peaks created by the flow reversals and has been derived using mass balances, rates of change, etc., and integrating the resulting partial differential equations under given initial and boundary conditions. It gives the concentration of the solute at the junction of the sampling and the diffusion column x=l′ or $z=0$ of Fig. 2, for different values of the time variable. The sampling equation predicts the sample peaks theoretically and its predictions coincide with the experimental sample peaks shown in Fig. 3, the only difference being that the peaks predicted are square, whereas those actually found are not square owing obviously to non-ideality. In fact, the experimental peaks can be made as narrow as we want, since the width at their half-height is equal to the duration of the carrier-gas flow reversal. The equation describing the height, H, of the sample peaks as a function of time, t, when each flow-reversal was made has the form:

$$H^{1/M} = g \cdot c(l',t) = \sum_i A_i \exp(B_i,t) \tag{1}$$

where i runs from 1−4, M is the response factor of the detector used (M=1 for a flame ionization detector), g is the calibration factor of the detector (in $cm^4 \cdot mol$), $c(l',t)$ is the measured sampling concentration of the gaseous analyte (in $mol \cdot cm^{-3}$) at $x=l'$ or $z=0$ of Fig. 2, and A_i, B_i are functions of the physicochemical quantities pertaining to the various phenomena occurring in the solid bed region. The detailed content of A_i and B_i, as found from a non-linear least-square analysis of the plot of $H^{1/M}$ versus time t, leads to the clear determination of the physicochemical quantities of the mathematical model used; for instance, catalytic reaction constants, adsorption-desorption rate constants, gas and surface diffusion coefficients, local adsorption isotherms, local adsorption energies, local adsorption energy probability density functions, local lateral molecular interactions and adsorption rates, as will be explained for the action of various gaseous pollutants on calcareous stones, marbles and statues in the next sections. In addition, a brief account of the general principles which construct the mathematical model of the method of RF-IGC will be given in the following section.

4.2.1 Mathematical model

The theoretical analysis for the measurement of the time resolved physicochemical parameters by RF-IGC is based on the following equations:

i. two mass balances for the gaseous concentration of the analyte in the regions y and z, c_y ($mol \cdot cm^{-3}$) and c_z ($mol \cdot cm^{-3}$), respectively;

$$\frac{\partial c_y}{\partial t} = D_2 \frac{\partial^2 c_y}{\partial y^2} - k_{-1} \frac{a_S}{a_y}(c_S^* - c_S) - k_{app} c_y \tag{2}$$

$$\frac{\partial c_Z}{\partial t} = D_1 \frac{\partial^2 c_Z}{\partial z^2} - k_{app} c_Z \tag{3}$$

ii. one rate of change of the adsorbed concentration c_S (mol · g^{-1}) of the analyte in the region y and

$$\frac{\partial c_S}{\partial t} = k_{-1}\left(c_S^* - c_S\right) - k_2 c_S \tag{4}$$

iii. one local adsorption isotherm, which correlates the adsorbed equilibrium concentration c_S^* (mol · g^{-1}) on the solid surface with the non-adsorbed concentration c_y:

$$c^*_S = \frac{m_S}{a_S}\delta\left(y - L_2\right) + \frac{a_y}{a_S} k_1 \int_0^t c_y(\tau) d\tau \tag{5}$$

$k_{app.}$: is the apparent rate constant of a first- or pseudofirst-order reaction of the gaseous adsorbate in the gas phase (in s^{-1}).
D_2 : is the diffusion coefficient of this gaseous adsorbate into the gas phase in section y (in cm^2 s^{-1}).
k_{-1} : is the rate constant for desorption of the solute from the solid bulk (in s^{-1}).
c_S : is the adsorbed concentration of the gaseous adsorbate adsorbed on the solid at time t (in mol g^{-1}).
k_2 : is the rate constant of a possible first-order or pseudofirst-order surface reaction of the adsorbed solute (in s^{-1}).

With the initial conditions $c_y(0,y) = \frac{m}{a_y}\delta(y - L_2)$, and $c_S(0,y) = 0$, m being the amount (mol) of the gaseous adsorbate introduced as a pulse at $y=L_2$, all the required adsorption parameters are calculated from the experimental data - pairs (H, t) and various geometrical characteristics of the diffusion column and the solid bed - on the basis of the following equations, by means of a suitable PC-program based on non-linear least-squares regression analysis (Agelakopoulou et al., 2009; Arvanitopoulou et al., 1994; Bakaoukas et al., 2005; Floropoulou et al., 2009; Katsanos et al., 1998,2003,2004; Metaxa et al., 2009a,2009b,2009c; Roubani-Kalantzopoulou, 2004,2009; Roubani-Kalantzopoulou et al., 1996; Sotiropoulou et al., 1995):

a. local adsorption energies, ε

$$\varepsilon = RT\left[\ln\left(KRT\right) - \ln\left(RT\right) - \ln K^0\right] \tag{6}$$

b. local adsorption equilibrium concentrations, c_S^* *and* c_y

$$c_s^* = \frac{\alpha_y}{\alpha_s} k_1 \frac{\upsilon L_1}{g D_z} \sum_{i=1}^{3} \frac{A_i}{B_i}\left[\exp\left(B_i t\right) - 1\right] \tag{7}$$

$$c_y = \frac{\upsilon L_1}{g D_1} \sum_{i=1}^{3} A_i \exp\left(B_i t\right) \tag{8}$$

c. local adsorption isotherm, θ_t

$$\theta_t = \frac{c_s^*}{c_{s\max}^*} \tag{9}$$

$$\theta = 1 - \frac{1}{c_{s\max}^*} \cdot \frac{1}{KRT} \cdot \frac{\vartheta c_s^*}{\vartheta c_y} \tag{10}$$

d. local monolayer capacity, $c_{s\max}^*$:

$$c_{s\max}^* = c_s^* + \frac{\partial c_s^* / \partial c_y}{KRT} \tag{11}$$

e. probability distribution function for adsorption energy, $\varphi(\varepsilon; t)$:

$$\varphi(\varepsilon;t) = \frac{\theta}{c_{s\max}^* RT}\left[\frac{KRT\left(\partial c_s^* / \partial t\right) + \partial^2 c_s^* / \partial c_y \partial t}{\partial (KRT) / \partial t} - \frac{\partial c_s^* / \partial c_y}{KRT}\right] \tag{12}$$

f. dimensionless parameter, β, for lateral interactions:

$$z\omega = \beta RT \tag{13}$$

where: ω is the lateral interaction energy, z the number of neighbors for each adsorption site and β is a dimensionless parameter. Thus, the "$\theta z \omega$" is the added to ε "differential energy of adsorption due to lateral interactions", namely:

$$\theta z \omega = \beta \theta RT \tag{14}$$

All these relations are based on the Jovanovic local isotherm Eq. (15):

$$\theta(p, T, \varepsilon) = 1 - \exp(-K p) \tag{15}$$

where

$$K = K^0(T) \exp(\varepsilon / RT) \tag{16}$$

R being the gas constant, and

$$K^0 = \frac{h^3}{(2\pi m)^{3/2} (kT)^{5/2}} \cdot \frac{\upsilon_S(T)}{b_g(T)} \tag{17}$$

where: m is the molecular mass of the adsorbate; k is the Boltzmann's constant; h the Planck's constant; and the ratio $u_s(T)/b_g(T)$ of two partition functions, namely that of the adsorbed molecule, $u_s(T)$, and that for rotations-vibrations in the gas phase $b_g(T)$. This ratio is taken as a unity, approximately, as was done before (Agelakopoulou et al., 2009; Arvanitopoulou et al., 1994; Bakaoukas et al., 2005; Floropoulou et al., 2009; Katsanos et al.,

1998,2003,2004; Metaxa et al., 2009a,2009b,2009c; Roubani-Kalantzopoulou, 2004,2009; Roubani-Kalantzopoulou et al., 1996; Sotiropoulou et al., 1995).

The contribution of lateral molecular interactions in the overall phenomenon of adsorption and desorption is taken into account, by correcting Eq. (16) to include this type of energy:

$$K' = K^0 \exp\left(\frac{\varepsilon}{RT} + \beta\theta_i\right) = K \exp(\beta\theta_i) \tag{18}$$

Thus, Jovanovic isotherm described in Eq. (15), is modified accordingly, as well as any other equation based on it.

4.2.2 Applications of RF-IGC in studying physicochemical phenomena being responsible for materials' decay

The differential time-resolved RF-IGC method provides a new pathway for solids characterization, by supplying us with experimental *local* values of important physicochemical quantities, such as adsorption energy, adsorption isotherm, monolayer capacity, non-adsorbed concentration of gaseous analyte in equilibrium with the solid surface, probability density function and energy owing to lateral molecular interactions, which pertain to the particular surfaces. It is reminded that the term "local" means "with respect to time *t*", namely it regards only adsorption sites active at time *t*.

In the following sections, some representative results are described, which have already published in high impact International Scientific Journals and announced in Scientific Symposiums, concerning cultural heritage monuments in Greece (Agelakopoulou et al., 2009; Arvanitopoulou et al., 1994; Bakaoukas et al., 2005; Floropoulou et al., 2009; Katsanos et al., 1998,2003,2004; Metaxa et al., 2009a,2009b,2009c; Roubani-Kalantzopoulou, 2004,2009; Roubani-Kalantzopoulou et al., 1996; Sotiropoulou et al., 1995).

4.2.2.1 The local character of adsorption in materials decay: chemisorption or physisorption?

Due to their different crystallographic properties, natural stones exhibit different types of structural surfaces–each one with its own adsorption energy distribution function–which determines the variations in the observed weathering, deterioration patterns and processes of these materials. The adsorption of atoms or molecules onto these heterogeneous surfaces may occur either by the formation of strong chemical bonds (chemisorption) or via weaker physical attachment (physisorption) of the adsorbed molecules. It should be noted that although the majority of the reported results regards experiments done at relatively low temperatures, where physisorption is more expected, the structural materials of samples used (e.g., statues from Museums) are energetically upgraded, as a consequence of the whole process they have been undergone from the moment of their natural formation until the moment of their use as materials for creating statues. Thus, in the presence of an aggressive environment a chemisorption process is favorable; by means of RF-IGC experiments, chemisorption is observed taking place in the beginning of the experiments (Agelakopoulou et al., 2009; Floropoulou et al., 2009; Katsanos et al., 1998,2003,2004; Metaxa et al., 2009a,2009b,2009c; Roubani-Kalantzopoulou, 2004,2009). Furthermore, any kind of adsorption process (chemisorption or/and physisorption) takes place in different extent, because of the development of molecular lateral interactions (attractive or/and repulsive) between the adsorbed species, in addition to the adsorbent-adsorbate interactions, the latter being the predominant type of forces characterizing chemisorption. In case of *chemisorption*, the lateral interactions between adsorbates are mostly *repulsive*, resulting in *random* topography. The adsorbent-adsorbate interactions leading to chemisorptions are chemical bond forces.

The active sites, where *chemisorption* takes place, are included in the region A and correspond on the minima of potential energy of the surface, namely the maxima of adsorption energy. The distribution of the adsorbed molecules on these active sites is *random*. These sites correspond to *high* values of coverage θ and adsorption energy ε. The regions B and C include active centers that correspond to adsorption at sites of *lower* energy, ε and surface coverage, θ. They are characterized by a *patchwise* or/and an *island* topography of the adsorbed molecules and created by weak forces of Van der Waals type between the admolecules. Adsorption sites in B and C are not on a free surface, but they are created through the *lateral* interactions with molecules have been already adsorbed. When chemisorption is taking place, high adsorption energy values are observed, since they correspond to the minima of the potential energy of the surface. Contrary to this, physisorption entails low adsorption energy values, since it usually corresponds to the maxima of the surface potential (saddle points) [Rudzinski & Everett, 1992]. The decrease of adsorption entropy values verifies that adsorption occurred. All the above are depicted in Figs. (4a)-(4d). The "topography" of the solid surface is described experimentally through the time-resolved analysis of the energy distribution function $\varphi(\varepsilon;t)$ versus t, where distinguished regions corresponding to different kinds of active sites produced at different times are observed, as the three types of active sites recorded in Figs. (4c) and (5c). The number of distinct peaks recorded corresponds to different kinds of active sites appear and the area under these peaks corresponds to the collection of adsorbed species onto active sites with a definite mean energy value; in fact, this reflects or/and explains the "local" character of adsorption. It is worthy of noting that these results coincide with those extracted with simulation methods for the submonolayer adsorption of argon on the surface of crystalline rutile (Bakaev& Steele, 1992).

In the absence of SO_2, an adsorption-induced surface reconstruction creates new adsorption sites, as it is indicated in Figs. 4 (a, b, d) and Figs. 5 (a, b, d). The above observed surface reconstruction [Agelakopoulou et al., 2009; Floropoulou, 2009; Metaxa et al., 2009a, 2009b] may be attributed to one or more of the following reasons:

i. An *entropy increase* owing to the close approach of adsorbates in various configurations which increase the surface entropy and unveil new active sites for adsorption on the solid surface [Jansen, 2008].
ii. The creation of sub-surface states which does not affect the surface free-energy [Christmann, 1995].
iii. A co-variation in chemisorption and physisorption so that an equilibrium has been established between adsorption and desorption.

The most possible reason in this case seems to be the entropic one, because of the momentary increment of the adsorption entropy observed in Fig. (4d) as a local maximum, which is followed by a new slight decrease in the adsorption entropy thus confirming the readsorption induced. Obviously, the third kind of active sites observed for the adsorption of C_2H_2 on the surface of the statues (L291 of Kavala and L1991 of Philippi), in the absence of SO_2, resulted from this surface-reconstruction.; in the presence of SO_2, only two different types of active sites are observed, as Fig. (4c) indicates. Although the above-mentioned explanation for this surface reconstruction and readsorption induced concerns, in fact, the formation of islands without attractive interactions [Jansen, 2008], another surface reconstruction process which favours the formation of islands due to attractive lateral interactions have been proposed by other researchers (Velasco & Rezzano, 1999). The latter explanation could be also accepted in our case, in the sense that this island-formation occurs in higher time-values and after θ=1, at saddle-points positions (Roubani-Kalantzopoulou, 2004, 2009; Agelakopoulou et al., 2009; Metaxa et al., 2009a).

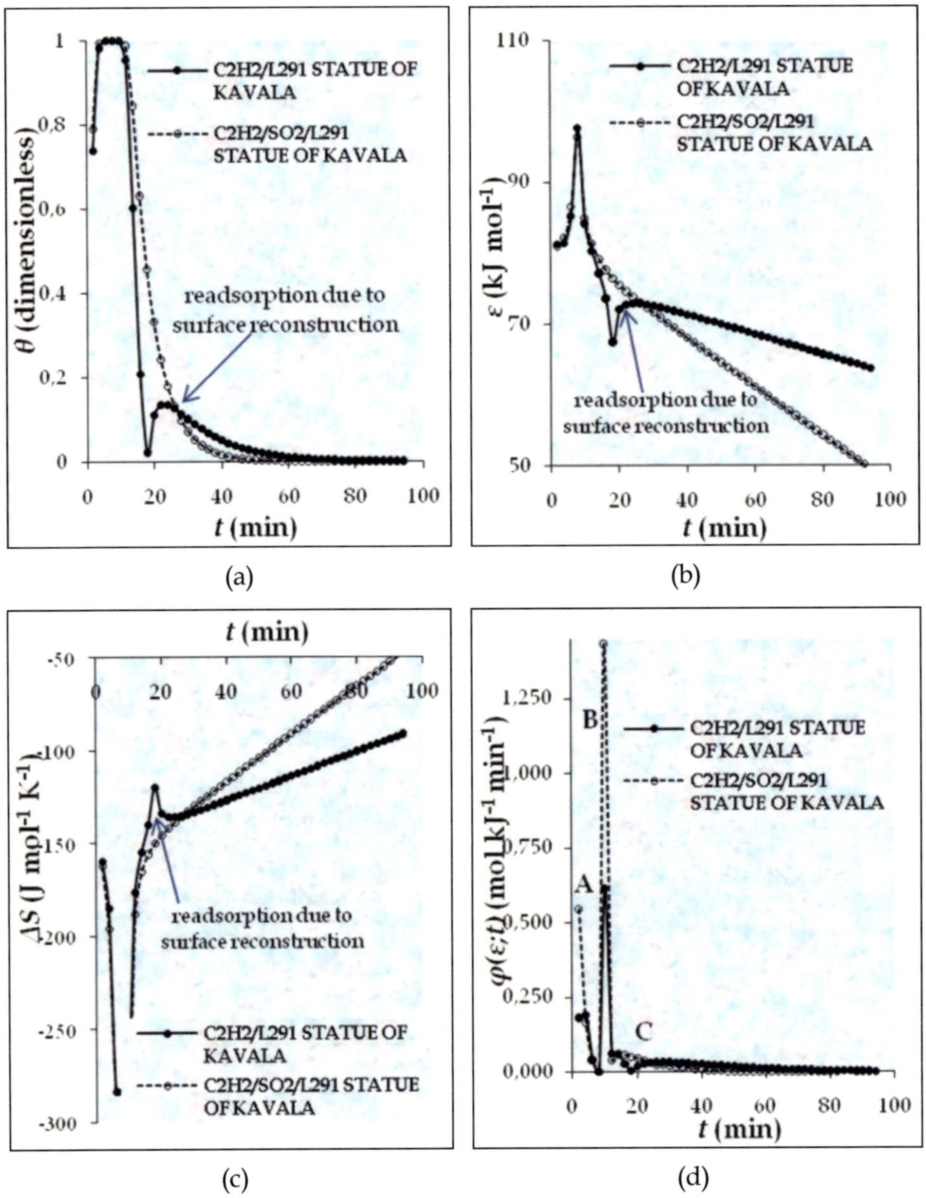

Fig. 4. Time-resolved analysis for the adsorption isotherms (4a), adsorption energies (4b), distribution energy functions (semi-logarithmic plot) (4c), and adsorption entropies (4d), concerning the systems $C_2H_{2(g)}$/ ($SO_{2(g)}$)/ L291 Statue of Kavala, Greece.

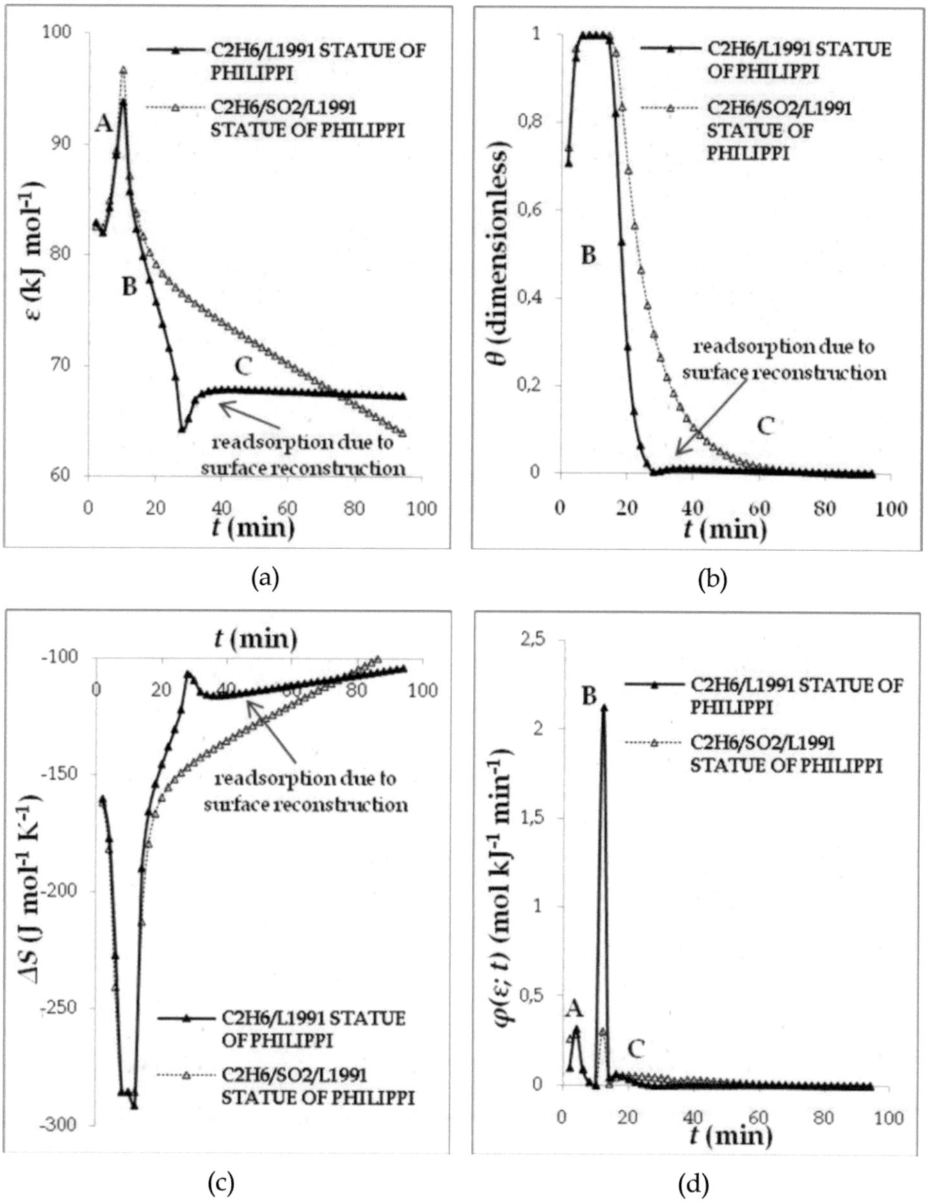

Fig. 5. Time-resolved analysis for the adsorption isotherms (4a), adsorption energies (4b), distribution energy functions (semi-logarithmic plot) (4c), and adsorption entropies (4d), concerning the systems $C_2H_{2(g)}/(SO_{2(g)})$/L1991 Statue of Philipi, Greece.

4.2.2.2 The role of synergy in the adsorption phenomena

The synergistic effect of a second pollutant has also been examined and is very obvious how it operates in each case. For example, in the presence of SO_2 lower values for c^*_{ssmax} are determined for the adsorption of ethane on the surface of the ancient statue L1991 taken from the interior of the Museum of Philippi, near Salonica, in Greece. This fact could be

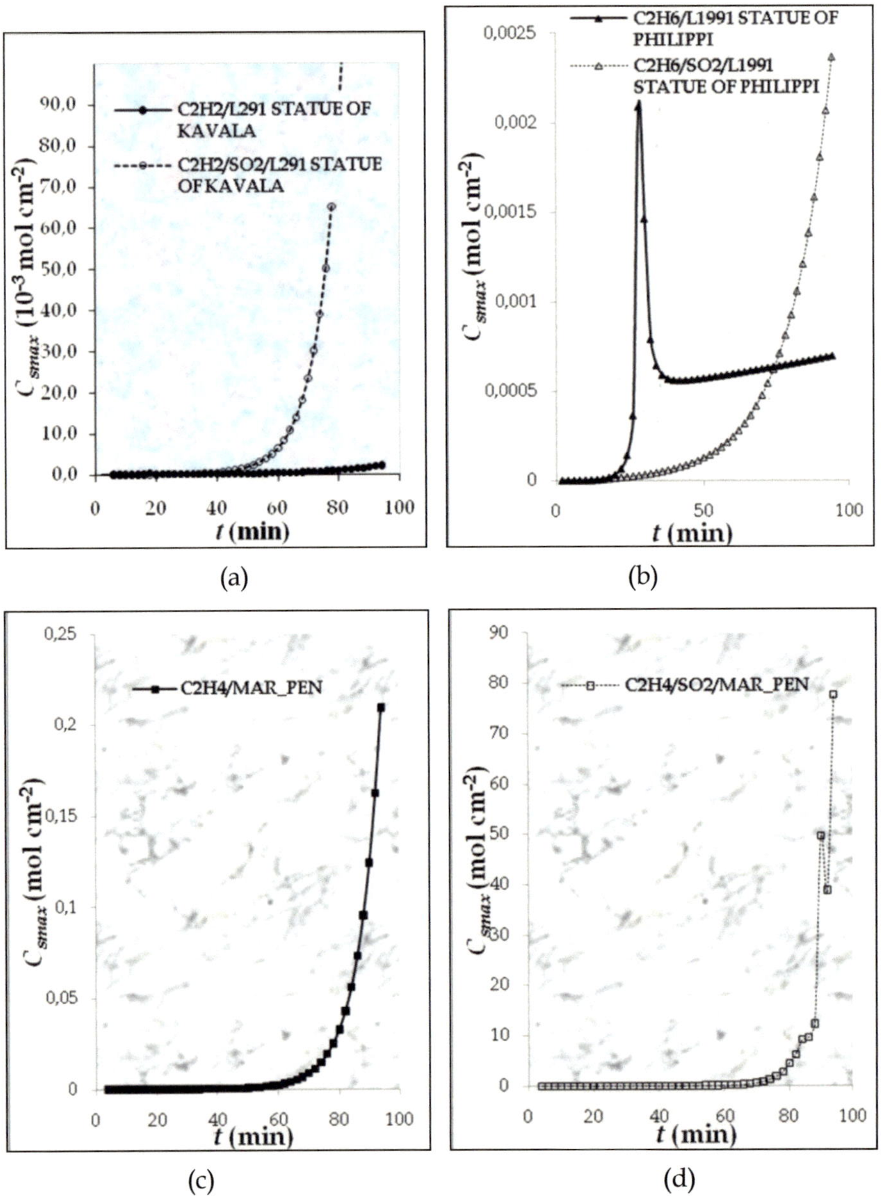

Fig. 6. Time-resolved analysis of local molecular capacity, c^*_{ssmax} , for the systems: (a) $C_2H_2/(SO_2)$/L291 statue of Kavala museum, (b) $C_2H_6/(SO_2)$/L1991 statue of Philippi museum and (c,d) $C_2H_4/(SO_2)$/Pentelic marble.

attributed either to an oxidation of the hydrocarbon molecules from SO_2 in the gaseous phase before adsorption takes place or to a competitive adsorption of SO_2-molecules towards ethane molecules on the active sites of the statue surface. This fact is confirmed by Fig. 5d, where the number of active sites on the statue surface dramatically decreases in the presence of sulphur dioxide. On the other hand, an opposite behavior is observed concerning the synergistic action of sulphur dioxide on the adsorption of acetylene on the surface of the ancient statue L291 - a pure calcite - from the exterior of Kavala Museum and on the adsorption of ethylene on the surface of a recently cut sample from Penteli mountain ore in Dionysos, Greece. The latter is also confirmed by Fig. 4d, where the number of active sites available for the adsorption of acetylene on the surface of L291 statue increases significantly in the presence of sulphur dioxide. Analogous observations have been drawn for the adsorption of ethane, ethene and ethyne (acetylene) on the surface of another ancient statue sample (L351), which was taken from the interior of the Kavala Museum and it was a pure dolomite, as an X-Ray diffraction analysis of this sample showed. Finally, the amounts of the hydrocarbons which totally adsorbed (c^*_s) on both of the statues from Museum of Kavala was calculated, with or without the presence of sulphur dioxide (Agelakopoulou et al, 2009). The results show that with the presence of SO_2 an increment of the estimated total adsorbed amount of acetylene is noted for the calcite's statue (L291) of Kavala, contrary to the other hydrocarbons where the presence of sulphur dioxide causes only a negligible or no effect. As regards the dolomite's statue (L351) of Kavala, the synergistic effect of SO_2 is more profound. First of all, the totally adsorbed amount of each hydrocarbon on L351 statue, either with or without the presence of sulphur dioxide, is higher than in case of L291 statue, a fact that is ascribable to the higher porosity of the former, which is dolomite, whereas the latter is calcite. Secondly, the higher adsorbed amount found for ethane in the *absence* of SO_2, something is reversed in the presence of SO_2 and concerns acetylene. In addition, the synergistic effect of SO_2 decreases from acetylene to ethene and ethane; obviously, the order of the bond of hydrocarbon has a significant role in the adsorption phenomenon.

4.2.2.3 The influence of the hydrocarbon's bond on the adsorption phenomena

For the same solid adsorbent (statue, pure oxide, etc.), the influence of the type of the bond in the molecule of the hydrocarbon is related with its molecular weight which reflects on the diffusion coefficient of the molecule. As the molecular weight of the hydrocarbon increases, the corresponding diffusion coefficient decreases. This becomes obvious in the following diagram (Fig. 7) which depicts the non-adsorbed concentration, c_y, of the hydrocarbon as a function of time t:

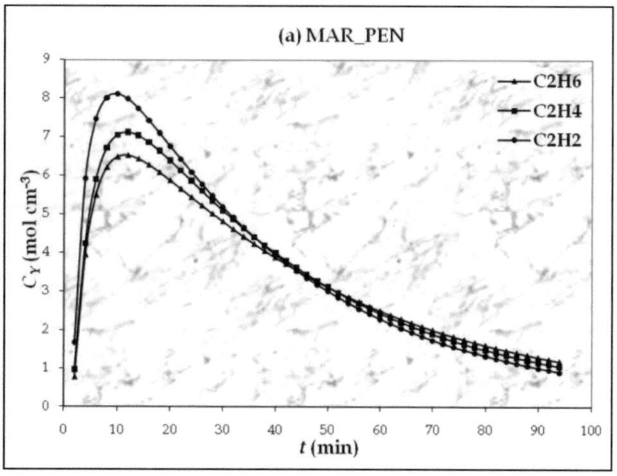

Fig. 7(a). Time-resolved analysis of the local non-adsorbed equilibrium concentration, c_y, of each hydrocarbon adsorbed on the various solid substrates: (a) marble of Penteli, (b) L1991 statue of Philippi, (c) L291 statue of Kavala and (d) L351 statue of Kavala.

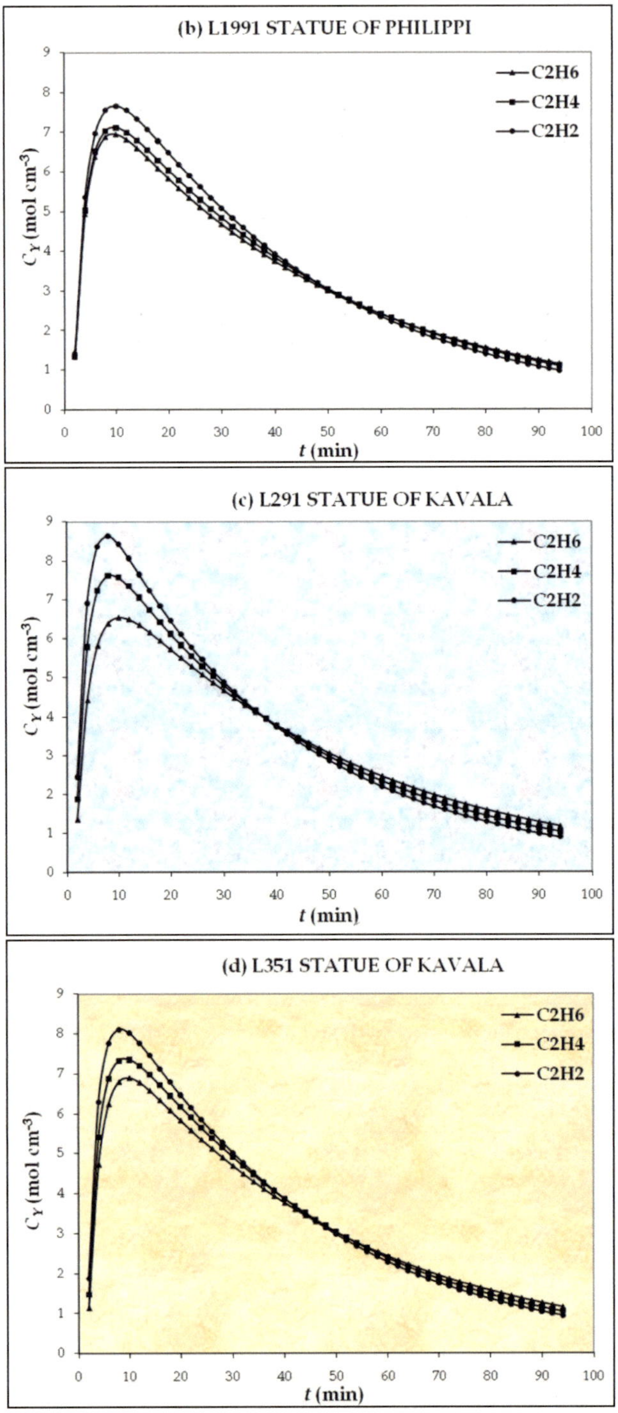

Fig. 7(b, c, d). Time-resolved analysis of the local non-adsorbed equilibrium concentration, c_y, of each hydrocarbon adsorbed on the various solid substrates: (a) marble of Penteli, (b) L1991 statue of Philippi, (c) L291 statue of Kavala and (d) L351 statue of Kavala.

4.2.2.4 Characterization of cultural heritage deterioration by means of XRD, SEM and Raman analysis

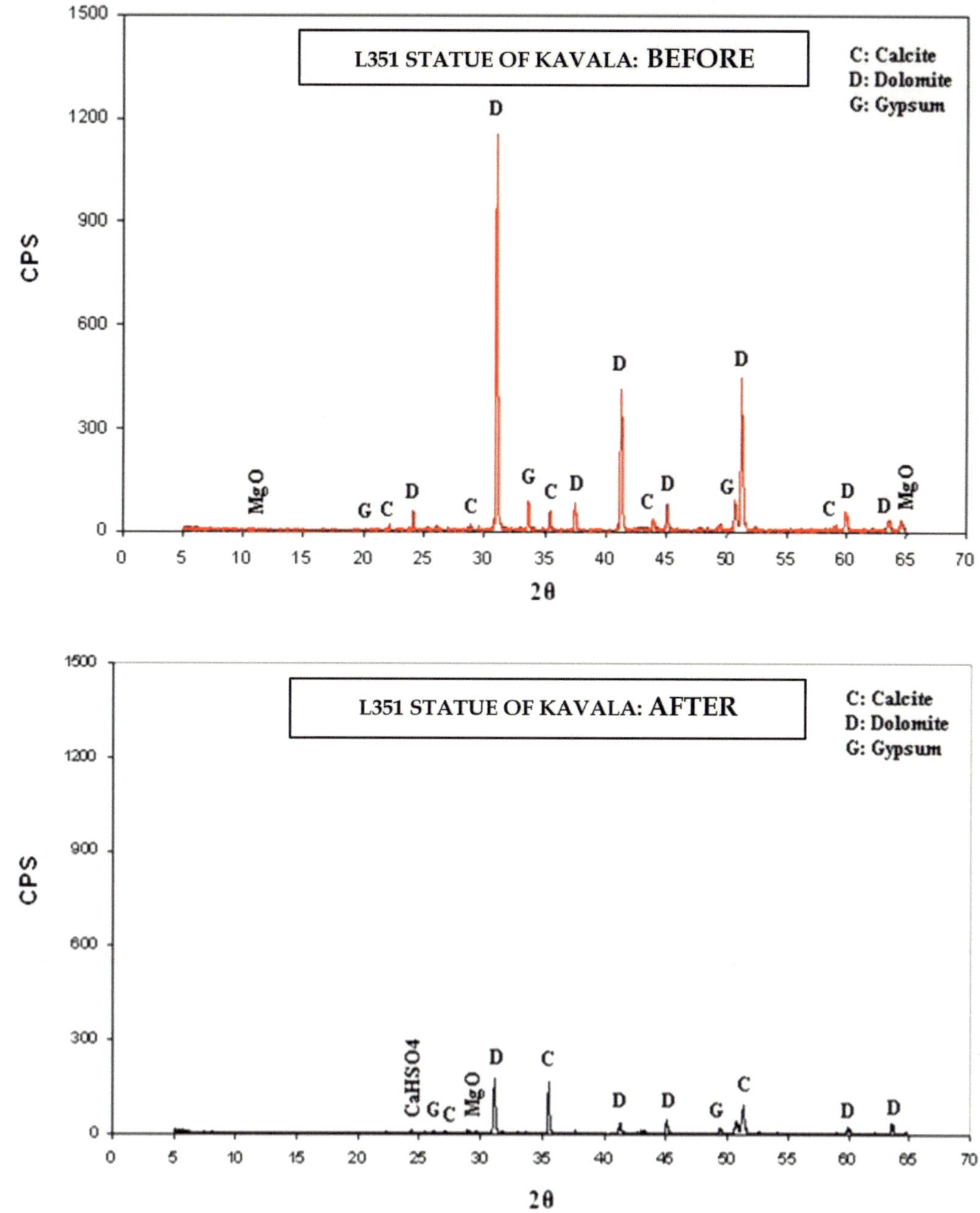

Fig. 8. XRD-diagram for the L351 statue from the interior of the Museum of Kavala: (on the top) before the injection of gaseous pollutants and (at the bottom) after the injection of gaseous pollutants.

The characterization of the above-mentioned cultural heritage materials through diffraction (XRD), spectroscopic (Raman) and microscopy techniques (SEM) was achieved following to the RF-IGC experiments. For example, the XRD-analysis for the L351 statue from the interior of the Museum of Kavala showed that it is a marble mostly composed of dolomite and quite less of calcite; the opposite stands for the L291 statue from the exterior of the same Museum. As it can also be revealed from the XRD-analysis (c.f. Fig. 8), some gypsum was detected before the exposure of the statue sample to the gaseous pollutants, so either the gypsum was a minor component of the marble or it was formed during a previous exposure of the statue to an atmosphere that permitted this formation. After the exposure to the pollutants, no gypsum was detected by any method used. This is rather expected because there was no humidity during the contact of the injected sulphur dioxide onto the marble. More, an improvement in the organization of the crystallites of the sample is observed after the injection of the gaseous pollutants. In the Raman spectra (c.f. Fig. 9) for the same statue after the exposure an accumulation of weak peaks at 1970-2200 cm^{-1} is distinguished owing to acetylene (C_2H_2). More, the main peak existed in 1095 cm^{-1}, before the exposure to the gaseous pollutants, which concerns the carbonate ion, is absent in the spectrum after the injection of the pollutants. The SEM-images for both statues (L351: dolomite and L1991: calcite) are shown in Fig. 10, where is evident the coarse and porous structure of dolomite. Calcite seems more tight and compact. The combination of Raman and SEM-EDAX analysis showed that SO_2 was adsorbed (Metaxa et al., 2009b).

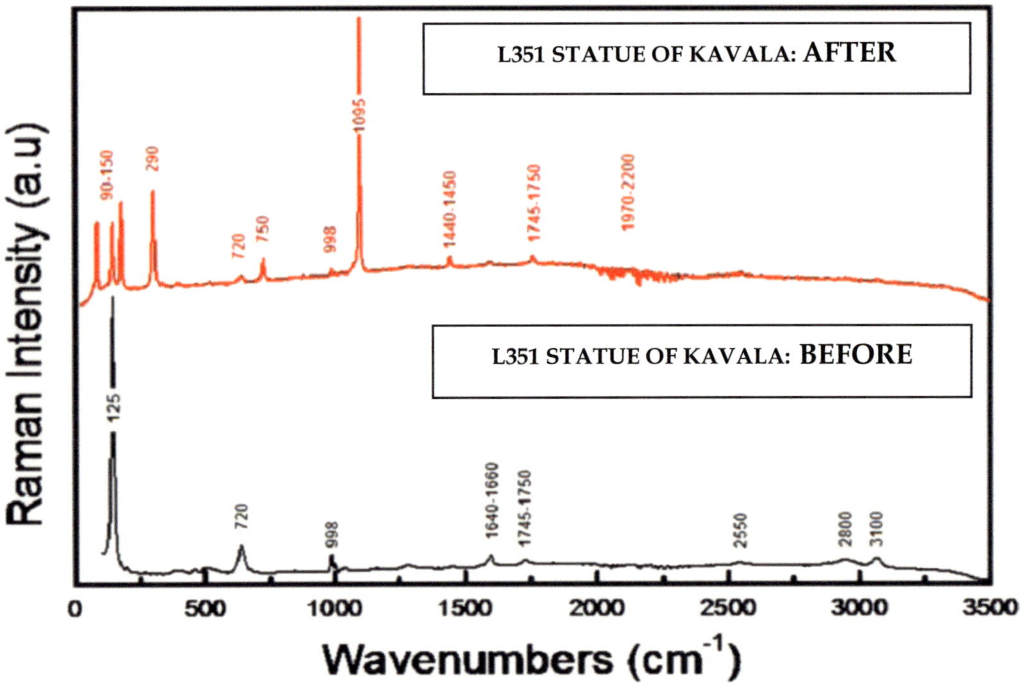

Fig. 9. Raman-spectra for the L351 statue from the interior of the Museum of Kavala: (on the top) before the injection of gaseous pollutants and (at the bottom) after the injection of gaseous pollutants.

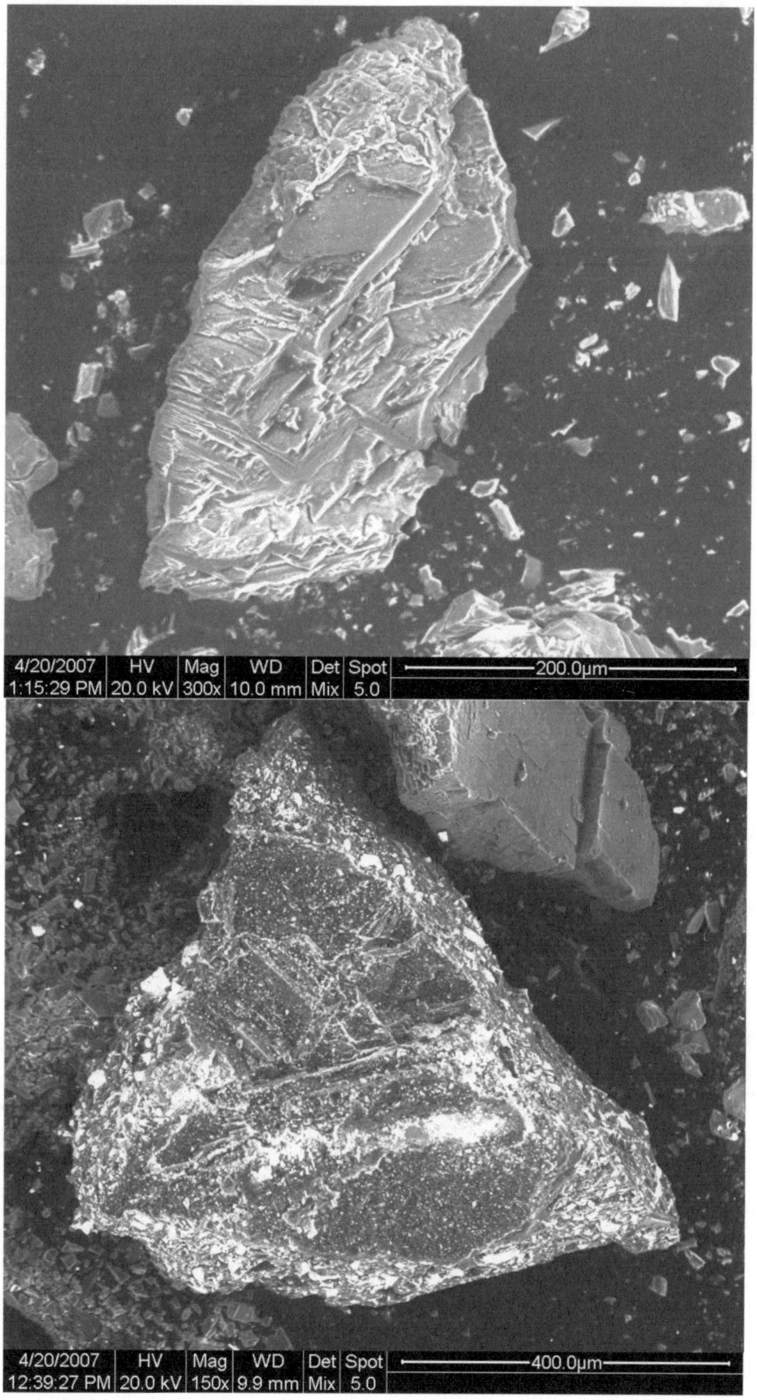

Fig. 10. SEM-images showing the shapes of calcite (on the top) and dolomite (at the bottom).

4.2.2.5 Extracting information about surface heterogeneity

The "non-ideal" behavior of a solid surface in the direction of the adsorption of a gaseous substance on it consists of an inherent energetic heterogeneity component and an adsorption-induced heterogeneity component originated from the lateral interactions between the adspecies. Surface heterogeneity is responsible for the time-variation in the activation of the various active sites onto the solid surface towards the adsorption process. The effects of surface heterogeneity on adsorption equilibrium have extensively studied, both experimentally and theoretically. The RF-IGC methodology constantly answers for the role of surface heterogeneity on the adsorption phenomena, through the local physicochemical quantities determined for various adsorption systems. The results obtained are in a good agreement with literature data as it has already been shown (Metaxa et al., 2009a).

4.2.2.6 Study of the action of sulfur dioxide on Penteli marble in the presence or the absence of protective materials-evaluation of the effectiveness of the protective materials for monuments against sulfur dioxide corrosion

Calcareous stones, such as marble, suffer from the attack of sulfur dioxide in polluted atmospheres because of the transformation of calcite-content ($CaCO_3$) into gypsum ($CaSO_4 \cdot 2H_2O$) as final product [Elgohary, 2008]. The choice of the repair materials is a crucial point in stone conservation. Such operations, in fact, are much critical, as they can alter the structure of the original material and create new textural heterogeneity at the natural stone-composite interface. In order to avoid unsuccessfully results, the characteristics of materials used in conservation works and their compatibility, including long-term effects, with existing materials should be fully established. Thus, new impregnation products, as well as those which already exist, must be viewed with caution and be subjected to laboratory research before they applied on historic buildings. The choice between "traditional" and "innovative" materials and techniques should be determined case-by-case with preference given to those that are least invasive and most compatible with heritage values, consistent with the need for safety and durability.

Protective materials, such as acrylic copolymers and siloxanes, have been largely used in conservation practice as coatings, consolidants and adhesives, because of their good adhesion, film forming properties and their environmental stability. These materials alter the physicostructural properties of the porous materials and change the physicochemical behavior of the interface between the work of art and the environment [Carreti & Dei, 2004]. Therefore, the characterization of the solid surfaces before and after the application of these materials is important for the evaluation of their ability to protect the historic monuments and buildings.

In order to study the action of SO_2 on Penteli marble, experiments were carried out by using the RF-IGC instrumentation and various physicochemical parameters (rate constants as well as equilibrium constants) were calculating by using a non-linear regression analysis PC-programs in GW-BASIC [Katsanos et al., 2003] for the experimental data. Afterwards, kinetic parameters such as adsorption rate constants k_1, adsorption/desorption k_R and surface reaction rate constants k_2, as well as surface diffusion coefficients D_y, deposition velocities V_d and reaction probabilities γ of SO_2 "on marble surfaces, at various temperatures," in the presence or in the absence of protective materials (an acrylic copolymer, Paraloid B-72 or a siloxane, CTS Silo 111) were calculated and discussed

[Bakaoukas et al., 2005]. The results showed that both materials are good enough at low temperatures (303.2-323.2K), while at high temperatures (333.2-353.2K) siloxane acts better as protective material than acrylic copolymer. More specifically, the values of surface reaction rate constant k_2 in all cases where the marble was coated by the acrylic copolymer were bigger than those for pure marble, while they were smaller in all cases where the siloxane was used as protective material. Probably that happens because SO_2 interacts with the acrylic copolymer and not with the siloxane and a mechanism was proposed for this interaction. On the other hand, the values of the adsorption rate constant k_1 for the system (SO_2 + marble) were bigger than those for the systems (SO_2 + coated marble), thus indicating that the adsorption of SO_2 is more difficult in the case of coated marble than in the pure marble. In addition, the values of V_d and γ in most cases where the marble was coated with anyone of the protective materials were smaller than those for pure marble, except from the higher temperature of 353.2K where the differences before and after coating with the acrylic copolymer were negligible; probably, because in this high temperature SO_2 interacts more rapidly with the acrylic copolymer. Concerning the effect of coatings on D_y-values, it was found that in the absence of protective materials, at extreme temperatures (>333.2K), the porous of the marble is being destroyed producing surface diffusion coefficients equivalent to diffusion coefficients in the gas phase. On the contrary, in the presence of protective materials, the later does not happen due to the shrink of the porous size of the marble[Bird et al., 2002; Carreti & Dei, 2004; Bakaoukas et al., 2005].

5. Conclusions

Among the various causes which are responsible for the destruction of cultural property, namely of historic artifacts and monuments, air pollution could be considered as an important one. Cultural goods are chiefly significant as a kind of evidence of past human activity. Conservation of cultural heritage allows this evidence to be consulted whenever new questions about the past are emerged. Thus it is well-understood that successful conservation has to be underpinned by a comprehensive understanding of the causes of decay and the factors controlling them.

In order to estimate the impacts of air pollution on the various solid surfaces, including them of cultural heritage, in a real scientific basis, theory and experiment needs to cooperate in a way close to real systems. From this aspect of view, the new dynamic version of classic inverse gas chromatography, the so-called Reversed-Flow Inverse Chromatography (RF-IGC), combining a powerful mathematical background with a very simple and smart experimental arrangement, attains this purpose by using time-resolved analysis in the way this method has developed it. Thus, by means of a simple PC-program, important local physicochemical quantities are determined, which characterize the main rate processes taking place at the phase boundaries, such as adsorption and desorption, aiming at their spreading in the bulk. Lately, this method has been successfully applied to the study of the impact of air pollutants on many cultural heritage objects, such as marbles, pigments of works of art, etc., by studying the topography of active sites in heterogeneous solid surfaces and their availability for adsorption, without using retention volume data, as it abandons the traditional role of the carrier gas in conventional chromatography and substitutes it with gaseous diffusion currents. The results of this method are based on a non-linear adsorption isotherm model and rate measurements over an extended period of time. All the chromatographic methods known up to now offer approximate functions for the probability density function for the adsorption energy, without any determination of local adsorbed parameters, as this technique does. All findings supplied from the application of this methodology in various systems gas/solid provide valuable information about the susceptibility of the examined artifact or statue to environmental gaseous pollutants.

6. References

Adriaens, A. (2005). Review: Non-destructive Analysis and Testing of Museum Objects: An Overview of 5 Years of Research. *Spectrochimica Acta Part B,* Vol. 60, pp. 1503-1516.

Agelakopoulou, T.; Metaxa, E.; Karagianni, Ch.-S. & Kalantzopoulou-Roubani, F. (2009). Air Pollution Effect of SO_2 and/or Aliphatic Hydrocarbons on Marble Statues in Archaeological Museums. *Journal of Hazardous Materials,* Vol. 169, pp. 182-189.

Arvanitopoulou, E.; Katsanos, N.A.; Metaxa, H. & Roubani-Kalantzopoulou, F. (1994). I. Simple Measurement of Deposition Velocities and Wall Reaction Probabilities in Denuder Tubes-II. High Deposition Velocities. *Atmospheric Environment,* Vol. 28, No. 15, pp. 2407-2412.

Bakaev, V.A. & Steele, W.A. (1992). Computer Simulation of the Adsorption of Argon on the Surface of Titanium Dioxide. 1. Crystalline Rutile. *Langmuir,* Vol. 8, No. 5, pp. 1372-1378.

Bakaoukas, N.; Kapolos, J.; Koliadima, A. & Karaiskakis, G. (2005). New Gas Chromatographic Instrumentation for Studying the Action of Sulfur Dioxide on Marbles. *Journal of Chromatography A,* Vol. 1087, pp. 1693-176.

Bhargava, R. & Levin, I.W. (2003). Time-Resolved Fourier Transform Infrared Spectroscopic Imaging. *Applied Spectroscopy,* Vol. 57, No. 4, pp. 357-366.

Bird, R.B.; Stewart, W.E. & Lightfoot, E.N. (2002). *Transport Phenomena* (2nd ed.), Wiley, ISBN:0471410772, New York.

Carreti, E. & Dei., L. (2004). Physicochemical Characterization of Acrylic Polymeric Resins Coating Porous Materials of Artistic Interest. *Progress in Organic Coatings,* Vol. 49, No. 3, pp. 282-289.

Cazes, J. (2009-10-12). *Encyclopedia of Chromatography* (3rd ed.), CRC Press, ISBN: 1420084593, U.S.A.

Christmann, K. (1995). Some General Aspects of Hydrogen Chemisorption on Metal Surfaces. *Progress in Surface Science,* Vol. 48, No. 1-4, pp. 14-26.

Doménech-Carbó, A.; Doménech-Carbó, M.T. & Costa, V. (2009). Chapter 1: Application of Instrumental Methods in the Analysis of Historic, Artistic and Archaeological Objects, In: *Electrochemical methods in archaeometry, conservation and restoration* (1st ed.), Scholtz Fritz (ed.), pp. 1-32, Springer-Verlag, ISBN: 978-3-540-92867-6, Berlin.

Elgohary, M.A. (2008). Air Pollution and Aspects of Stone Degradation "Umayyed Liwān-Amman Citadel as a Case Study". *Journal of Applied Sciences Research,* Vol. 4, No. 6, pp. 669-682.

Giakoumaki, A.; Melessanaki, K. & Anglos, D. (2007). Review: Laser-Induced Breakdown Spectroscopy (LIBS) in Archaeological Science-Applications and Prospects. *Analytical & Bioanalytical Chemistry,* Vol. 387, No. 3, pp. 749-760.

Isnard, O. (2006). In Situ and/or Time Resolved Powder Neutron Scattering for Materials Science. *Journal of Optoelectronics and Advanced Materials,* Vol. 8, No. 2, pp. 411-417.

Jansen, A.P.J. (2008). Island Formation without Attractive Interactions. *Physical Reviews B*, Vol. 77, No. 7, article: 0732408, 4 pages.

Jenkins, R. (2000). X-Ray Techniques: Overview, In: *Encyclopedia of analytical chemistry*, Meyers, R.A. (eds.), pp. 1-20, John Wiley & Sons Ltd, ISBN: 9780470027318, U.S.A.

Katsanos, N.A., (1988). *Flow Perturbation Gas Chromatography* (1st ed.), Marcel Dekker, New York-Basel.

Katsanos, N.A.; Thede, R. & Roubani-Kalantzopoulou, F. (1998). Review: Diffusion, Adsorption and Catalytic Studies by Gas Chromatography. *Journal of Chromatography A*, Vol. 795, pp. 133-184.

Katsanos, N.A.; Gavril, D. & Karaiskakis, G. (2003). Time-Resolved Determination of Surface Diffusion Coefficients for Physically Adsorbed or Chemisorbed Species on Heterogeneous Surfaces, by Inverse Gas Chromatography. *Journal of Chromatography A*, Vol. 983, No. 1, pp. 177-193.

Katsanos, N.A. & Karaiskakis, G. (2004). *Time-Resolved Inverse Gas Chromatography and its Practical Applications* (1st ed.), HNB Publishing, ISBN: 0-9728061-0-5, New York.

La Russa, M.F.; Ruffolo, S.A.; Barone, G.; Crisci, G.M.; Mazzoleni, P. & Pezzino, A. (2009). The Use of FTIR and micro-FTIR Spectroscopy: an Example of Application to Cultural Heritage. *International Journal of Spectroscopy*, Vol. 2009, pp. 1-5.

Metaxa, E.; Kolliopoulos, A.; Agelakopoulou, T. & Kalantzopoulou-Roubani, F. (2009a). The Role of Surface Heterogeneity and Lateral Interactions in the Adsorption of Volatile Organic Compounds on Rutile Surface. *Applied Surface Science*, Vol. 255, pp. 6468-6478.

Metaxa, E.; Agelakopoulou, T.; Bassiotis, I.; Karagianni, Ch. & Kalantzopoulou-Roubani, F. (2009b). Gas Chromatographic Study of Degradation Phenomena Concerning Building and Cultural Heritage Materials. *Journal of Hazardous Materials*, Vol. 164, pp. 592-599.

Metaxa, E.; Agelakopoulou, T.; Karagianni, Ch.-S. & Kalantzopoulou-Roubani, F. (2009c). Study of the Adsorption of Ozone on the Surface of Ferric Oxide by Revered-Flow Inverse Gas Chromatography. *Instrumentation Science & Technology*, Vol. 37, No. 5, pp. 584-606.

Miliani, C.; Rosi, F.; Brunetti, B.G. & Sgamelloti, A. (2010). In Situ Noninvasive Study of Artworks: the MOLAB Multitechnique Approach. *Accounts of Chemical Research*, Vol. 43, No. 6, pp. 728-738.

Osticioli, I.; Mendes, N.F.C.; Porcinai, S.; Cagnini, A. & Castellucci, E. (2009). Spectroscopic Analysis of Works of Art Using a Single LIBS and Pulsed Raman Setup. *Analytical & Bioanalytical Chemistry*, Vol. 394, No. 4, pp. 1033-1041.

Putzig, C.L.; Leugers, M.A.; McKelvy, M.L.; Mitchell, G.E.; Nyquist, R.A.; Papenfuss, R.R. & Yurga, L. (1994). Infrared spectroscopy. *Analytical Chemistry*, Vol. 66, No. 12, pp. 26R-66R.

Quellette, J. (2004). Time-Resolved Spectroscopy Comes of Age. *The Industrial Physicist*, 2, pp. 16-19.

Roubani-Kalantzopoulou, F.; Metaxa, H.; Kalantzopoulos, A.; Kalogirou, E.; Sotiropoulou, V. & Katsanos, N.A. (1996). Contribution to the Mechanism of Marble Deterioration by Gas Chromatographic Studies, Koutsoukos, P. & Kontoyiannis, Ch. (Eds.), *Eurocare-Euromarble EV496 Workshop 7*, Patras, Greece, October 1996, pp. 33-38.

Roubani-Kalantzopoulou, F. (2004). Review: Determination of Isotherms by Gas-Solid Chromatography Applications. *Journal of Chromatography A*, Vol. 1037, No. 1-2, pp. 191-221.

Roubani-Kalantzopoulou, F. (2009). Review: Time-Resolved Chromatographic Analysis and Mechanisms in Adsorption and Catalysis. *Journal of Chromatography A*, Vol. 1216, No. 10, pp. 1567-1606.

Sotiropoulou, V.; Vassilev, G.P.; Katsanos, N.A.; Metaxa, H. & Roubani-Kalantzopoulou, F. (1995). Simple Determination of Experimental Isotherms Using Diffusion Denuder Tubes. *Journal of the Chemical Society Faraday Transactions,* Vol. 91, pp. 485-492.

Thielmann, F. (2004). Review: Introduction into the Characterization of Porous Materials by Inverse Gas Chromatography. *Journal of Chromatography A*, Vol. 1037, No. 1-2, pp. 115-123.

12

Effect of Air Pollution on Archaeological Buildings in Cairo

Mohamed Kamal Khallaf
Restoration Department, Faculty of Archaeology, Fayoum Universiy, Egypt

1. Introduction

Islamic Cairo is a part of central Cairo noted for its historically important mosques and other Islamic monuments. It is overlooked by the Cairo Citadel.Islamic Cairo was founded in 969 AD as the royal enclosure for the Fatimid caliphs, while the actual economic and administrative capital was in nearby Fustat. Fustat was established by Arab military commander 'Amr ibn al-'As following the conquest of Egypt in 641 AD, and took over as the capital which previously was located in Alexandria. Al-Askar, located in what is now Old Cairo, was the capital of Egypt from 750 AD to 868 AD. [1] Ahmad Ibn Tulun established Al-Qatta'i as the new capital of Egypt, and remained the capital until 905 AD, when the Fustat once again became the capital. After Fustat was destroyed in 1168 AD /1169 AD to prevent its capture by the Crusaders, the administrative capital of Egypt moved to Cairo, where it has remained ever since. [2]It took four years for the General Jawhar Al Sikilli (the Sicilian) to build Cairo and for the Fatimid Calif Al Muizz to leave his old Mahdia in Tunisia and settle in the new Capital of Fatimids in Egypt. Fustat became a regional center of Islam during the Umayyad period. Later, during the Fatimid era, Al-Qahira (Cairo) was officially founded in 969 AD as an imperial capital just to the north of Fustat. [3] Over the centuries, Cairo grew to absorb other local cities such as Fustat, but the year 969 AD is considered the "founding year" of the modern city. In 1250 AD, the slave soldiers or Mamluks seized Egypt and ruled from their capital at Cairo until 1517 AD, when they were defeated by the Ottomans. [4] By the 16th century, Cairo had high-rise apartment buildings where the two lower floors were for commercial and storage purposes and the multiple stories above them were rented out to tenants. Napoleon's French army briefly occupied Egypt from 1798 AD to 1801 AD, after which an Albanian officer in the Ottoman army named Muhammad Ali Pasha made Cairo the capital of an independent empire that lasted from 1805 AD to 1882 AD. [5] The city then came under British control until Egypt was granted its independence in 1922 AD. Cairo is a world heritage city. It contains possibly the finest collection of monuments in the Islamic world. It contains some of the best surviving monuments of the medieval period in the Islamic world. [6]. The wealth, prosperity, and power of Cairo are reflected in the grand architecture of the monuments that are crowded together into the Fatimid city and just beyond, Fig. (1). [7] Cairo's Islamic monuments are part of an uninterrupted tradition that spans over a thousand years of building activity. No other Islamic city can equal Cairo's spectacular heritage, nor trace its historical and architectural development with such clarity. [8] Cairo contains the greatest concentration of Islamic monuments in the world, and its mosques, mausoleums, religious schools, baths, and caravanserais, built by prominent patrons between the seventh and nineteenth centuries, are

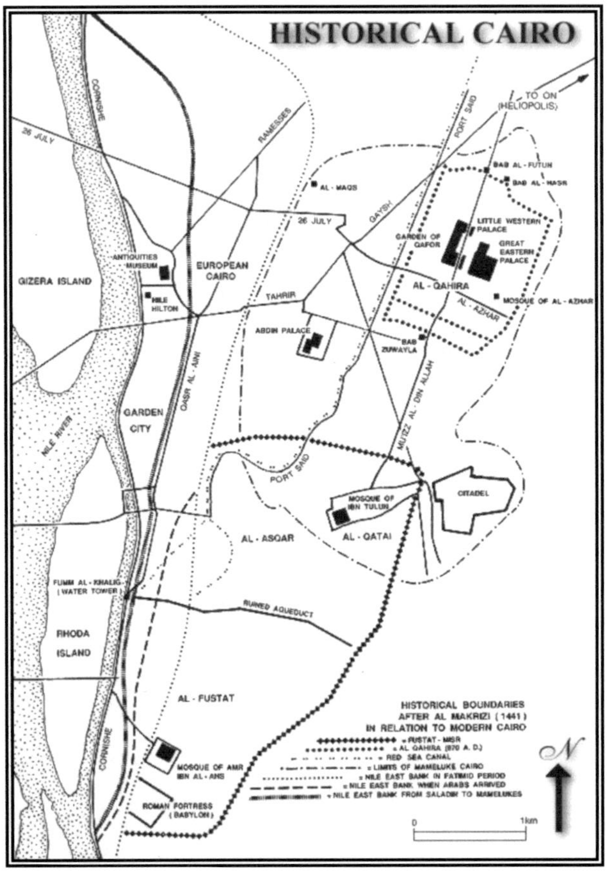

Fig. 1. Shows a map of historical Cairo.
http://www.touregypt.net/Map08.htm.

among the finest in existence [9] , fig. (2) Shows some of Islamic archaeological buildings in Cairo. The air pollution in Cairo is a matter of serious concern. Greater Cairo's volatile aromatic hydrocarbon levels are higher than many other similar cities. Air quality measurements in Cairo have also been recording dangerous levels of lead, carbon dioxide, sulphur dioxide, and suspended particulate matter concentrations due to decades of unregulated vehicle emissions, urban industrial operations, and chaff and trash burning. There are over 4,500,000 cars on the streets of Cairo, 60% of which are over 10 years old, and therefore lack modern emission cutting features like catalytic converters. Cairo has a very poor dispersion factor because of lack of rain and its layout of tall buildings and narrow streets, which create a bowl effect. Cairo also has many unregistered lead and copper smelters which heavily pollute the city. The results of this have been a permanent haze over the city with particulate matter in the air reaching over three times normal levels. [10] Pollutants are deposited on the surface of stone from the air. Where the surface of the stone is totally dry, the stone is discolored as the deposits increase. Where the surface of the stone is moist, the pollutants are converted to acids that eat away the surface of the stone by dissolving the binder in the stone causing the stone particles or grains to separate and erode away easily. [11] Carbon dioxides, Nitric oxides, and Sulphur oxides product mineral acids in humid conditions. They dissolve the calcium and magnesium carbonates in limestone, marble, lime mortars, and plasters in archaeological buildings. Archaeological buildings in Cairo suffer from different deterioration phenomena for example, black crust formation, chemical alterations, disintegration between surface mineral grains, pitting, cracks, missing parts, erosion, and white stains. [12]. This chapter aims to study deterioration and decay of building materials in archaeological buildings in Cairo because of air pollution, Discussion and explanation of deterioration phenomena which forming in archaeological building in Cairo according to air pollution and Discussion of different methods and materials of treatment, restoration and conservation of building material in archaeological buildings from deterioration phenomena related to air pollution.

2. Sources of air pollution in Cairo

Air pollution plays a major role in the deterioration of building materials used in historic buildings. Industrial facilities such as factories and plants emit toxic gases into the atmosphere. Another major source of toxic emissions in Egypt is the widespread open-air burning of trash and waste. Waste landfills also give off methane, which, although not toxic, is highly flammable and can react in the air with other pollutants to become explosive. [13] There are numerous sources to air pollution in Egypt, as in other countries. However, the formation and levels of dust, small particles and soot are more characteristic in Egypt than presently found in industrialized countries. Some of the sources for these pollutants, such as industries, open-air waste burning and transportation, were also well known problems in most countries only 10 to 20 years ago. Another important source for particulate matter is the wind blown dust from the arid areas. Suspended dust (measured as PM10 and TSP) can be seen to be a major air pollution problem in Egypt. PM10 concentrations can exceed daily average concentrations during 98% of the measurement period. On the other hand it seems that the natural background of PM10 in Egypt may be close to or around the Air Quality Limit value. These levels can be found also in areas where local anthropogenic sources do not impact the measurements. Further measurements may be used in the future to quantify the relative importance of the different sources relative to a background level that varies dependent upon the area characteristics. In addition to particles, also SO_2 in urban areas and in industrial areas, as well as NO_2 and CO in the streets may exceed the Air Quality Limit value. Major industrial pollutants include sulphur oxides, nitrogen oxides, carbon monoxide and carbon dioxide. [14] For instance, Cairo is surrounded by various industrial sites. Thirty Kilometers to the south of Cairo is Helwan, where different factories produce iron, steel, coke, chemicals, automobiles, and cement. To the north of Cairo are Shubra Al - Khayma, Musturud, and Abu Zabal. In this area factories produce dyes, textiles, glass, ceramics, and chemical products. All of these factories emit different pollutants (gaseous, liquid, solid), which are carried by the dominant winds (north and northeast, and west or south, southwest) down to Cairo, many of the historic buildings are located. Every day Cairo receives a high dose of pollutants composed of 52 percent monocarbon oxide (CO), 14 percent sulphur dioxide (SO_2), 21 percent hydrocarbons, 10 percent dust, solid materials, and 2 percent (NO_x) nitrogen oxides, The dust particles from the Muqattam hills to the east of Cairo was 27 gm / m^2 / month in 1962. This increased to more than 60 gm / m^2 / month in 1988, with a particularly high a mount in the summer when the aerosols of dust in the air were more than 500 gm / m^2 / month [15]. Many Egyptians rely upon extremely old vehicles for transportation. These inefficient vehicles cause the carbon present in fuel to ineffectively react with oxygen during combustion, producing carbon monoxide or condensing to form particles of soot. The hydrocarbons do not combust completely and are released as gaseous hydrocarbons or absorbed by particles, increasing the particulate mass

Fig. 2. Shows some of Islamic Archaeological buildings in Cairo: (A) El- Sultan Hassan Madrassa (1362 AD / 764 AH) and El - Refae Mosque. (B) El-Mosabeh Mosque(1792 AD / 1192 AH). (C) El- Ghouri Mosque in El - Sayeda Aisha Square (1504 AD / 909 AH). (D) El - Mahmoudya Mosque (1568 AD / 975 AH). (E) Qaitbay Sabil (1479 AD / 884 AH). (F) Taghri Bardi Mosque (1439 AD / 843 AH). (G) Azbak El- Yusufi Mosque. (H) Singer and Slar Mosque (1303 AD / 703 AH). (I) Lagen El - Sayfi Mosque (1296 AD / 696 AH).

in the air. The speed at which pollutants disperse in the air is determined by meteorological conditions such as wind, air temperature and rain. Egypt and Cairo, particularly, have a very poor dispersion factor due to lack of rain and the layout of streets and buildings, which are not conducive to air flow. [16] Emissions that arise from the combustion of solid fossil fuels are of prime concern. Coal and oil both contain sulphur in varying amounts, and both therefore produce sulphur dioxide when burnt. There are a number of nitrogen oxides (NOx), but the one of principal interest as an air pollutant likely to have adverse effects on human health and soiling properties is nitrogen dioxide (NO_2). Nitrogen compounds are also contributors to the wet and dry deposition of acidic compounds on vegetation and buildings. Particulate matter is a term that represents a wide range of chemically and physically diverse substances that can be described by size, formation mechanism, origin, chemical composition, atmospheric behavior and method of measurement. The concentration of particles in the atmosphere varies across space and time and as a function of the source of the particles and the transformations that occur to them as they age and travel. Particles less than 10 mm in diameter (PM10) are often measured that include both fine and coarse dust particles. [17].

3. Materials and methods

Limestone and marble samples of original stones and crusts were collected from different deteriorated parts of Archaeological buildings, according to the decay and the damage levels fig. (3) as follows: - Limestone samples from El- Ghouri Mosque, El - Mahmoudya Mosque, Taghri Bardi Mosque and Lagen El - Sayfi Mosque. - Marble samples from Qaitbay Sabil, Taghri Bardi Mosque and Azbak El- Yusufi Mosque. Analytical study have been carried to selected samples by Polarizing Microscope [PLM], Scanning Electron Microscope [SEM], Energy dispersive X-ray analysis [EDX], X-ray diffraction (XRD) and FTIR analysis.

Fig. 3. Shows some details of Islamic Archaeological buildings, (A) , (B) , (G) El - Mahmoudya Mosque (1568 AD / 975 AH). (C) , (D) Qanibay Al - Ramah Mosque (1503 AD / 908 AH). (E) Qaitbay Sabil (1479 AD / 884 AH) (F) , (H) , (I) , (L) Taghri Bardi Mosque (1439 AD / 843 AH). (J) Lagen El - Sayfi Mosque (1296 AD / 696 AH). (K) Azbak El- Yusufi Mosque (1494 AD / 900 AH).

The x-ray diffraction analysis of samples was carried out using philips x-ray diffractometer. The operating conditions were: Generator applied on a Cu kα radiation (1.5418 A°) with Ni filter, 40 KV, 20 mA° target tube. Scavenging velocity 2° per minute and chart velocity 5 mm per minute were applied in Bulk sample powder. Fragments of crusts collected were prepared for observation using scanning electron microscope (SEM), operated at accelerating voltage of 30 kV. Infrared spectra were recorded employing a Nicolet Nexus 870 FTIR spectrometer. A small amount of samples were mixed with KBr and pressed into pellets, then scanned from 4000 to 400 cm−1.

4. Results and discussion

Limestone samples were examined by Polarizing Microscope (PM) and, it is found that: Samples consist mainly of fine-grained calcite besides presence of iron oxides, quartz, clay minerals and fossils include nummulite fossils these components increase the rate of stone decay [18], fig. (4 , A-F). On the other hand the thin section of fragments taken from marble objects shows that the major mineral is calcite, The crystals appeared in mosaic texture, The crystals have irregular faces and highly variable grain size, the cleavage planes of the calcite crystals and the presence of rare and very little amount of opaque minerals [19] fig. (4, G-I). When the limestone samples were examined by [SEM], it is found disintegration between calcite crystals and the stone lost the binding materials between grains by the effect of salts crystallization fig. (5, A-D). Examination of marble samples by [SEM] shows that, erosion of calcite crystals, presence of salts because of chemical reaction with climatic conditions, alteration of calcite into gypsum because of air pollution effect, voids and disintegration between grains by crystallization of salts stresses and lose of binding material [18] fig. (5, E-F). XRD data fig. (6) (A-D) shows that, the examined limestone samples consist of Calcite $CaCO_3$, Card No. (5-0586) in addition to Gypsum $CaSO_4.2H_2O$ Card No. (6-0046), Quartz SiO_2 ,Card No. (5-0490), Halite Card No.(5-0628) and Dolomite Card No.(11-078) . XRD data of the marble samples shows that, they consist of Calcite $CaCO_3$, Card No. (5-0586) in addition to Dolomite $Ca,Mg(CO_3)_2$, Card No. (11-078), and Halite, NaCl Card No. (5-0628) fig (5-a). and Anhydrite, Card No (6-0226) fig (5-c). The surface of the marble is covered by a crust of Hydrated Calcium Sulphate (Gypsum) related to reaction with air pollution in presence of moisture. Gypsum crusts are the most common type of growth found on building surfaces. Gypsum is Calcium Sulphate Dihydrate, with the chemical formula $CaSO_4.2H_2O$. Gypsum crusts are formed on calcareous stones following SO_2 deposition to the surface in the presence of moisture, followed by the dissolution of Calcite and the precipitation of Gypsum. The black color of gypsum crusts is the result of the accumulation of particulate matter within the crust [20]. When the water evaporates from soluble salts as chlorides, it leaves behind concentrations of salt solutions which crystallize on the stone surfaces and between mineral grains of stone, this process cause disintegration and deterioration of stone [21]. Energy Dispersive X-ray analysis (EDX) of Limestone samples shows that it consists of calcium element (Ca), sulphur element (S) , Silicon element (Si) and Sodium element (Na) in addition to traces from other elements. The relative enrichment of Si, Al and Fe might be derived from the deposition of wind-borne articles since the archeological stone buildings in Cairo are near a road with much traffic [22] fig. (2). Rich-S is originated from SO_2 emitted by anthropogenic sources like combustion of fuels, automobile emissions, foundries and smelters. EDX data shows high content of calcium related to calcite mineral, silicon and aluminum due to clay minerals, silicon due to quartz mineral, iron related to iron oxides, sodium and chlorine due to presence of halite salt fig (7)(A-D). Rich-Ti-Mn is associated with industrial and urban emissions. The fly-ash particles play an active role in the damage processes affecting stone, since the content of transition metal oxides contribute to the catalytic oxidation of atmospheric gaseous SO_2 and to the sulphation of calcium carbonate. XRD, SEM, EDX results show that black crusts are essentially composed by gypsum crystals, fly ashes and soot, including some limestone and marble materials. Fly ashes usually are rich in Si and Al with higher or lower amounts of K, Fe, Ca, Ti and Cl. Combustion of fuel and natural gas in car engines and house heating originates carbon rich soot about one hundred times smaller than fly ashes. [23]. The term "atmospheric particulate material" refers to all airborne particles, so it is by definition non-specific. It includes material from such diverse sources as, for example, vehicle emissions, the resuspension of surface dusts and soils and chemical reactions between vapours and gases in the atmosphere, which result in the formation of secondary particles [24]. Therefore emission inventories of PM relate to primary sources of PM only (not secondary sources) [25]. The principal types of primary particulate material are Petrol and diesel vehicles, the latter being the source of most black smoke [26]. Controlled emissions from chimney stacks. Fugitive emissions. These are diverse and mostly uncontrolled and include The resuspension of soil by wind and mechanical disturbance [27]. The resuspension of surface dust from roads and urban surfaces by wind, vehicle movements and other local air disturbance [28]. Emissions from activities such as quarrying, road and building construction, and the loading and unloading of dusty materials,[29]. Secondary particles are those arising when two gases or vapours react to form a substance that condenses onto a nucleation particle, [30]. The main sources of secondary particles are the atmospheric oxidation of sulphur dioxide to sulphuric acid and the oxidation of nitrogen dioxide to nitric acid; the sulphuric acid is present in air as droplets, the nitric acid as a vapour, [31]. Hydrochloric acid vapour (arising mainly from refuse incineration and coal combustion) is also present in the atmosphere, and both this and nitric acid vapour react reversibly with ammonia to form ammonium salts,[32]. Sulphuric acid reacts irreversibly in two stages to form either ammonium hydrogen sulphate or ammonium sulphate. These ammonium salts are formed continuously as sulphur dioxide and nitrogen dioxide are oxidised, and ammonia becomes available for neutralization, [33]. FTIR spectra of a limestone sample fig. (8) shows that the characteristic absorption peaks of $CaCO_3$ is at 1798, 1424, 874, 711 cm−1, the characteristic absorption peaks of, CO_3-apatite $[(Ca_5(PO_4)_3)2CO_3]$ is at 565, 604, 1040 cm−1 and the characteristic absorption peaks of gypsum is at 672, 1623, 3408 cm−1 .The results of infrared spectroscopy are also confirmed by XRD analysis which provides information on the crystalline components. The limestone contains Calcite and Quartz, and Gypsum. In consideration of the high average relative humidity and rainwater in the environmental conditions in Cairo, the most probable process of crust formation on stone substrate is the absorption of sulphur dioxide in rainwater, liquid atmospheric aerosols or moist film supported on a stone surface, [34]where it is oxidized to form a sulphuric acid solution that dissolves the Calcium Carbonate by Gypsum formation. Kaolinite has been identified on a stone flake collected from a washed-out surface [35]. Its presence can be related to Calcite dissolution, which is strongly enhanced by its exposition to rainwater and winds[36]. The deposition of wind-born soil dust on the surface may also be a source of kaolinite. [37] The mineralogical, textural and physicochemical differences of the examined crusts suggest that it is unlikely that they have the same origin or the same pattern of development [38]. In Cairo, high relative humidity, frequent fogs, sulphur, nitrogen pollutants, carbonaceous and deposition of airborne particles either on exposed or sheltered areas of Cairo archaeological buildings[39]. In consequence of these processes, these deterioration products grow on sheltered areas leading to thick encrustations, which are washed-out on surfaces exposed to rainwater[40]. On the unsheltered surfaces, newly formed soluble salts, washed-out by water and percolated through the bedding planes of the stone substrate, create a network of parallel and deep fissures, which increase the stone susceptibility to further deterioration [41]. On the other hand archaeological buildings in Cairo suffer from soiling, fig. (3). Soiling is a visual effect resulting from the darkening of exposed surfaces following the deposition and accumulation of atmospheric particles [42]. Deposition, removal and accumulation processes are numerous and complex, [43] depending on the physical and chemical properties of the particles, the nature of the surface, the local meteorology and the pathways followed by rainwater after it hits the building surface[44]. As a result of these complex interactions, there can be substantial variations in the level of soiling observed on building surfaces. It is one of the effects of air [45].

Fig. 4. (A-F) Thin section photomicrographs showing iron oxides, clay minerals, fossil and grains of quartz in a mass ground of fine- grained calcite. 60X (C.N). (G-I) shows that it is a mosaic texture; the calcite crystals have irregular faces and cleavage planes ,120X (C.N).

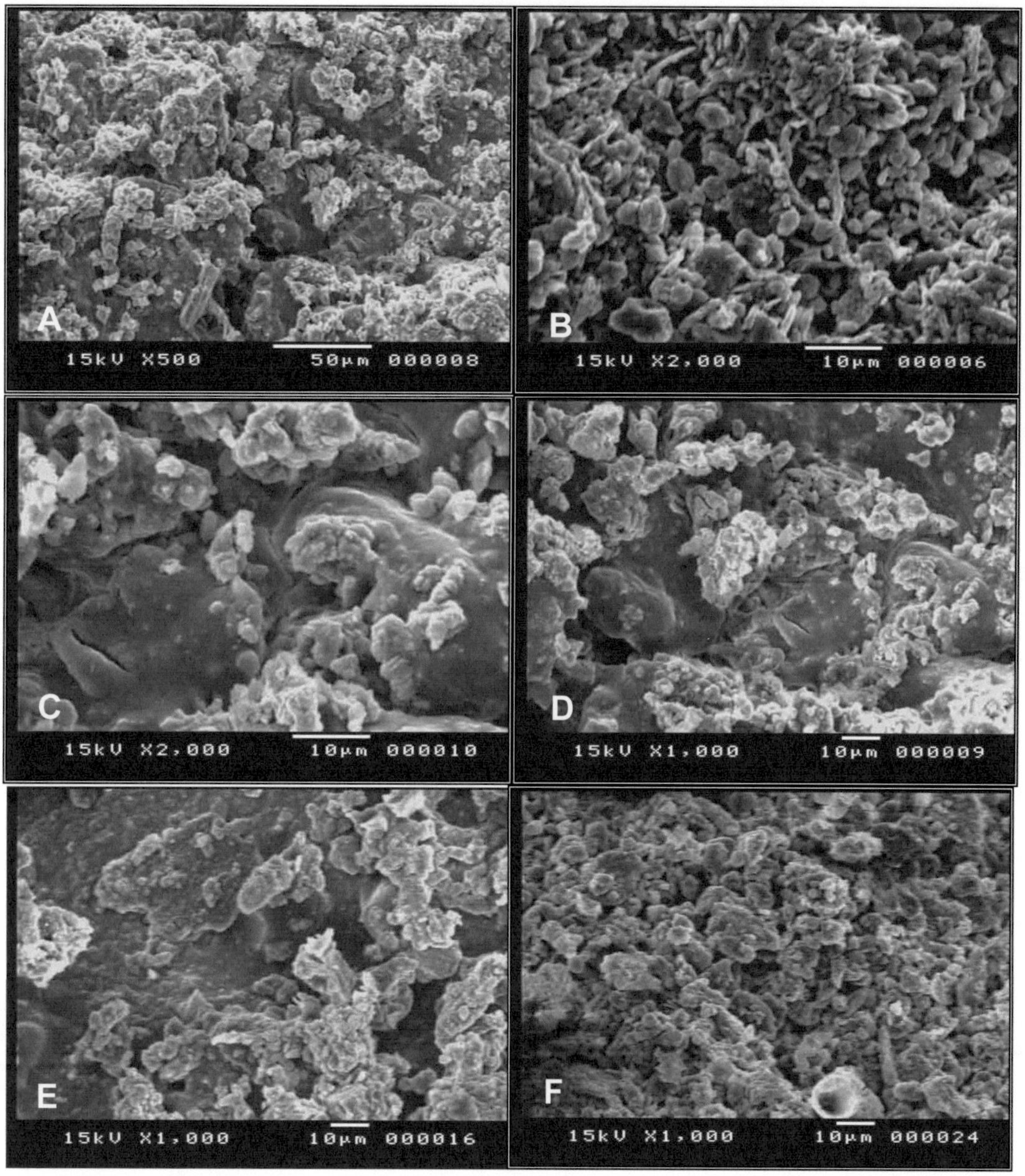

Fig. 5. (A-D) SEM photomicrographs of limestone samples showing the collapse of internal structure, salts crystallization between grains of limestone ornaments.
(E-F) photomicrographs of Marble samples showing voids due to lose of binding material erosion, discoloration, a coat of Carbon (C-D), chipping, fly ashes in a black crust and particles from the combustion of fuel oil and coal, containing a quantity of Carbon, Iron, Manganese and Sulphur.

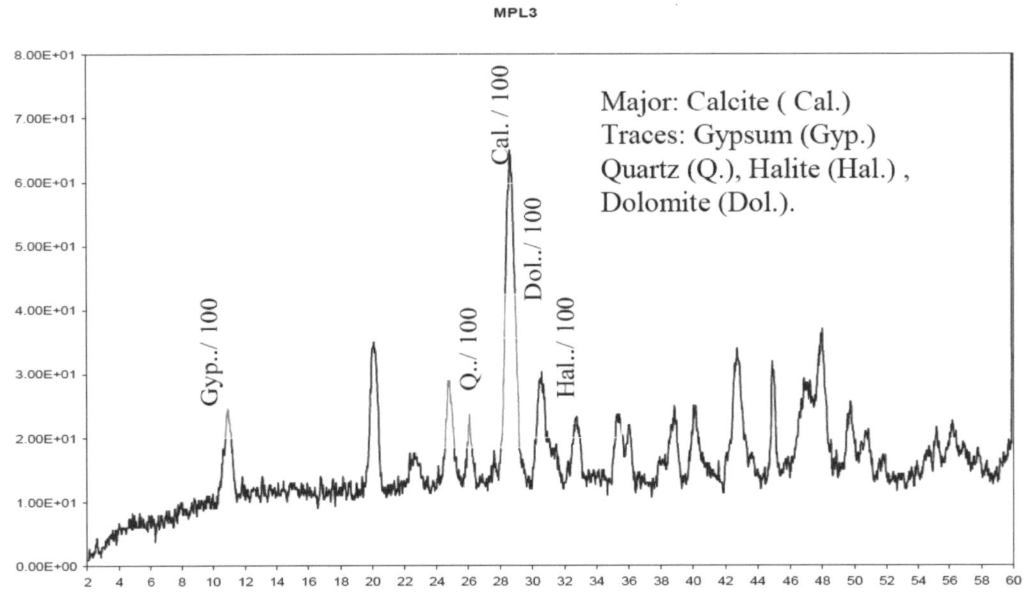

Fig. 6. (A) Shows XRD patterns of Limestone sample from El- Ghouri Mosque.

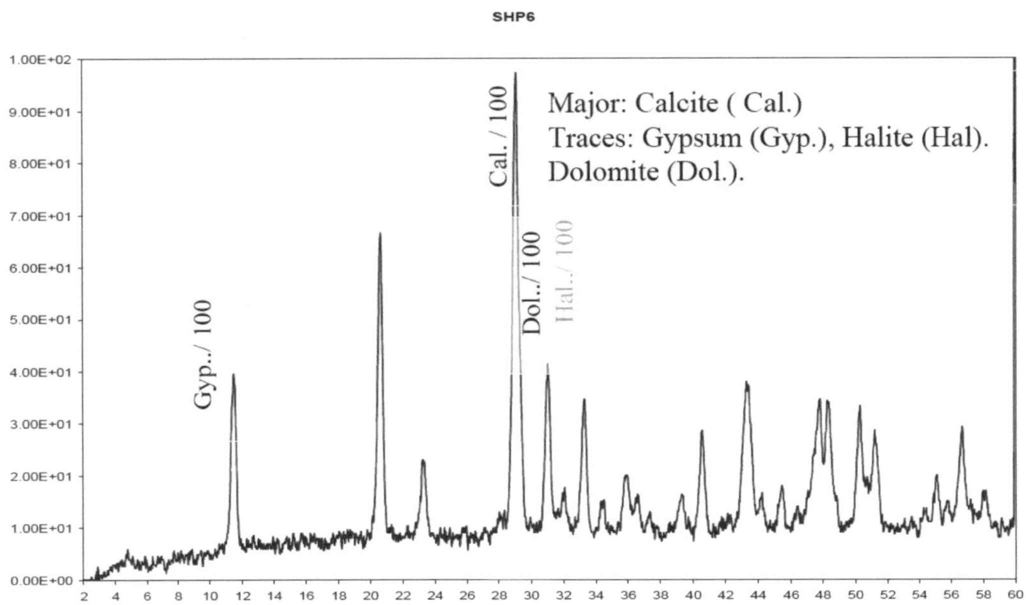

Fig. 6. (B) Shows XRD patterns of Limestone sample from El – Mahmoudya Mosque.

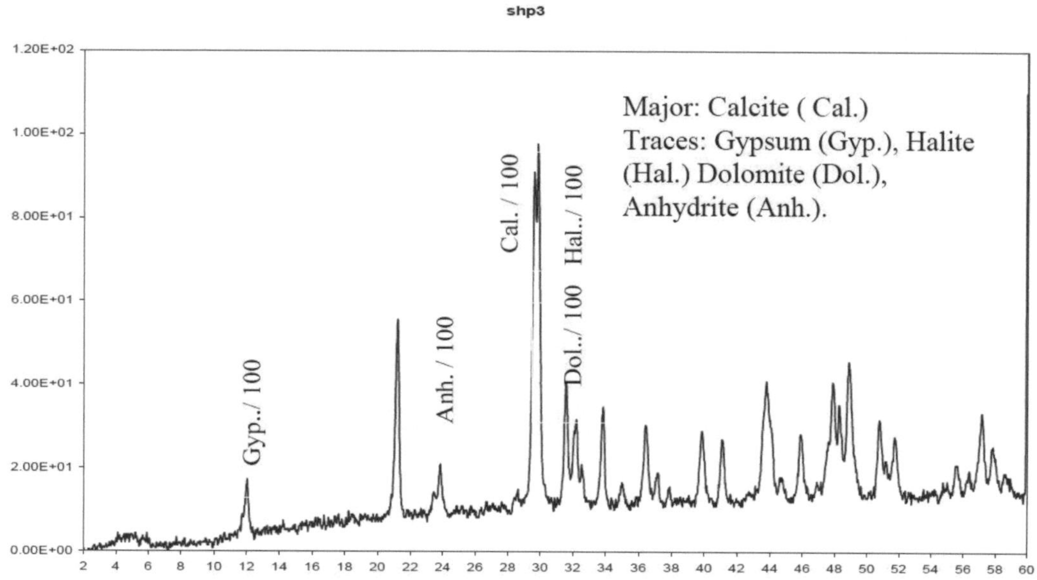

Fig. 6. (C) Shows XRD patterns of Marble sample from Taghri Bardi Mosque.

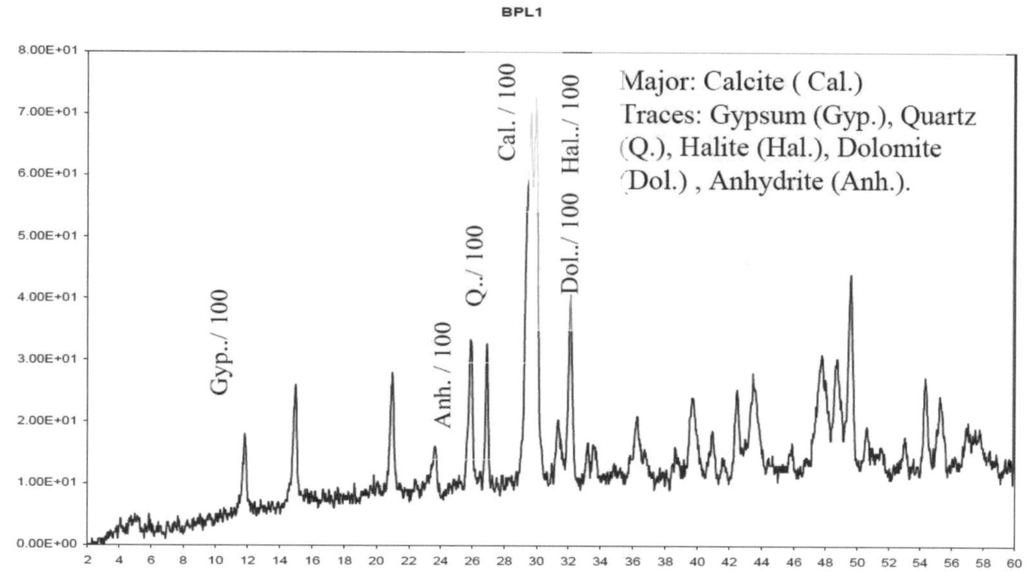

Fig. 6. (D) Shows XRD patterns of Marble sample from Qaitbay Sabil.

Atomic%	Weight%	Element
69.16	45.02	O K
1.98	2.17	Al K
5.40	6.17	Si K
2.27	2.96	S K
0.62	0.99	K K
1.69	2.75	Ca K
9.20	17.92	Ti K
0.45	1.00	Mn K
9.25	21.01	Fe K
	100.0	Total

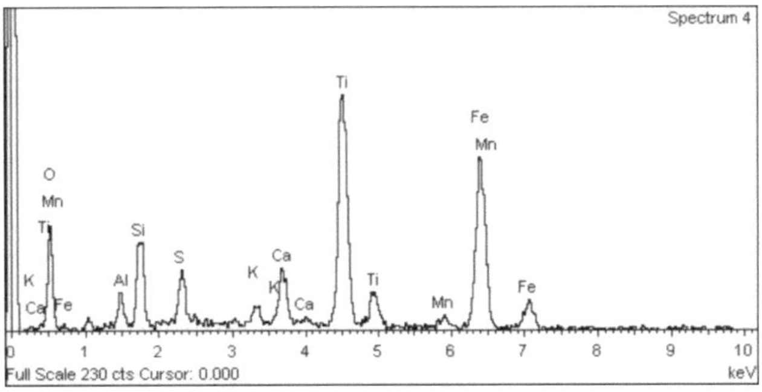

Fig. 7. (A) shows EDX patterns of limestone sample from El- Ghouri Mosque.

Atomic%	Weight%	Element
49.75	36.59	C K
38.35	37.58	O K
0.46	0.65	Na K
0.21	0.41	P K
5.26	10.32	S K
0.52	1.14	Cl K
0.63	1.50	K K
4.81	11.81	Ca K
	100.0	Total

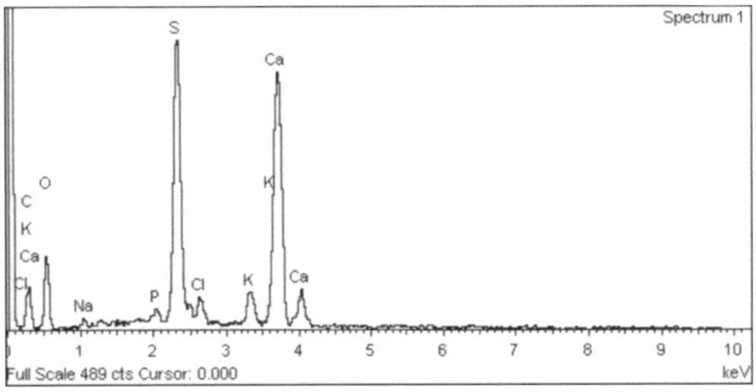

Fig. 7. (B) shows EDX patterns of limestone sample from El – Mahmoudya Mosque.

Atomic%	Weight%	Element
75.34	58.45	O K
1.35	1.51	Na K
0.72	0.85	Mg K
0.63	0.83	Al K
2.10	2.85	Si K
0.14	0.21	P K
7.27	11.30	S K
1.32	2.27	Cl K
1.35	2.57	K K
9.57	18.60	Ca K
0.21	0.56	Fe K
	100.0	Total

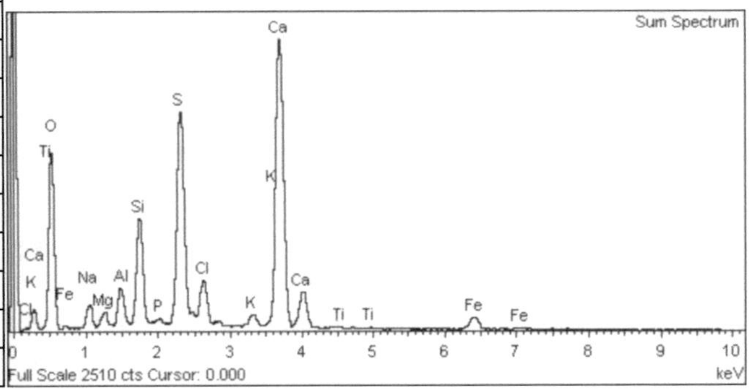

Fig. 7. (C) shows EDX patterns of Marble sample from Taghri Bardi Mosque.

Atomic%	Weight%	Element
42.72	31.40	C K
46.38	45.40	O K
0.86	1.21	Na K
0.17	0.33	P K
4.99	9.79	S K
0.22	0.49	Cl K
0.47	1.14	K K
4.18	10.25	Ca K
	100.0	Total

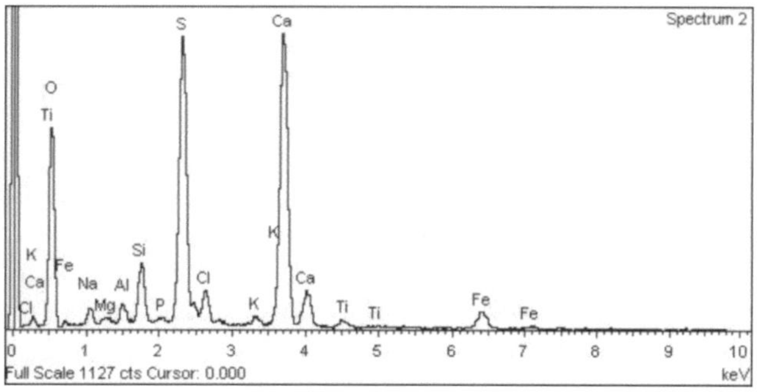

Fig. 7. (D) shows EDX patterns of Marble sample from Qaitbay Sabil.

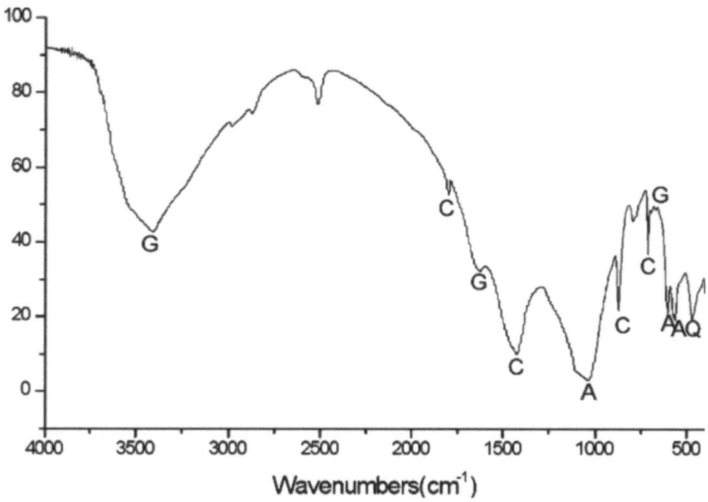

Fig. 8. FTIR spectra of limestone sample from El - Mahmoudya Mosque C, calcite (1798, 1424, 874, 711 cm−1); G, gypsum (672, 1623, 3408 cm−1); A, apatite (565, 604, 1040 cm−1); Q, quartz (469 cm−1).

5. Treatment and conservation processes

There are many methods and materials of treatment, restoration and conservation of building material in archaeological buildings from deterioration phenomena related to air pollution Cleaning Methods and Materials. These methods include cleaning, extraction of salts, consolidation and Water-Repellent Coatings.

5.1 Cleaning

Masonry cleaning methods generally are divided into three major groups: water, chemical, and abrasive. Water methods soften the dirt or soiling material and rinse the deposits from the masonry surface [46]. Chemical cleaners react with dirt, soiling material or paint to effect their removal, after which the cleaning effluent is rinsed off the masonry surface with water. Abrasive methods include blasting with grit, and the use of grinders and sanding discs, all of which mechanically remove the dirt, soiling material or paint (and, usually, some of the masonry surface). Abrasive cleaning is also often followed with a water rinse. Laser cleaning, although not discussed here in detail, is another technique that is used sometimes by conservators to clean small areas of historic masonry. It can be quite effective for cleaning limited areas, but it is expensive and generally not practical for most historic masonry cleaning projects. Although it may seem contrary to common sense, masonry cleaning projects should be carried out starting at the bottom and proceeding to the top of the building always keeping all surfaces wet below the area being cleaned, [47]. The rationale for this approach is based on the principle that dirty water or cleaning effluent dripping from cleaning in progress above will leave streaks on a dirty surface but will not streak a clean surface as long as it is kept wet and rinsed frequently.

5.2 Removal and extraction of salts

The notion of the poultice has been adapted for the cleaning of historic buildings and a true poultice is intended to draw out deep-seated contaminants and staining from the surface of masonry and sculpture. In current practice the word poultice is extended to a wide range of cleaning materials and techniques, not all of which achieve a true poultice effect on the substrate. What might be termed the true or plain poultice contains water and the poultice medium only, relying on these ingredients to achieve the mobilisation and removal of the contaminant. The most common poultice medium is clay, although paper and cotton fabrics are also used, and talc, chalk and even flour are traditional poultice materials. A mixture of clay and paper fabric produces an absorbent and plastic mixture that is often favoured by conservators of stone sculpture.This plain or true poultice is normally used for desalination, to draw out soluble salts, or as a cleaning method on substrates such as limestone that respond to water cleaning. In these cases the poultice is allowed to dry out and the soiling and/or salts are drawn into the poultice by capillary action with the moisture. Multiple applications may be necessary to draw the salts from within the surface pores, [48]. Whatever the medium, the poultice is mixed with water to form a material that will adhere to the substrate. Clay forms a sticky mass that adheres well to stone and other surfaces. These plain poultices can be conveniently mixed by hand as required on site with the addition of water to the poultice medium. Alkaline poultice cleaners and strippers are commonly used for cleaning or degreasing masonry surfaces and for paint removal. Sodium hydroxide is the most common alkaline cleaning agent in proprietary cleaners for a range of masonry substrates, including limestone, sandstone, brick and terracotta and is the most common ingredient in proprietary paint removers. Care must be taken in the use of sodium hydroxide based cleaners to minimise risks to the building and the user. Sodium hydroxide based cleaners and strippers must be neutralised with acid afterwash. Adjacent, dissimilar building surfaces must be protected and personal protective equipment worn by the cleaning operative. In the field of stone conservation ammonium carbonate is added to clay and clay/paper poultices to remove soiling from limestone. Ammonium carbonate is a less alkaline cleaner than sodium hydroxide. It works by reacting with calcium sulphate on the soiled surface to form calcium carbonate and soluble ammonium sulphate that can be rinsed off with water. These 'active' or 'chemical' poultices are all applied to a pre-wetted surface to minimize penetration of the chemical into the masonry surface and covered with plastic film to prevent the poultice drying out. The cleaning additives in these mixtures chemically dissolve the soiling or staining which is held to the surface of the poultice, and then both the cleaning agent and the contaminant are removed with the clay. Rinsing with water and, where necessary neutralization, follows to remove any soiling that remains on the surface and also to remove residues of the chemical cleaners. Strictly speaking these materials are clay-based cleaning packs rather than true poultices, but the word poultice is now widely used in the building cleaning industry [49].

5.3 Consolidation

Stone strengtheners based on ethyl silicates are generally applied plied by spraying or flooding. It is usually also possible to treat moveable parts by immersing them in a bath. Compresses can serve as an alternative to immersion .they ensure maximum length of con – tact between the stone strengthener and the stone[48]. Equipment employed for flooding includes electrical pumps air- less sprays and simple hoses. The pressure has to be kept as low as possible since the aim is to apply the material to the surface so that it will be absorbed naturally by the stones capillary system the excess will run off and be absorbed immediately by untreated areas below[49]. Several wet – on – wet treatments are generally needed applied at intervals of 20 to 30 minutes .the exact number of treatments quantity of material and desired minimum penetration depth have to be ascertained by preliminary tests and trials .the construction materials must be dry since the active in- gradient in the stone strengthener, ie, the ethyl silicates reacts with moisture. The moisture required by the stone strengthener for chemical deposition of the silica gel is sup- plied by the construction material which always has a certain sorption moisture content varying in equilibrium with the atmospheric humidity [50]. The best working conditions are a relative humidity of 40 to 70 % and a surface temperature on the construction material of 10 to 25 c each coating operation be so arranged that the entire surface can be covered in one working day . otherwise there is the danger that gel which gas been deposited in the pore system will prevent the strengthener from penetrating further this in turn might cause gel to be deposited in the surface regions of the stone and to gloss or crust formation . Very often instead of the whole object only small sections are treated such as a precious or- nametag detail or areas that are severely damaged in such cases it is advisable to follow up the last treatment with a solvent wash suitable solvents are hydrocarbons methyl ethyl ketene and ethyl alcohol [47]. Freshly treated surfaces must be covered for 2 to 3 days against the rain. Considerable loss of the active ingredient by evaporation may occur at temperatures exceeding 25 c at such temperatures the freshly consolidated surfaces have to be protected against direct sunlight. Temperatures below 5 c cause the stone strengthener to react very slowly this may result very slowly this may result in discoloration or glaze on the surface [51]. The total time needed for the stone strengthener to deposit the silica gel depends on the relative humidity and the temperature. it varies form one to at most three weeks therefore before any further restoration work is carried out on period of roughly one week should elapse this will allow 90 to 95 % of the silica gel to be de -posited . On no account should water be added to the ethyl silicate preparation in an attempt to speed up the reaction this can result in extensive glazing of the surface that is extremely difficult to remove if indeed this is at all possible.

5.4 Water-repellent coatings

Water-repellent coatings are formulated to be vapor permeable, or "breathable". They do not seal the surface completely to water vapor so it can enter the masonry wall as well as leave the wall. While the first water-repellent coatings to be developed were primarily acrylic or silicone resins in organic solvents, now most water-repellent coatings are water-based and formulated from modified siloxanes, silanes and other alkoxysilanes, or metallic stearates [49]. While some of these products are shipped from the factory ready to use, other water-borne water repellents must be diluted at the job site. Unlike earlier water-repellent coatings which tended to form a "film" on the masonry surface, modern water-repellent coatings actually penetrate into the masonry substrate slightly and, generally, are almost invisible if properly applied to the masonry. They are also more vapor permeable than the old coatings, yet they still reduce the vapor permeability of the masonry [48]. Once inside the wall, water vapor can condense at cold spots producing liquid water which, unlike water vapor, cannot escape through a water-repellent coating. The liquid water within the wall, whether from condensation, leaking gutters, or other sources, can cause considerable damage. Water-repellent coatings are not consolidants. Although modern water-repellents may penetrate slightly beneath the masonry surface, instead of just "sitting" on top of it, they do not perform the same function as a consolidant which is to "consolidate" and replace lost binder to strengthen deteriorating masonry. Even after many years of laboratory study and testing, few consolidants have proven very effective. The composition of fired products such as brick and architectural terra cotta, as well as many types of building stone, does not lend itself to consolidation. Some modern water-repellent coatings which contain a binder intended to replace the natural binders in stone that have been lost through weathering and natural erosion are described in product literature as both a water repellent and a consolidant The fact that the newer water-repellent coatings penetrate beneath the masonry surface instead of just forming a layer on top of the surface may indeed convey at least some consolidating properties to certain stones. However, a water-repellent coating cannot be considered a consolidant. In some instances, a water-repellent or "preservative" coating, if applied to already damaged or spalling stone, may form a surface crust which, if it fails, may exacerbate the deterioration by pulling off even more of the stone [52].

6. Achievements and planned activities for improvement of air quality in Cairo

Air quality represents a major priority for the Egyptian Ministry of state for Environmental Affairs, Egyptian Environmental Affairs Agency as it has dangerous impacts on the public health and it is effect on archaeological buildings. This concern encompasses a number of trends [53]:

6.1 Alleviating the vehicles' emissions

Through the coordination and effective cooperation between the Ministry of Environment and the Ministry of Interior, the decree of the Minister of Interior was issued:

a. To link between the issuance of the licenses of the Vehicles and its emissions testing, and the start of the implementation of this decree in the Qaluibia and Giza governorates. Such decree provides a new hope of the improvement of air quality and the first step of overcoming the problem of the vehicles' emissions, to be applied in many other governorates. This decree is essential for the reinforcement of Law No. 4 for 1994 on the protection of Environment. [53].
b. The Ministry of State has already, in collaboration with USAID through the Cairo Air Improvement Project, delivered the traffic departments in Giza and Qaluibia

governorates 38 devices for vehicles' emission testing, in addition to training those who are designated to the technical inspection of vehicles using diesel and benzene. It is worth mentioning that the application of the issuance of the vehicles' licenses in both governorates has started from June 1, 2003 on vehicles' emissions testing to combat the emissions of Carbon monoxide and Hydrocarbons.

c. The cooperation between the Ministry of Environment and the Ministry of Interior has resulted in the establishment of the environment police: the first police stations to be inaugurated will be in the Regional Branch of the Egyptian Environmental Affairs Agency in Greater Cairo and El-Fayoum as well as in Beni Siweif.

6.2 Relocation of the heavily polluting activities outside the populated areas

Due to the variety of pollution sources especially within Greater Cairo, the Ministry of Environment has formulated a plan of the relocation of the polluting activities outside the populated areas, among them the smelters, quarries, potteries, crackers, brick factories and coal and lime facilities as well as 1206 mining factories and 6000 textiles factories.

This funding plan is based on the contribution of the owners of these activities, applying the principle "Polluter pays". The estimated budget of this plan is L.E. 1745 million; the share of the government is about 15% of its total. In addition to that, the government provides soft loans for the relocation of these polluting activities to the desert. The owners of these activities contribute to the remainder of cost for 4 years starting from July 1, 2003 to June 30, 2007.

The Ministry has, in cooperation with the competent governorates, identified the places of relocation of these polluting activities in El-Amal region in the Ain El Sokhna Road for all Cairo smelters and in Akrasha region for Qaluibia smelters, in addition to relocation of coal facilities to the industrial zone in Belbeis as well as the brick factories to Arab Abu Saad region. [53].

6.3 Combating the industrial pollution

As for the plan of the Ministry of Environment for pollution sources control in the big factories, it has prepared a plan in two phases as follows:

The first phase: factories in need of limited funds for approximately L.E. 23.13 million to combat pollution discharged from them.

The second phase: factories in need to huge funds for about L.E. 545.9 million. In this respect, the EEAA is implementing some of the projects that make available the funding and technical support for the industrial establishments, such as the Industrial Pollution Abatement Projects providing grants and soft loans offered by the World Bank as well as technical assistance as a grant from the Finnish Government. In addition, there is the Environment Protection Fund for the Public Sector Industries funded by the German Construction Bank that provides Euro 25.56 million as a grant from the German Government, representing a partial funding of 50 % of the necessary investments for the implementation of industrial waste treatment projects as well as soft loans presented from the Egyptian banks participating in this project[54].

6.4 The environmental inspection on the establishments

Since the start of the practical application of Law No. 4 for 1994 and after the termination of the grace period provided by the Law and its Executive Regulation, the Ministry has established the Environmental Inspection Unit at the central level and prepared a manual of the policies and procedures for the inspection unit, which is considered the first manual in this field. This manual has specified the role and authorities of the environmental inspection in comparison to the other supervisory agencies concerned with the inspection on the establishments. This manual reemphasizes that the periodical follow-up and inspection are the effective means of the non-replication of violations.

6.5 The safe use of the treated sewage water in the irrigation of forests

For further improvement of air quality and the reduction of dust and sand rates, arising from Al-Khamaseen wind, the Ministry of State for Environmental Affairs is implementing the Green Belt Project around Greater Cairo (Cairo-Giza-Qaluibia) along 100 km on the sides of the circular road with a width of 10-25 m, [55].cultivating it with Acacia and Cypress trees.

This project aims at protecting the citizens of the Greater Cairo from dust and sands and conserving their health. In addition, it provides job opportunities to the graduated youth whether in the implementation of the project or its maintenance, besides using the treated sewage water for the irrigation of these trees to be economically made use of.

This project is implemented in four phases in the three governorates: starting from Cairo Governorate in the region from El-Moneib Bridge to Misr Ismailia Desert Road, in Qaluibia Governorate to El-Kanater establishment, and in Giza Governorate to El-Moneib Bridge. The total cost of the project is L.E. 13.7 million. [53].

The Green Belt is not the last project implemented by the Ministry for the improvement of air quality but there is also the National Programme for the Safe Use of Treated Sewage Water, in collaboration with the Agriculture, Irrigation, Housing, Local Development and Environment Ministries as well as the different governorates.

The concept of this project depends on the investment of treated sewage water since Egypt produces about 3 billions m3 annually at the cost of 14 Piastres/meter with a total of approximately L.E 14 million, and turns this problem into a social, environmental and economical value. Instead of disposing this treated water into water channels and contaminating it, it can be used in afforestation.

This project achieves several social, economic and environmental benefits as it basically improves air quality through the plantation of trees that are the source of Oxygen for they intake Carbondioxide and produce Oxygen[53]. In addition, it helps in combating desertification, protecting water resources and soil from pollution, building green belts and wind obstructers, to be used in producing woods instead of importing them. It also helps in providing job opportunities for the youth and establishing the new urban communities side by side with these forests. [54]. There are successful attepts for this project in Serabuim area in Ismailia, Sadat City, Asuit, Sohag, Luxor, Qena, New Valley, Tour Sinai, El-Saaf , and Aswan.

This national project is carried out at several phases. The first phase is executed in an area of 82940 thousand Fedan around 72 Sewage stations in the different governorates all over the Republic at the cost of L.E. 5 thousand/ Fedan, providing a collective revenue during the lifetime of the forest, i.e. 12 years. The implementation is carried out for 8 thousands Fedan annually.

6.6 Manufacturing the construction materials from rice straw using unconventional technology

There is no doubt that success that will come out of the real partnership between the Government and the Private Sector in the relocation of polluting activities outside the residential regions, based on environmental principles and standards supported by environment friendly technologies will directly assist in combating the Black Cloud phenomenon that we suffer from in October annually. Scientists from the Scientific Research Academy, the National Research Center, the Environment Research Council, the Meteorology Organization and the Specialized National Councils have a consensus that the real reasons for the Cloud are confined to a climate phenomenon, namely, the existence of high pressure that appears every year at the same time, accompanied by a thermal change and stability of wind, which all lead to the accumulation of pollution in Cairo air. [53].

7. Conclusion

The danger to archaeological buildings from air pollution comes from two main sources - gases that increase the corrosivity of the atmosphere and black particles that dirty light-colored surfaces. Acid rain comes from oxides of sulphur and nitrogen, largely products of domestic and industrial fuel burning and related to two strong acids: sulphuric acid and nitric acid. Sulphur dioxide (SO_2) and nitrogen oxides (NOx) released from power stations and other sources form acids where the weather is wet, which fall to the Earth as precipitation and damage both heritage materials and human health. In dry areas, the acid chemicals may become incorporated into dust or smoke, which can deposit on buildings and also cause corrosion when later wetted. Atmospheric chemistry is, of course, far more complex than this and a variety of reactions occur that may form secondary pollutants that also attack materials. Particulate matter is much more complicated because it is a mixture rather than a single substance - it includes dust, soot and other tiny bits of solid materials produced by many sources, including burning of diesel fuel by trucks and buses, incineration of garbage, construction, industrial processes and domestic use of fireplaces and woodstoves. Particulate pollution can cause increased corrosion by involvement in a number of chemical reactions and, often more importantly, it is the source of the black matter that makes buildings dirty. The influence of heavily polluted atmosphere in the urban environment results in different weathering patterns, mainly in the form of crusts. It might be assumed that the analytical results of Polarizing Microscope, XRD, SEM, EDX and IR. alone are not sufficient to clarify and interpret the growth mechanisms of crusts. However, they do provide valuable information about changes in compositions of crusts and original rock, and the relationship between crusts composition and air pollution. The compositions of the crusts collected from areas on the archaeological stone buildings with different decay patterns show that the deterioration is mainly due to the atmospheric pollutants and its extent is strongly dependent on the surface exposition to the environment. According to the obtained results, an appropriate conservation plan will be developed, that includes the steps of cleaning and consolidation, in order to identify the most suitable materials and methodologies to remove the deterioration crusts avoiding the loss of original substrate and ensuring an increased cohesion to deteriorated stone.

8. References

[1] Creswell, (1959), The Muslim Architecture of Egypt, Oxford, P.112.
[2] Beattie, Andrew (2005). Cairo: A Cultural History (illustrated ed.). New York: Oxford University Press.

[3]Butler, Alfred J. (2008). The Arab Conquest of Egypt - And the Last Thirty Years of the Roman Dominion. Portland, Ore: Butler Press.

[4] Behrens-Abouseif, Doris (1992). Islamic Architecture in Cairo (2nd ed.). Brill. Daly, M. W.; Petry, Carl F. (1998). The Cambridge History of Egypt: Islamic Egypt, 640-1517. Cambridge, UK: Cambridge University Press.

[5] Glassé, Cyril; Smith, Huston (2003). The New Encyclopedia of Islam (2nd revised ed.). Singapore: Tien Wah Press.

[6] Rose, Christopher; Linda Boxberger (1995). "Ottoman Cairo". Cairo: Living Past, Living Future. The University of Texas Center for Middle Eastern Studies.

[7] Mortada, Hisham (2003). Traditional Islamic principles of built environment. Routledge. p. viii.

[8] Williams,C., Islamic Monuments in Cairo: The Practical Guide, American University in Cairo Press,2004.

[9] Anoniou, J. "Historic Cairo - A Walk through the Islamic City", American University in Cairo Press,1999.

[10] Watt, J. et al.Creighton NP et al., (1990) Soiling by atmospheric aerosols in an urban industrial area. J Air Waste Manag Assoc. 40, 1285–1289.

[11]Aksu R, Horvath H, Kaller W, Lahounik S, Pesava P and Toprak S (1996) Measurements of the deposition velocity of particulate matter to building surfaces in the atmosphere. J Aerosol Sci. 27, S675–676

[12] Davidson CI, Tang F, Finger S, Etyemezian V and Sherwood SI (2000) Soiling patterns on a tall limestone building: changes over 60 years. Environ Sci Technol. 34, 560–565.

[13] Air Pollution Levels Measured in Egypt Exceeds Air Quality Limit Values.(2002), EEAA/EIMP, Ministry of state for Environmental Affairs, Egyptian Environmental Affairs Agency.

[14] Hopkins; N. 2003"The Environmentalist: Living with Pollution in Egypt".

[15] Abo El-Ela, A., The Impact of Environmental pollution on the Mosque of Al-Azhar and the complex of Al-Ghuri, In : [16] The Restoration and Conservation of Islamic Monuments in Egypt, The American University in Cairo Press,1995.Pp. 99-114.

[17] Del Monte M, Sabbioni C and Vittori O (1981) Airborne carbon particles and marble. deterioration, Atmos Env. 16, 2253–2257.

[18] Mohamed ,K. Khallaf, (2006), Role of Investigation and Analytical Methods in study of Archaeological Stone Ornaments's Deterioration and Its Treatment,The Seventh International Symposium on New Trends in Chemistry " Analytical Chemistry for a better Life " Egyptian Journal of Analytical Chemistry - Volume (15), January .

[19] Mohamed ,K. Khallaf , (2006) Analysis and Preservation of Marble in Archaeological Buildings, The Seventh International Symposium on New Trends in Chemistry " Analytical Chemistry for a better Life " Egyptian Journal of Analytical Chemistry - Volume (15), January.

[20] Mohamed ,K. Khallaf, (2008), Degradation and Conservation of Marble Floors in Archaeological Buildings, 5th Symposium of the Hellenic Society for Archaeometry, 8-12 October, 2008, Athens, Greece. October.

[21] Mohamed ,K. Khallaf ,(2006), Environmental Detrioration and Conservation Studies of Building Materials of Qaitbay Citadel, Rosetta City, Egypt, Civil Engineering

Research Magazine, Faculty of Engineering, Al - Azhar University, Volume(28) No.(1), January.
[22] Saiz-Jimenez C. (editor) 2004, Air pollution and cultural heritage. A.A. Balkema Publishers, Taylor & Francis Group plc, London.
[23] Parker A. (1955) The destructive effect of air pollution on materials. National Smoke Abatement Society. London. pp 3–15.
[24] Watt J. (1998) Automated Characterisation of Individual Carbonaceous Fly Ash Particles by Computer Controlled Scanning Electron Microscopy -Analytical Methods and Critical Review of Alternative Techniques. Water AirSoil Pollut. 106, 309–327.
[25] Pesava P, Aksu R, Toprak S, Horvath H and Seidl S (1999) Dry deposition of particles to building surfaces and soiling. Sci Total Env. 235, 25–35.
[26] Pio CA et al., (1998) Atmospheric aerosol and soiling of external surfaces in an urban environment. Atmos Env. 32, 1979–1989.
[27] Brimblecombe P 2003, The effects of air pollution on the built environment. Imperial College Press, London.
[28] Hamilton RS and Mansfield TA 1991, Airborne particulate elemental carbon: its sources, transport and contribution to dark smoke and soiling, Atmos. Env. 25, 715–723.
[29] Hinds WC (1999), Aerosol Technology: properties, behaviour and measurements of airborne particles, 2nd edition. Wiley ISBN 978-0-471-19410-1.
[30] Tidblad, J., Mikhailov, A.,& Kucera, V. (2000). Acid deposition effects on materials in subtropical and tropical climates. Data compilation and temperate climate comparison. SCI Report 2000:8E, Swedish Corrosion Institute, Stockholm, Sweden.
[31] Kucera, V., Tidblad, J. (2005). Comparison of environmental parameters and their effects on atmospheric corrosion in Europe and in South Asia and Africa. Proc. 16th Int. Corrosion Congress, Beijing.
[32] Cole, I. S. (2000). Mechanisms of atmospheric corrosion in tropical environments. ASTM STP 1399. In S. W. Dean, G. Hernandez-Duque Delgadillo & J. B. Bushman (Eds), American Society of Testing and Materials. West Conshohocken, PA.
[33] Maeda, Y., Moriocka, J., et al., (2001). Materials damage caused by acidic air pollution in East Asia. Water, Air and Soil Pollution, 130, 141–150.
[34] Beloin NJ and Haynie FH (1975) Soiling of building materials. J Air Pollut Control Ass. 25, 393–403.
[35] Hamilton RS and Mansfield TA (1992) The soiling of materials in the ambient atmosphere. Atmos Env. Part A - Gen Topics. 26, 3291–3296.
[36] Lanting RW (1986). Black smoke and soiling in aerosols: research, risk assessment and control strategies. In Proceedings of the Second US-Dutch Symposium, Ed. Lee SD, Lewis Publishers, Williamsburg, VA.
[37] Mansfield TA and Hamilton RS (1989). The soiling of materials: models and measurements.
[38] Parker A. (1955) The destructive effect of air pollution on materials. National Smoke Abatement Society. London. pp 3–15.
[39] K.L. Gauri, G.C. Holdren, (1981) Pollutant effects on stone monuments, Environ. Sci. Technol. 15 (4), 386–390.
[40] F. Delalieux, C.P. Cardell, V. Todorov, (2001) Environmental conditions controlling the chemical weathering of the Madara [41] Horseman monument, NE, J. Cult. Herit. 2, 43–54.

[42] A. Moropoulou, K. Bisbikou, K. Torfs, (1998) Origin and growth of weathering crusts on ancient marbles in industrial atmosphere, Atmos. Environ. 32 (6), 967–982.
[43] P. Maravelaki-Kalaitzaki, (2005) Black crusts and patinas on Pentelic marble from the Parthenon and Erechtheum (Acropolis, Athens): characterization and origin, Anal. Chim. Acta 532, 187–198.
[44] C. Vazquez-Calvo, M. Buergo, R. Fort, (2007),Characterization of patinas by means of microscopic techniques, Mater. Charact. 58, 1119–1132.
[45] Brimblecombe P and Grossi CM (2005) Aesthetic thresholds and blackening of stone buildings. Sci Total Env. 349, 175–198. Fig. 4.15 Variation of soiling with PM10 concentration (white painted steel) 124
[46] Mack, R.C. and Grimmer, F.A.: Assessing Cleaning and Water-Repellent Treatments for Historic Masonry Buildings, Washington DC, (2003).
[47] Mohamed ,K. Khallaf, (2008), Interfacial Characteristics of Polymeric Coatings for Archaeological Stones Conservation., Sixth International Conference: Science and Technology in Archaeology and Conservation, Rome, Italy, December 9th - 13th.
[48] Ana Luque, Giuseppe Cultrone and Eduardo Sebastián: (2010) The Use of Lime Mortars in Restoration,Work on Architectural Heritage, In book: Materials, Technologies and Practice in HistoricHeritage Structures, Edited by, Maria Bostenaru Dan, Springer , New York .
[49] Clifton, J.R.: (2005) Stone Consolidating Materials, A Status Report, Cool, Documents, August.
[50] Gansicke, S., and J. Hirx. (1997) Mortars as A filling materials for the compensation of losses in objects. Journal of the American Institute for Conservation 36:17-29.
[51] Wheeler, G.: (2005) Alkoxysilanes and the Consolidation of Stone, Columbia University Press, U.S.A.
[52] Noll ,W., Chemistry and technology of silicones. Academic Press, New York, (1986).
[53] The Cairo air improvement project (2004), Final Report | March, Prepared by: Chemonics International Inc.1133 20t h Street, NW, Washington, DC 20036 / USA, Prepared for: USAID/Egypt, Office of Environment , Contract 263-C-00-97-00090-00.
[54] Khoder, M.I. (2007). " levels of volatile organic compounds in the atmosphere of Greater Cairo". Atmospheric Environment (Air Pollution Research Department, National Research Centre, Dokki, Giza) 41 (3): 554–566.
[55] Hopkins N., (2003) "The Environmentalist: Living with Pollution in Egypt" A.A. Balkema Publishers, Taylor & Francis Group plc, London.

13

Influence of Air Pollution on Degradation of Historic Buildings at the Urban Tropical Atmosphere of San Francisco de Campeche City, México

Javier Reyes[1] et al.,*
[1]*Autonomous University of Campeche,*
[2]*National Center for Scientific Research,*
[3]*CINVESTAV-Mérida,*
[1,3]*Mexico*
[2]*Cuba*

1. Introduction

The role of atmospheric pollution in degradation of historic building has been studied for long time along the world because it increases stone decay and the lost of historic materials (Massey, 1999; Graedel, 2000; Monna, 2008). The preservation of Cultural Heritage is considered a strategic factor in countries integration because of their economical, social and cultural implications (Cassar et al., 2004; Sessa 2004, Moropoulou and Konstanti, 2004).
Latin-american countries have an important building legacy from prehispanic, colonial and modern periods. This is the case of México which currently count with 15 sites included in UNESCO´s cultural heritage list. Most of them are located in urban areas like Mexico City, Morelia, Guanajuato, San Miguel de Allende and San Francisco de Campeche, between others.
San Francisco de Campeche is a small City located at the south east of Mexican Republic, just in the occidental coast the Peninsula of Yucatan, inside the Gulf of Mexico Basin (Fig. 1). The City was founded in 1527, by Spanish colonizers leaded by Francisco de Montejo, "el Mozo". During the XVII century, it was the only point for exportation of goodness from Yucatan to Europe. Because of these conditions, French, Netherlanders and British pirates considered the city a legitimate target.
At that time, authorities designed an impressive military defensive system to protect the City and their inhabitants. Forts, batteries and a rampart surrounding San Francisco de Campeche urban core were built by using calcareous materials based on masonry structures made with limestone quarry blocks joined and covered with mortars made with slike lime and sahacab, a typical calcareous clay material used since prehispanic period for building construction. Nowadays, about 1500 buildings are located into the historic and architectonic complex included in 1999 in the UNESCO`s Cultural Heritage List.

* Francisco Corvo[1], Yolanda Espinosa-Morales[1], Brisvey Dzul[1], Tezozomoc Perez[1], Cecilia Valdes[2], Daniel Aguilar[3] and Patricia Quintana[3]

Since their construction, these buildings have been exposed to the action of environmental agents that induce their deterioration. For long time, natural parameters like high relative humidity, extended rainfall periods and the effects of marine aerosols were the principal factors related with buildings degradation (Zendri, 2001, Cardell et al., 2003). Karstification, crust formation, lost of components and biodegradation are typical pathologies of degradation observed in the buildings. Nevertheless, in the last decade, the City has been under a dynamic development. As a consequence, a sensible increase in automobile units has been registered in specific areas of the city, including the historical centre.

Fig. 1. The State of Campeche located at the South East of Mexican Republic. Red dots indicate the location of San Francisco de Campeche City and Iturbide town, current environmental monitoring sites operated by the Corrosion Research Center (CICORR).

Automotive emissions generate atmospheric particles and corrosive gases like sulphur dioxide (SO_2) and nitrogen oxides (NO_X) that, in contact with environmental humidity produce acid precipitation that dissolve calcareous materials, or induce black crust formation (Lipfert, 1989; Gobi, et al., 1998, Kucera, 2007). Systematic studies related to atmospheric pollution and their effects in historic building degradation at San Francisco de Campeche City are scarce.

2. Stone decay

Stone materials have a natural tendency to degradation as a consequence of change in their chemical stability when they are extracted from the quarry and submitted to the building fabric, atmospheric action and change in air quality.

Before industrial revolution, natural agents were the main cause of stone buildings degradation, sometimes through suddenly destructive actions as earthquakes, volcanic eruptions or hurricanes. Most of the times, acting in slow weathering process. Nevertheless, with the appearance of the industrial society, atmospheric pollution got a major role in building deterioration.

In natural conditions atmospheric water is the main agent associated to stone degradation. Its influence is especially important in tropical climates, where high relative humidity and large rain forest period along the year guarantee water availability to lead chemical reaction over stone substrata or to produce secondary pollutant's potentially harmful for stone materials.

In San Francisco de Campeche City, historic buildings were constructed using calcareous stone materials including quarry blocks and mortars. Calcareous stone and traditional mortars used during buildings construction or restitution works usually show a wide interval of porosity (Reyes et al., 2010; Torres, 2009). It is well known that water circulation in porous stones and their exchange with atmosphere or ground, affect their behavior and durability.

The flux of water across porous structure of stones and mortars is consequence of wet- to dry- cycles, that induce chemical reactions and salts crystallization leading materials lost and decreasing their mechanical capabilities. Furthermore, direct impact of rainfall is cause of erosion on stone surface and the appearance of run-off inside of masonry structures. On the other hand, when water table level is high, a capillary effect could appear. Then, a continuous flux of soluble salts inside and outside materials stone structure is established. Water presence also facilitates the development of microorganism colonies and the growth of superior plants. In both cases, their consequences on stone materials are chemical and mechanical damage.

In urban environments, decay of historic buildings is strongly influenced by the presence of atmospheric pollutants like SO_2, NO_X, atmospheric particles and acid rain.

In the atmosphere water drops incorporates carbon dioxide **(CO_2)** to produce the weak carbonic acid (HCO_3^-), which is partially dissociated according to the next reaction:

$$CO_2 + H_2O \leftrightarrow H^+ + HCO_3^- \qquad (1)$$

As a consequence, water acquires a pH of 5.65. It means that in unpolluted atmosphere water tends toward acid. Under this condition, dissolution of calcareous materials is possible. Dissolution of carbonates inside walls as their migration and deposition to the evaporation front lead the formation of crusts, as is demonstrated in the next reaction:

$$CaCO_3 + H_2O + CO_2 \rightarrow Ca(HCO_3)_2 \qquad (2)$$

Soluble calcium bicarbonate ($Ca(HCO_3)_2$) is transported by water to the surface of stone and mortars across porous system of built and decorative elements. When water evaporates CO_2 drags (equation 3).

$$CaCO_3 + H_2O + CO_2 \rightarrow Ca(HCO_3)_2 \qquad (3)$$

Formation rate of $Ca(HCO_3)_2$ depends on CO_2 levels, that is the reason why in urban environments with high levels of this gas, carbonation of calcareous materials is most important than in rural environment. On the other hand, recrystallized calcite is bigger in size and more porous than the microcrystalline original calcite. The increase in size is extreme harmful for stone materials, because it creates conditions for a deep penetration of acidic solutions (like acid rain), insoluble salts and gases like SO_2 and NO_X.

Acid rain is produced when gases like SO_2 or NO_X reacts with water drops, increasing their acidity under pH value of 5.65 to form the so called acid rain. Acid rain is a global phenomena and its effect can be observed at long distances from their precursor sources (Bravo et al., 2000).

The presence of a minimum water amount is enough to oxidize SO_2 to sulphuric acid (H_2SO_4) according to the next reactions:

$$2SO_{2(g)} + O_{2(g)} \rightarrow 2SO_3 \quad (4)$$

$$SO_{3(g)} + H_2O_{(l)} \rightarrow H_2SO_4 \quad (5)$$

H_2SO_4 can easily react with calcareous materials to form gypsum ($CaSO_4{\cdot}2H_2O$) as is indicated in equation (6)

$$CaCO_3 + H_2SO_4 \rightarrow CaSO_4{\cdot}2H_2O + CO_2\,. \quad (6)$$

Gypsum formation is a serious problem because when it crystallizes gradually expands up to 30 % of their original size (Feddema, et al., 1987). $CaSO_4.2H_2O$. It is highly soluble at predominant temperatures in tropical regions, so it requires a minimum water amount to dissolve and lead a fast migration to evaporation front by capilar mechanisms. When gypsum lost humidity, it can recrystallize into porous, where induce the formation of microcracks and fatigue of materials. In urban environments, gypsum incorporates into their mineral structure atmospheric particles, dust and biomass to form the so called black crust.
NO_X formation depends on environmental conditions and the kind of pollutant present in the atmosphere. It is expressed in the next reactions:

$$2NO + O_2 \rightarrow 2NO_2 \quad (7)$$

$$2NO_2 + H_2O \rightarrow HNO_3 + HNO_2 \quad (8)$$

Ozone (O_3), also can also react with nitrogen oxide (NO) and nitrogen dioxide (NO_2):

$$NO_2 + O_3 \rightarrow NO_3 + O_2 \quad (9)$$

$$NO + O_3 \rightarrow NO_2 + O_2 \quad (10)$$

The products of these reactions establishes an equilibrium with dinitrogen pentoxide, which react with water to form nitric acid (HNO_3):

$$NO_2 + NO_3 \leftrightarrow N_2O_5 \quad (11)$$

$$N_2O_5 + H_2O \rightarrow 2HNO_3 \quad (12)$$

In urban zones, O_3 and NO also react with water to form HNO_3

$$2NO + O_3 + H_2O \rightarrow 2HNO_3 \quad (13)$$

Nitric acid dissolves calcareous stone to produce calcium nitrate $(CaNO_3)_2$:

$$CaCO_3 + 2HNO_3 \rightarrow Ca(NO_3)_2 + H_2O + CO_2 \quad (14)$$

$Ca(NO_3)_2$ is more water soluble than $CaCO_3$. If it is present, is transported across porous capillars to finally crystallize on monuments surface to be washed during rainy events (Allen et al, 2000). Deposition mechanisms also play an active role in historic building deterioration. Atmospheric particles and aerosols are transported by wind toward monumental structures.

Here, they are incorporated into neo-mineral matrix of degradation products or participate in oxidation reactions induced by carbonaceous particles or metals like iron (Fe), vanadium (V), and nickel (Ni) content in dust.

In coastal zones, marine aerosols also contribute to deterioration of stone. It is primary composed by sea water along with particles naturally generated by the action of the wind on the seawater surface to introduce ionic species into the atmosphere, principally chlorides and sulfates (Stefanis et al., 2009). Chlorides are a destructive agent of porous materials.

Because of its high solubility, it penetrates into porous network, and crystallizes inside the material. Its crystallization produces disruptive pressure forces that lead to microcracks formation (Cardell et al., 2003).

On the other hand, suspended particles are also natural substrata for oxidation reactions (Primerano et al., 2000). New products eventually reach stone surface were they originates physical, chemical and aesthetic changes (Fig. 2).

Once stone materials have been sensitized by physical or chemical factors it is more sensible to the action of biological agent causing biodegradation. Biodegradation is an undesirable change in materials properties caused by the action of microorganisms, animals and plants. The presence of microorganisms causes the formation of biofilms.

Biofilms are sessile communities adhered to substrate enclosed in a polymeric matrix producing metabolites with capabilities to initiate, promote or magnify stone degradation through modification in pH levels, ionic concentrations, and redox conditions at the interfase between substrate and surrounding media to produce chemical and physical alterations (Gorbushina *et al.*, 2002; Ortega- Morales, 2003; Guiamet *et al.*, 2005; Little y Ray, 2005).

a b

Fig. 2. General aspect of degradation at Forts San Pedro (a) and San Carlos (b), historic buildings of San Francisco de Campeche City.

3. Degradation of historic buildings: the case of San Francisco de Campeche

3.1 Meteorological conditions

San Francisco de Campeche City is located under a gently slooping flood plain. The City is limited at the Norwest by the Gulf of México and at the South, Southeast and Southwest by a group of softened hills. Under these conditions, the natural expansion of the city follows to South and Southeast direction. The City presents a tropical summer rain forest climate (Aw) (Castro Mora, 2002). Table 1 concentrate the annual average value of meteorological parameters registered during 1992 to 2002 period at National Meteorological Service station (SMN), located into the installation of the aeronaval airport of the City.

Year	Meteorological parameter					
	Precipitation (mm)	Temperature (° C)	Relative humidity (%)	Atmospheric pressure (mb)	Predominant Wind direction	Wind velocity ($m.s^{-1}$)
1992	1224.30	27.20	73	12198.50	SE	3.40
1993	1294.30	27.60	71	12168.70	E	3.60
1994	1084.80	27.30	73	12166.10	E	2.90
1995	1688.40	26.90	74	12149.10	SE	3.10
1996	938.50	26.40	74	12159.20	ESE	3.10
1997	1115.60	27.30	74	12153.40	SE	2.60
1998	815.40	27.80	71	12143.50	ESE	2.70
1999	1227.70	26.70	73	12164.20	ESE	2.80
2000	927.00	26.70	72	12168.20	E-SE	2.60
2001	1004.70	27.00	73	12159.20	E	3.10
2002	1297.20	27.10	71	12157.50	E	2.80

Table 1. Annual average value of meteorological parameters registered during 1992 to 2002 period at National Meteorological Service station (SMN), at San Francisco de Campeche City.

The existence of high relative humidity values along the year can be observed and the persistent sum of rainfall covering an extended period from June to November. Those conditions guarantee water availability for occurrence of chemical and physic processes able to deteriorate calcareous stone materials through binder dissolution mechanisms, including also the penetration of soluble salts and atmospheric pollutants (Corvo et al., 2010).

Two characteristics regional meteorological phenomena affecting coastal zones are related to inland humidity penetration from the sea: along the autumn season, tropical storms (hurricanes) carry on humidity from Caribbean Sea raising up rainfall precipitation levels. It is especially worthily in September and October. During winter, cool dry fronts come in from North America, drag humidity when they cross the Gulf of Mexico warm water, increasing haze episodes in the coast and eventually the rainfall events. In this period rainfall events tend to minimum, and an extended dry season from November to May begin. In spite of those situations, during this period, San Francisco de Campeche City temperature rise up to its maximum levels, while relative humidity falls to the lowest value.

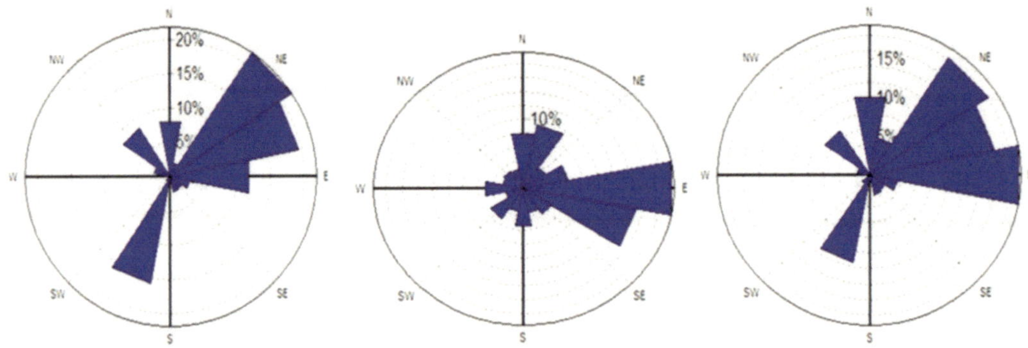

(a) (b) (c)

Fig. 3. Characteristics wind pattern observed at San Francisco de Campeche City during 2007. (a). Dry season, (b) rainy season, (c) polar front season. (Miss, 2008).

Along the year, three wind patterns can be observed in dependence of meteorological conditions at San Francisco de Campeche City (Fig. 3). Dry season is characterized by winds from E-ENE and SW directions that increase dust level at the atmosphere. At the rainy season, the wind pattern is dominated by E-ESE and a small contribution from N-ENE, due to, eventually, strong tropical storms hitting the city. In winter, when polar front reaches the coast of Campeche, winds from E-NE, N NW and SW are more frequent.

3.2 Environmental conditions

San Francisco de Campeche City is an emerging place located at the occidental coast of the Peninsula of Yucatan. At the present time, it has about 235,000 inhabitants. Until the last decade of the XX century the City was considered a place of scarce economical and industrial development, since the main productive activities were administration, fishing, and processing of food. There are no installed heavy industries were installed, except by a power plant located at Lerma town, about 6 km SE from downtown.

Nevertheless in 1999, the historic and architectonic complex of the City was included into UNESCO´s Cultural Heritage List. It considered the city as a historic and cultural reference in Mexico and other countries. As a consequence, an intense urban and economical development occurs, mainly due to the raising of cultural sector. Also an increasing of infrastructure needs because of the parallel demographic expansion that is occurring in the city. Environmental problematic like water supply, solid residues management, residual water disposition and atmospheric pollution are associated with urban development.

Studies about the number of existing automotive units ordered in 2003 by the Government of Campeche State demonstrated that during 1996 to 2003 period, San Francisco de Campeche City suffered an increasing of 8% the vehicular units, while between 2002 to 2003 the increasing was of 13.13 %, this situation cause serious vial problems. According to this study, projections for 2010 indicates an increase of 69,130 units (Government of the State of Campeche, 2004). Under these conditions, it is expected an increase in atmospheric pollution level.

Atmospheric pollution is mainly related with industrial and vehicle exhaust emissions. Gases like ozone (O_3), carbon oxides (CO, CO_2), nitrogen oxides (NO_2, NO_3), sulfur dioxide (SO_2), and atmospheric particles (PST, PM_{10}, $PM_{2.5}$), have been used to indicate air quality in urban areas. Those pollutants can be the origin of health diseases, changes in environmental conditions and degradation of materials. In this sense, they are precursors of acid rain and the blackening of stone materials in historic buildings and monuments (Reyes et al., 2009; Corvo et al., 2009).

It is interesting to report that since 1992, atmospheric corrosion under structural materials aluminium (Al), carbon steel (Fe), copper (Cu) and zinc (Zn), was monitored in five sites distributed across urban city area (Fig. 4) (Reyes, 1998; Cook, et al., 2000; Corvo et al., 2008).

These studies were performed according to criteria established by the program ISO CORRAG (Tidblad et al., 2000). During the study, an estimation of corrosion rates was carried out considering deposition rate of corrosive parameters like chloride ions (Cl^-), SO_2, and their correlation with the temperature-humidity complex represented as time of wetness (TOW) (Tables 2 and 3).

It was established that the corrosion rate at exposures sites depends strongly of their distance to the coast, since Cl^- levels decrease when distance increases (Corvo et al., 2008). Here, SO_2 deposition rate was very low, except for Technological Institue of Campeche (ITC) site located at Lerma town, closer to the Power Station (Table 3).

Fig. 4. Map of San Francisco de Campeche City. Red dots indicate the location of selected monitoring sites for atmospheric corrosion research. ITC, Technological Institute of Campeche¸ CRIP, Regional Center for Fisheries Research; CICORR, Corrosion Research Center; INAH, National Institute for Anthropology and History; SMN, National Meteorological Service.

It has the highest SO_2 content between all exposure sites. Nevertheless, it does not appear as decisive as chloride and TOW in prediction of corrosion rate as usually occurs in Regional Center for Fisheries Research (CRIP) and CICORR stations.

The results of the study also shows that in San Francisco de Campeche atmospheric corrosion rates are lower than those located in Mexican and Cuban coastal stations, where industrial and marine influence was more important.

The only exception to this rule was the station located at the CRIP, located at 4 meters from the coastline. It is an interesting data in order to consider the possible effects of atmospheric condition on stone materials decay.

Station	Atmospheric corrosivity			
	Al	Cu	Fe	Zn
SMN	Medium to high	Medium	Medium	Medium
CICORR	Medium to high	High	High	High
INAH	Medium to high	Medium	Medium	Medium
CRIP	Medium to high	High	High	High
ITC	Medium to high	Medium	Medium	Medium

Table 2. Estimation of atmospheric corrosivity at selected monitoring sites in San Francisco de Campeche City.

Station	Distance to the coast (m)	Corrosive parameters		
		SO_2 mg.m^{-2}.day^{-1}	Cl^- mg.m^{-2}.dia^{-1}	TOW annual hours
SMN	4.000	2.42	19.98	4576
CICORR	0.300	2.61	70.50	4894
INAH	0.615	1.47	18.08	3271
CRIP	0.004	2.64	76.20	4572
ITC	0.300	15.83	29.50	3380

Table 3. Corrosive parameters registered at San Francisco de Campeche City selected monitoring sites.

3.3 Atmospheric pollution

During the dry season of 1998, exceptional natural fires were declared along the south east of Mexican Republic. It was especially worthily at Campeche State, where several health problems like skin, respiratory and ocular diseases were observed in people. For the first time, atmospheric particles considering the fraction of atmospheric particles with diameter below 10 μm (PM_{10}) fraction was measured at San Francisco de Campeche following procedures established by Official Mexican Standards. The study (carried out in May 21th, 1998), yielded average value of 40 μg.m^{-3}, that was considered below health risk levels for inhabitants.

This study was the only reference of atmospheric pollution measured at San Francisco de Campeche City until 2005, when an initiative to study the effects of the environment on degradation of Cultural Heritage was driven by Autonomous University of Campeche (Reyes, 2005a, 2005b). Atmospheric parameters like SO_2, atmospheric particles (TSP and PM_{10} fractions) and acid rain were measured in different periods during 2005 to 2009 following Mexican and International Standards (NOM, US-EPA, ISO and UNE).

Until the beginning of this project, there was no additional information on air pollutants measured using standard methods in the City of San Francisco de Campeche.

3.3.1 Present atmospheric pollution levels at San Francisco de Campeche City

We proceed to determine the levels of air pollutants in the city of San Francisco de Campeche, considering two important aspects: its effect on materials and the possibility of using standardized methods to generate a database that could be used as a reference on air quality in the city (Reyes 2005a, 2005b; Miss 2008, Villaseñor, 2008., Dzul 2010, Góngora 2010, Quirarte, 2010).

Two atmospheric pollution stations were placed at "home of Lieutenant of the King" and San Pablo Buildings, historic buildings belonging to Centro INAH-Campeche (INAH-National Institute of Anthropology and History).

Another station was installed on the Corrosion Research Center (CICORR), main Campus of the Autonomous University of Campeche (Fig.4). Passive (SO_2, NO_X, Cl^-), active (Total Suspended Particles –TSP and PM_{10} fraction) and automatic (SO_2) samplers were employed. Also, wet precipitation was sampled by using a wet/dry rain sampler. The results of the sampling are condensed on Table 4.

Table 4, shows medium, maximum and minimum deposition rates and concentrations of the different types of pollutants determined using standardized methods at selected atmospheric monitoring stations.

Pollutant	Method	Standard	Medium value	Maximun value	Minimun value	Station
SO_2 ($mg.m^{-2}$) Feb. 2007 to Feb. 2009	Sulphation plate (passive)	ISO 9225:1992	1.31	3.52	0.58	INAH
Cl^{-1} ($mg.m^{-2}$) Feb. 2007 to Feb. 2009	Wet Candle (passive)	ISO 9225:1992	20.28	31.90	3.53	INAH
NO_X ($mg.m^{-2}$) Feb. 2007 to Feb. 2009	Diffusion tubes (passive)	UNE EN 13528	9.82	13.48	4.71	INAH
TSP ($mg.m^{-2}$) Nov. 2006 to Dec. 2008	High-volume sampler (Active)	NOM-035-SEMARNAT-1993	47.23 48.71	101.30 106.67	15.26 23.89	INAH CICORR
PM_{10} ($mg.m^{-2}$) May to August 2007	Low volume sampler (Active)	US-EPA standard	3.54 3.30	8.69 9.72	1.49 1.34	INAH CICORR
SO_2 ($mg.m^{-3}$) Jan. 2007 to Jan. 2008	Fluorescence (Automatic)	NOM-038-SEMARNAT-1993	6.95	74.70	1.30	INAH

Table 4. Average deposition rate and concentration of the different types of pollutants determined by standard methods at monitoring stations in San Francisco de Campeche City during 2006 to 2009.

3.3.1.1 Passive methods

Passive methods consist of an absorbent substrate that reacts with a specific chemical compound in the atmosphere. Afterwards, the samplers are removed and analyzed quantitatively in the laboratory. These devices work by principles of deposition or diffusion, but they are not considered appropriated for air quality studies; however, they provide trends on the spatial-temporal distribution.

Fig. 5 shows the values of SO_2, NO_X and Cl- determined by passive methods in the historic center of the city of San Francisco de Campeche (urban-marine atmosphere). These data can be compared with results obtained in the rural monitoring station installed at Iturbide town (Fig. 6), about 100 km E far away from the City (Fig. 1).

As can be seen, the values of all pollutants are higher in the city of San Francisco de Campeche in comparison with Iturbide, due to the urban nature of the city. Pollutants such as SO_2 and NO_X are usually produced during combustion of fossil fuels and emitted into the atmosphere by motor vehicles.

These gaseous pollutants are considered acid contaminants because they corrode metals and stone materials due to its ability to form acid solutions in contact with environmental humidity on the surface of materials (Tercer 1998; Massey, 1999; Zappia et al., 1998; Allen et al., 2000).

The levels of airborne salinity in a particular site depend upon the geographical position and the existence of orographic accidents. Its marine origin causes a preferential distribution in coastal areas.

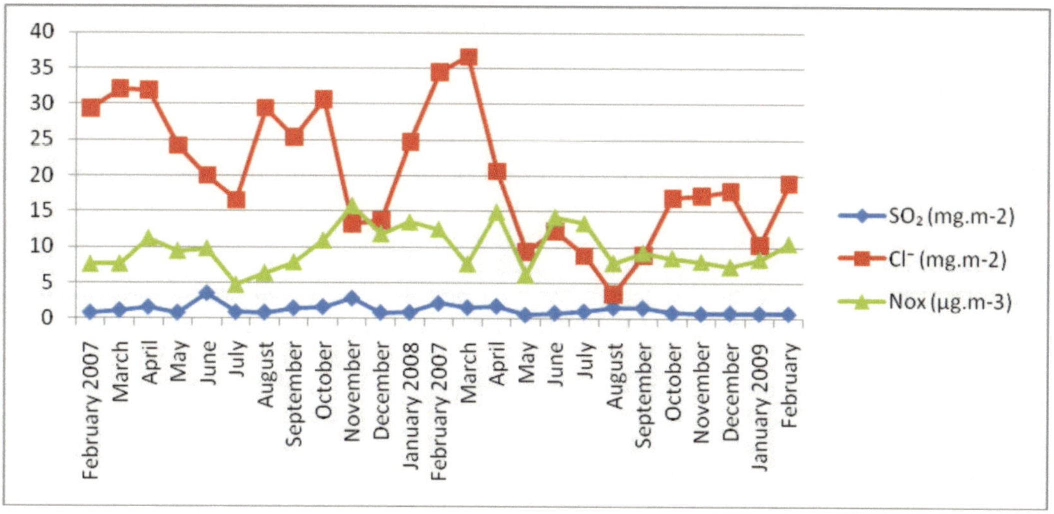

Fig. 5. SO_2, NO_X and Cl-, levels determined by passive methods in the urban marine atmosphere of San Francisco de Campeche City (INAH station).

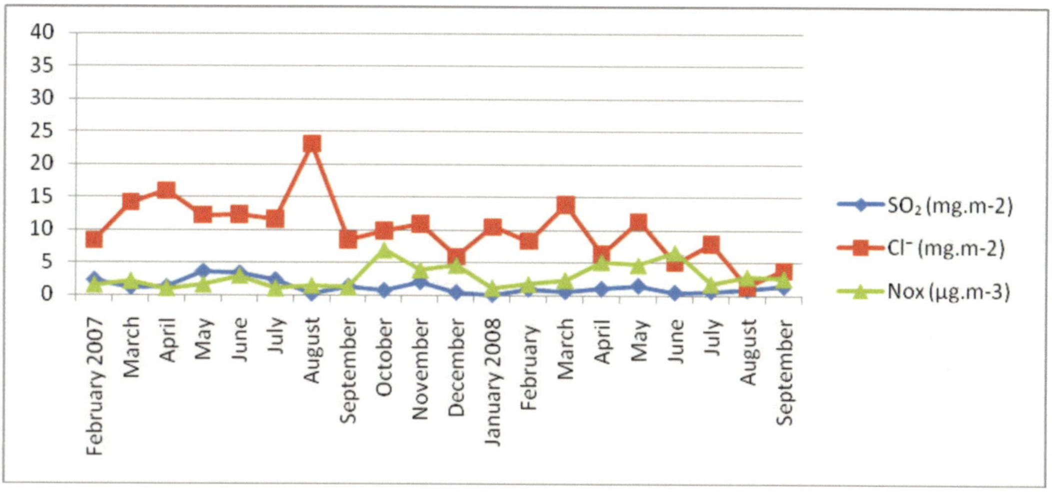

Fig. 6. SO_2, NO_X and Cl- levels determined by passive methods in the rural monitoring station installed at Iturbide Town.

Its concentration decreases when the distance to the coastline increases. This distribution also depends on the speed and wind direction. Higher levels of airborne salinity are expected near the coastline

It is appropriate to mention that despite the proximity to the coast of INAH station (600 m), marine aerosol levels are relatively lower than those observed in Boca del Río coastal stations (600 m from shore line) or Coatzacoalcos petrochemical complex (1000 m from shore line) (Carpio et al., 1996; Reyes 1998; Cook et al., 2000).

It occurs because the prevailing wind patterns in Campeche is most of the year from the E (they are called offshore winds) (Fig. 3), opposing the entry of masses of moisture from the Gulf of Mexico (Reyes, 1998; Cook et al 2000). This wind regime, suffers slight modifications relatively constant during winter, since the winds from the N increases in intensity and frequency, so that marine aerosol levels tend to rise, increasing the potential corrosivity of the atmosphere.

3.3.1.2 Atmospheric particles

On the other hand, active methods involve a flow of air through an absorbent medium or a physical collecting medium. A suction pump is used. Samples thus obtained are quantitatively analyzed in the laboratory. Two types of samplers are used: high volume and low volume. Two sampling sites were selected: the "Home of Lieutenant King", central building of INAH-Campeche, located in the historic center of the city of San Francisco de Campeche, and the building of the CICORR in the main campus of Autonomous University of Campeche.

The level of total suspended particles (TSP) was determined at both sites during the period August 2006-October 2008. PM_{10} fraction of airborne particles was recorded during the period May to August 2007. Table 4 display the average, maximum and minimum values determined for the corresponding sampling periods.

Table 4 shows statistics for data sets obtained for PST in both sampling stations. In all cases the maximum, minimum and average values were higher for CICORR related to INAH station although a "t" test performed showed no significant differences between the average obtained in both sampling sites ($t = 1.57225$ $p > 0.05$). Moreover, during the sampling period, none of the stations exceeded the maximum permissible limit for Mexican Standard (210 $\mu g.m^{-3}$), as shown in Fig. 7.

Higher average values of TSP were monitored during the month of July coinciding with the end of the dry season and beginning of summer rainfall season. Average TSP values were found to be 47.23 and 48.71 $\mu g.m^{-3}$ for INAH and CICORR monitoring stations respectively. Several authors suggest that in drought periods, atmospheric particles concentration is higher than in rain periods, those because of the lack of washing of the atmosphere caused by rainfall (Muñoz et al. 2001; Miss 2008).

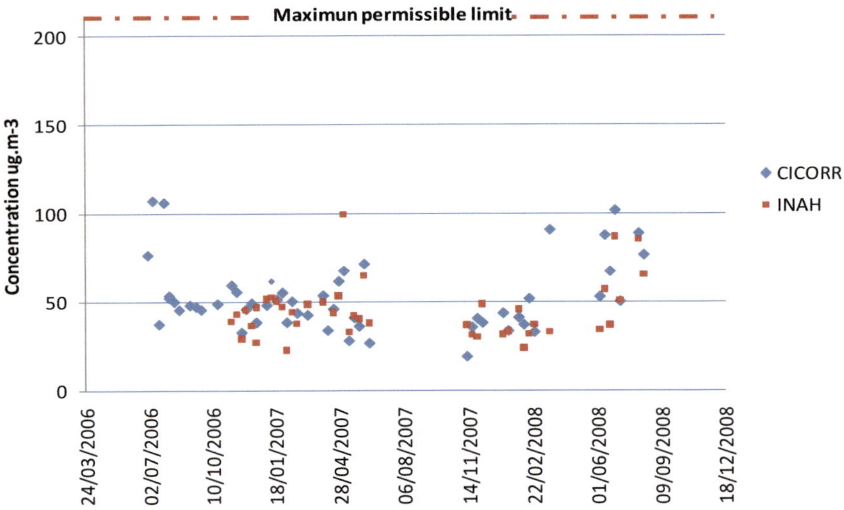

Fig. 7. TPS at San Francisco de Campeche monitoring sites during the period August 2006-September 2008. Red dotted line represents the maximum permissible limit of 240 $\mu g.m^{-3}$ According to Mexican Legislation.

The city of San Francisco de Campeche is located in the middle of a small valley, surrounded at N, S and E by hills, with elevations not higher than 150 meters. The Bay of Campeche is located in the W. Many of these hills are suffering continuous erosion and clearing of land for the construction of living houses or are employed by construction companies as sources of construction materials. These activities give rise to soil erosion and constant dust storms, which in times of drought contribute to increased levels of local TSP.

In a regional scale, the prevailing winds in the dry season (April to July), converge towards the sea ground by the E-NE quadrant and an important component S-SW (Fig. 3a). It contributes to the transport of atmospheric particles, originated in farming areas, eroded land and cattle ranches in the state, which add to the locally originated TSP.

The role of rainfall in the levels of TSP is evident in urban and industrialized areas, since water acts as a purifier of particles in the atmosphere (Muñoz et al., 200; Sosa et al., 2006; Miss, 2008), also the wind disperse atmospheric particles and reduce their content at the atmosphere.

It is confirmed by the minimum average value of 15.26 $ug.m^{-3}$ recorded during the month of March 2007 at INAH, when cool fronts introduce strong wind velocities and eventually rain episodes. During the period from August to November there is a significant decrease in the levels of TSP on both stations as a result of purifying effect of seasonal rains which masses are originated in the Caribbean Sea (Fig. 3b).

The presence of polar fronts in the Gulf of Mexico during the period from December to March becomes a factor of atmospheric instability that contributes to the dispersion of pollutants and the introduction of humidity from the ocean in coastal areas (Reyes 1998).

It coincides with the monthly average minimum of 35.25 $\mu g.m^{-3}$ registered at CICORR during December 2006, precisely at the end of the rainy season and early winter seasonal fronts when the wind increases in strength and components N-NE direction (Fig. 3c).

Atmospheric particulate matter PM_{10} fraction was determined during the end of the dry season and the beginning of the rainy season (May-August 2007). A Student "t" test to compare the arithmetic means of data sets collected at stations CICORR and INAH was used. The test results indicated no significant difference between values observed in the testing sites ($t = 0.612$, $p > 0.5$).

At both monitoring sites, the concentration of PM_{10} follows the same tendency being the maximum concentration of 9.72 $mg.m^{-3}$ and minimum concentration of 1.34 $mg.m^{-3}$ for CICORR, while for INAH, maximum and minimum concentrations were 8.69 and 1.49 $mg.m^{-3}$, respectively (Table 4). Regarding the maximum concentrations obtained during evaluation, values of 8.69 and 9.72 $mg.m^{-3}$ for CICORR and INAH were determined, respectively. These values represent no health risk to people and the environment because do not exceed the average value of 120 $ug.m^{-3}$ in 24 hours established by the Mexican Standard (Dzul, 2010).

Respecting the average values, a concentration of 3.54 $mg.m^{-3}$ and 3.30 $mg.m^{-3}$ was determined for INAH and CICORR, respectively, indicating a slight difference in concentration between both sites which follow the same behavior. According to the results, a higher concentration of PM_{10} particles in the CICORR station was found with respect to INAH. This behavior coincides with that observed previously for TSP in both seasons, given the prevalence of similar environmental conditions (Miss, 2008).

CICORR station is surrounded by trees and by the athletic field of the Autonomous University of Campeche. In the West side of CICORR is located Juan de la Barrera Street, showing steady traffic during the morning and tends to diminish in the evening during the class activities, a period which coincided with the sampling. INAH station is located in the center of the city of San Francisco de Campeche in an urban area with heavy traffic flow during most of the day.

3.3.1.3 Sulfur dioxide

SO_2 is considered as an indicator of atmospheric pollution in urban sites. It has been included in air quality indexes in several cities along the world (Valeroso, et al., 1992, Shifer et al., 2000; Raavindra et al., 2003). Industrial emissions and vehicle exhaust are the mains source of this pollutant which is precursor of acid rain and black crust formation (Mala, 1999; Primerano et al., 2000; Reyes 2004; Reyes et al., 2004).
This parameter was monitored during January 2007 to January 2008 in the historic center of San Francisco de Campeche City (INAH station), by using a visible fluorescence automatic equipment (NOM-038-SEMARNAT-1993). Fig. 8 shows the behavior of SO_2 during the sampling. Maximum, minimum and medium values are reported in Table 4. According to the results, both 24 hours maxima and annual arithmetic average were reported below maximum limits established by Mexican Standard. It means that its effects in health are limited. Nevertheless, the behavior of SO_2 during sampling period indicates a continue increase in their atmospheric concentration.

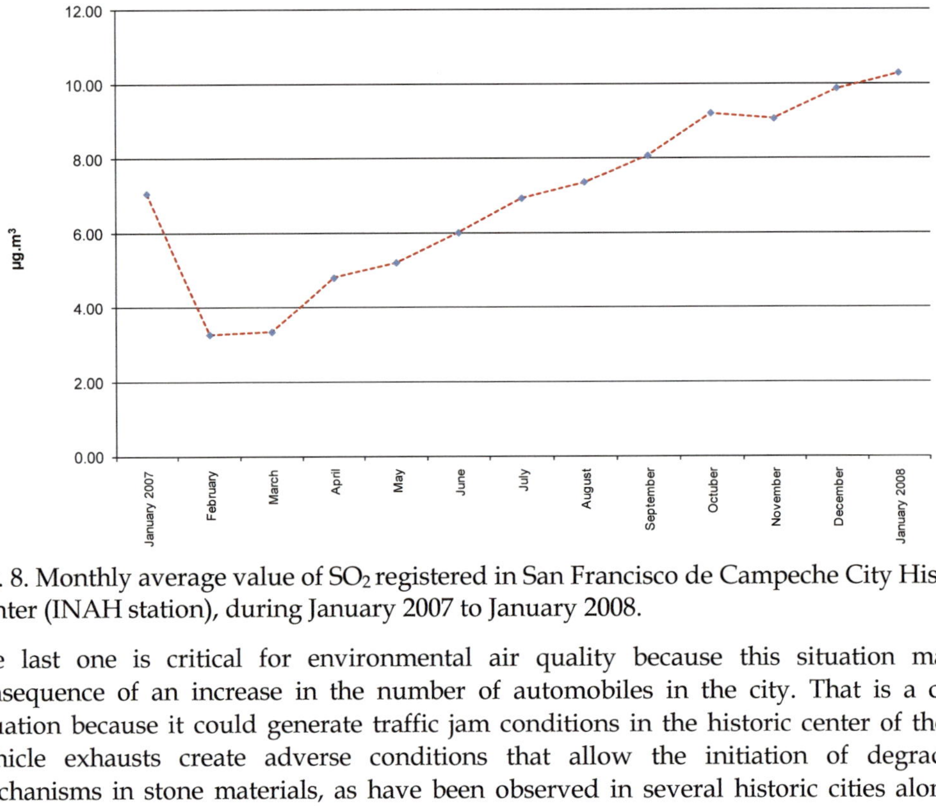

Fig. 8. Monthly average value of SO_2 registered in San Francisco de Campeche City Historic Center (INAH station), during January 2007 to January 2008.

The last one is critical for environmental air quality because this situation may be consequence of an increase in the number of automobiles in the city. That is a critical situation because it could generate traffic jam conditions in the historic center of the city. Vehicle exhausts create adverse conditions that allow the initiation of degradation mechanisms in stone materials, as have been observed in several historic cities along the world (Primerano, et al., 2000).

3.3.1.4 Acid rain

During the years of 2006 and 2007, a wet sample collecting campaign was carried on by using an automatic wet/dry sampler (US-EPA, 1994) installed at the INAH station

(Quirarte, 2010). A total of 147 samples were obtained. Table 5 shows the maximum, minimum and average weighted pH registered during the campaign. Fig. 9, represents the tendency in change of pH value along the rainy period.

It is important to note that in both years, a natural tendency to alkalinity exists in rain water pH. During the months corresponding to dry season (from December to June) rain water pH are usually higher than 6. This general tendency changes from July to November, period in which the atmosphere has been washed of dust particles by the rainy season. Then, the minimum values of pH are reached and eventually, sporadic acid rain events can be observed, probably as a consequence of atmospheric transport (Quirarte, 2010).

Year	Number of samples	PH			% of acid samples
		maximum	minimum	average	
2006	83	7.54	5.19	6.04	12
2007	73	7.80	4.97	6.39	5

Table 5. Maximum, minimum and average ponderated pH registered at San Francisco de Campeche City.

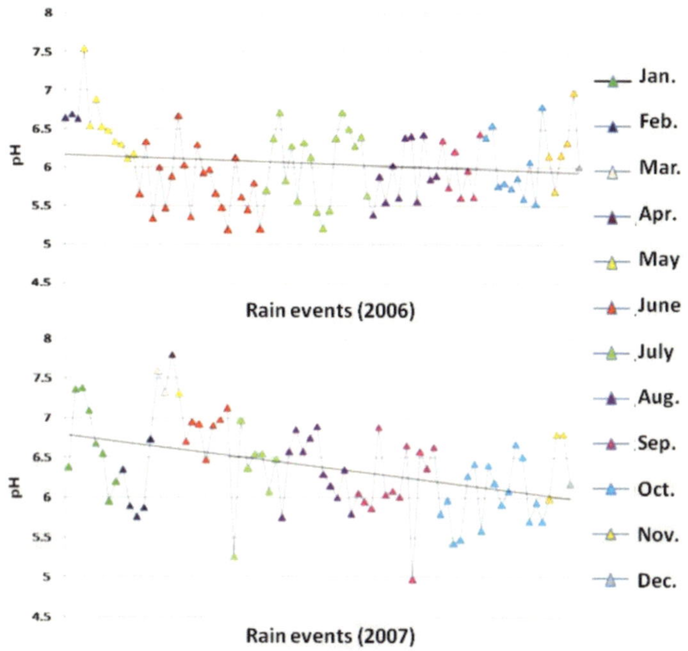

Fig. 9. Tendencies of rain pH during 2006 and 2007 at San Francisco de Campeche City.

Torres (2009), studied the ionic enrichment in rain samples collected at INAH station during 2007. The study indicate an enrichment on sulphates ($SO_4^{=}$), nitrates (NO_3^-), calcium (Ca^{2+}) and Cl^- ions. $SO_4^{=}$ and NO_3^- are acidic compounds present as a consequence of human activity, while Ca^{2+} is dragged from alkaline soils of Peninsula of Yucatan, because it is transported by the wind and incorporated to the rain drops in the atmosphere, contributing to the neutralization of acidic compounds.

Under this condition, rain acidity is not a determinant factor in recession rates of calcareous materials, since volume and intensity of precipitation seems like key factor in deterioration of the historic building at San Francisco de Campeche City.

3.4 Degradation of historic buildings in San Francico de Campeche City

Two representative building from the old military complex of the City were studied in order to analyze the influence of environmental condition on degradation of their mansory structure: Forts San Carlos and San Pedro (Fig. 10). Both buildings were constructed in masonry base structure made by calcareous stone quarry blocks and mortars, made with slike lime and stone dust named sahacab.

Fort of San Carlos is a pentagonal-shaped structure located at the city´s bastions-and-rampart system´s northwestern corner, in front of the south of Gulf of Mexico shoreline. Until mid of the XX century, when, state government, public works reclaimed some portion of land from the original previous shorefront, three walls suffered direct wave impact and tidal movements. At present, the State and Municipal Government office buildings as well as the State Congressional offices and Legislature auditorium are located adjacent to Fort San Carlos.

Continuous vehicular movements flow through this immediate area, which houses peripheral urban core lanes and formal entrance into the 8th Street downtown historic district.

Fort San Pedro crowns the city´s bastion- and- rampart system´s southeaster sector located at the Southeastern sector. While functioning as a bastion again possible inland attacks and "watchtower" for surrounding neighborhoods located to the south, southeast and southwest, this structure does not receive direct marine aerosols and tidal movements as noted in the case Fort San Carlos located in the northern parapet perimeter.

These factors suggest that deterioration followed a slow natural process over a long time period. However, at present this Colonial construction is surrounded by traffic jammed streets, municipal bus terminals and intense anthropogenic activity in the immediate area.

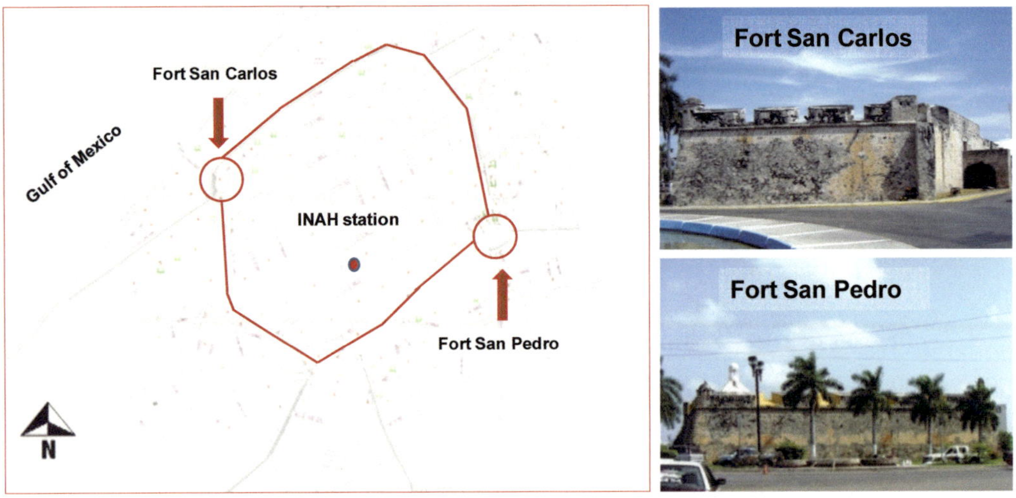

Fig. 10. Location of Forts San Carlos and San Pedro at the historic center of San Francisco de Campeche City. Also location of INAH station is showed.

In spite of consider the effects of environment in degradation of historic buildings, samples were collected from their walls and mineral alteration was investigated by XRD analysis in a Bragg-Brentano geometry X-ray diffractometer (Siemens D5000), and analyzed under the following conditions: Cu Kα radiation (λ=1.5416 Å) and operational conditions of 25 mA and 35 kV at a step size of 2°/2θ/min in the 2–60° range 2θ.

Table 6, shows the mineral phases identified during the analysis in crusts samples from both Forts. Calcite ($CaCO_3$), a rhombohedric form of calcium carbonate, seems like the major compounds in all the samples. As have been described before, tropical climate guarantee the water availability to lead dissolution of calcium carbonate content in calcareous materials and their later recrystallization to form crusts. Also minerals like, aragonite ($CaCO_3$), sodium silicate ($Na_2Si_4O_9$), quartz (SiO_2), dolomite ($CaMg(CO_3)_2$) and portlandite ($CaOH_2$) were present.

There are mineral components of limestone and traditional mortars employed during the construction of the Forts or the utilization of cements to make modern mortars during recent preservation works. Aragonite ($CaCO_3$), is a polymorphous of calcium carbonate and is present in bioclastic limestones.

The identification of neomineral phases like whewellite ($C_2CaO_4.H_2O$), and wheddellite ($C_2CaO_4.2H_2O$) keep relation with bio-deterioration phenomena. Calcium oxalates are formed during oxalic acid dissolution of calcareous materials (Arocena et al., 2007). Oxalic acid is produced by metabolic activity of microorganisms like cyanobacteria and lichens (Del monte y Sabbioni, 1985; Rampazi et al., 2004). In the walls of Forts San Carlos and San Pedro, was evident the colonization by abundant microbial communities (Fig. 11).

a b

Fig. 11. Aspect of the biodeterioration in the historic buildings of San Francisco de Campeche City. (a) Fort San Carlos. (b). Microbial community at West wall of Fort San Pedro.

On the other hand, it is important to note the presence of gypsum in Fort San Pedro samples while it was absent in Fort San Carlos ones. Gypsum is a neomineral product formed as a consequence of SO_2 reaction with $CaCO_3$ in urban environments (Graedel et al., 2000; Reyes et al., 2010b). It is an indicator of the certain pollution level in specific areas submitted to the pressure of vehicular and industrial emissions. San Pedro Fort is localized in the east area of the historic centre of the city. All their walls (except the west), are bordered by heavy traffic jams avenues, while south and southwest walls are very close to a bus station from Municipal Urban System.

Mineral phase	Sample code																
	Fort of San Carlos	1	2	3	4	5	6	7	8	Fort of San Pedro	9	10	11	12	Convent of San Francisco de Asís	CNC	PCC
Calcite		+	+	+	+	+	+	+	+		+	+	+	+		+	+
Aragonite		+	-	+	+	-	+	+	+		-	+	+	+		-	-
Sodium fedespar		+	-	+	+	+	+	+	+		+	+	+	+		-	-
Quartz		+	+	+	+	+	-	-	-		-	-	-	+		+	+
Orthoclase		-	-	-	-	-	-	-	-		-	-	+	-		-	-
Dolomite		-	-	-	+	+	+	-	-		-	+	+	-		-	-
[1]Goethite		-	-	-	-	-	-	-	-		-	+	-	-		-	-
[2]Iron hydroxide carbonate		-	-	-	-	-	-	-	-		-	+	-	-		-	-
Clay minerals		-	-	-	-	-	+	-	-		-	+	-	-		-	-
Whewellite		+	+	+	-	-	+	+	+		-	-	+	-		-	-
Bassanite		-	-	-	-	+	-	-	-		-	-	-	-		-	-
Weddellite		-	+	-	-	-	-	-	-		+	-	-			-	-
Portlandite		-	-	-	-	-	-	-	+		-	-	+	+		-	-
Hidroxiapathite		-	-	-	-	-	-	-	-		+	-	-	-		-	-
Gypsum		-	-	-	-	-	-	-	-		+	-	+	+		+	+

Table 6. Mineral phases identified in samples from Forts San Carlos and San Pedro by XRD. Samples CNC and PCC correspond to the Convent of San Francisco de Asís (Havana City). (+) present, (-) not present. [1]ICD card number 29-0713. [2] ICD card number 33-650.

3.5 City of Havana: A comparison of air pollution and stone degradation

3.5.1 The City of Havana

The City of Havana was founded on November 16, 1519 by Spanish conquest Diego Velázquez de Cuellar. Its historical center was declared a World Heritage Site by UNESCO in 1982. Havana was strengthened in the XVII century by order of the Spanish kings who signed as "Key to the New World and bulwark of the West Indies".

In 1763 construction began on the fortress of San Carlos de la Cabaña, the largest built by Spain in the New World, which shored up the defensive system of Havana after the British occupation.

The port of Havana was considered one of the most important of the region during the colonial era and one of the strategic points for Spain, which is why the bay was protected with a very important network of fortifications, including the Tower of San Lazarus, El Morro de La Habana, the Fortress of San Carlos de la Cabaña, the Castle of "La Fuerza" and other fortress dedicated to protecting the harbor and the city.

During the colony, Havana was also the major transshipment point between the New World and Europe. As a result Havana was the most fortified City in the Americas. Most examples of early architecture can be seen in military fortifications such as Fortress San Carlos de la Cabaña (1558 - 1577) and the Morro Castle (1589 -1630).

The Convent of San Francisco de Asis, is a religious building of Baroque architecture located in the plaza of the same name in the Old Havana (Figure 12). Construction began in 1548 until 1591, although it opened in 1575, fully completed nearly 200 years later, with a series of structural reforms that occurred from 1731 to 1738. It has a tower of 48 yards high, which in colonial times was the tallest structure in the city for several centuries.

a

b

Fig. 12. The Convent of San Francisco de Asis (a). Location of the Convent into the Historic Center of Havana City.

3.5.2 Degradation of historic buildings: a comparative Havana vs San Francisco de Campeche

Nowadays Havana is a City having about 2.2 million inhabitants and different types of industries, particularly around the Bay, a different situation respecting the Mexican City of San Francisco de Campeche. At Havana, air pollution levels are higher than those observed in the Mexican City (Corvo et al., 2010). In this order, a comparison of the influence of air pollution on stone buildings degradation can be made between both cities located in tropical climate.

San Francisco de Campeche City shows a tendency to alkaline rain water with percent of acid rain event of 12% and 5% during 2006 and 2007 respectively (Quirarte, 2010); however, in Havana City, during the period 1981-1994, rain having a pH lower than 5,6 oscillated between 25% and 75% of the samples. It indicates a general tendency to acid rain in Havana.

On the other hand, Table 7 shows the results of atmospheric contamination measured in San Francisco Convent and the Basilica. It can be noted that there is an evident difference in the deposition level of sulfur compounds between Havana and San Francisco de Campeche sites (Table 4). Havana sites show a significant higher deposition of sulfur compounds respecting San Francisco de Campeche. The two selected monitoring sites were located inside San Francisco de Asis Convent and Basilica Minor.

This building is located at less than 200 m from Havana Bay shoreline. Under indoor conditions, deposition rate is usually lower than outdoors. One of the monitoring sites was located inside the Basilica building, in the concert hall at about 3 m from the floor. The second monitoring site was located in the Chorus, in the same Basilica Building, at about 10 m from the floor. Evaluation was carried out beginning September 2006 up to March 2007. Chloride deposition rate was negligible because it was determined in indoor conditions, it is very well known that chloride aerosol significantly decreases in indoor conditions; however, in outdoor conditions, in sites near Havana Bay, an average chloride deposition around 10-20 $mg.m^{-2}d^{-1}$ has been measured. It is important to note that even under indoor conditions, values of sulphur and nitrogen compounds inside the Convent are higher than those reported for San Francisco de Campeche outdoors. It confirms that air pollution in Havana City is significantly higher (Corvo et al., 2010; Reyes et al., 2010).

City	Site	Sulphur compounds deposition rate ($mg.m^{-2}d^{-1}$)			Chloride deposition rate ($mg.m^{-2}d^{-1}$)			NO_2 concentration ($\mu g.m^{-3}$)		
Havana		Ave.	Max	Min	Ave.	Max	Min	Ave.	Max.	Min.
Indoor	Basilica	10.50	12.50	6.51	Neg.	Neg.	Neg.	16.35	26.08	6.23
Indoor	Chorus	11.60	14.65	7.60	Neg.	Neg.	Neg.	16.29	24.49	11.50

Table 7. Air pollution levels inside San Francisco de Asis Convent and Basilica Minor in Havana, Cuba. Neg: Negligible.

a b

Fig. 13. Main Façade of San Francisco de Asis Convent and Basilica Minor in Havana, Cuba (a). Black crust deposits (b).

Crust representative samples were taken from the façade of the Convent of San Francisco de Asis and analyzed according the same procedure previously described by Forts San Carlos and San Pedro samples (Fig. 13). Mineral composition of Cuban samples is included in Table 6 (CNC and PCC samples).

Black crusts formed at the Basilica façade (outdoors) in Obispo Street show gypsum as a predominant phase with small amounts of calcite and quartz. It means that black crust composition is almost completely gypsum due to contamination by atmospheric SO_2. No presence of nitrogen degradation product was detected.

Crust formed at Forts San Carlos and San Pedro (San Francisco de Campeche) is mainly formed by calcite, the original main content of the stone. Different minor phases are: aragonite, dolomite, and quartz. The presence of whewellite and weddelite in the samples is an index of the influence of biological activity in stone deterioration, although the presence of bassanite in sample 5 from Fort San Carlos, shows the influence of environmental SO_2. It is important to note that gypsum was identified in samples 9, 11 and 13 corresponding to Fort San Pedro, but not at Fort San Carlos. Gypsum is produced by the action of SO_2 over calcareous materials.

The comparison between crust composition in Campeche and Havana is a demonstration of the role of air pollution in deterioration of stone buildings. According to the present results, the influence of sulphur contamination is higher than nitrogen contamination, because degradation products do not show nitrogen compounds in its composition.

Sulphur dioxide is highly soluble in water; however, nitrogen dioxide is not significantly soluble, it could be a cause for a higher influence of sulphur compounds in stone degradation. In addition, nitrogen degradation compounds are more soluble than sulphur degradation compounds, so the first are easily eliminated by rain.

4. Conclusions

The present contribution, showed a general description of the current air quality conditions at San Francisco de Campeche City. From the health point of view, SO_2, TSP and PM_{10} fraction are below the limits of risk considered by Mexican Legislation. The creation of a local air monitoring program in order to prevent an increase of atmospheric pollution levels is necessary as a consequence of the recent economical, demographic and urban expansion suffered by the City. In this order, although SO_2 concentration was always below critical risk levels, it suffered a continuous increase during the monitoring period.

From the materials point of view, the tropical climate and the presence of natural and anthropogenic pollutants create conditions for degradation of both, metals and stony materials. In this order, the degradation of historic building in San Francisco de Campeche City shows a closer relationship with the effect of natural environmental factors, led by water actions that induce mechanisms of salt dissolution and recrystallization across wet to-dry cycles.

The majority presence of calcium carbonates in crust formed on walls of Forts of San Carlos and San Pedro seems to confirm this fact. On the other hand, in spite of the low levels of atmospheric pollutants observed in the City, the presence of gypsum (Fort San Pedro) and bassanite (Fort San Carlos), is an indicator of a growing influence that the anthropogenic pollution could have on deterioration mechanisms. The last one result clear in the case of Fort San Pedro, which actually is under high environmental pressure.

In contraposition, samples from San Francisco de Asis Convent (Havana), show gypsum as a majority neomineral phase. Gypsum is produced in urban environments with high content of SO2, which agrees with the higher levels of atmospheric pollution detected at Havana City. In case of increase of pollution levels at San Francisco de Campeche City a similar situation will be found.

5. Acknowledgements

The realization of this contribution was possible thanks to the support of FOMIX CAMP2005-C01-025 Project (Urban Environmental Influence on degradation of colonial military and religious buildings at Campeche City) financed by the Government of State of Campeche and the Council of Science and Technology of México. Also thanks to Centro INAH-Campeche for their giving facilities to the development of the project.

6. References

Allen, G., El-Turki, A., Hallam, K. R., McLLaughlin, D., Stacey, M. (2000). The role of NO_2 and SO_2 in degradation of limestone. British Corrosion Journal. 35, 35-38.

Arocena J., Siddique T., Thring R., Kapur S. (2007) Investigation of lichens using molecular techniques and associated mineral accumulations on a basaltic flow in a Mediterranean environment. Catena 70, 356-365.

Castro-Mora, J. (2002). Geological and mineral monography of the State of Campeche. Council for Mineral Resources. Secretary of Economy. Mexico. ISBN 968-6710-95-7.

Bravo, H., Soto, R., Sanchez, P., Torres, R., Granada, L. (2000) Chemical composition of precipitation in a Mexican Maya región. Atmospheric Environment. 34, 1197-1204.

Cardell, C., Delalieoux, F., Roumpopoulos, K., Moropoulou, A., Auger, F., Van Grieken, R. (2003). Salt-induced decay in calcareous stone monuments and buildings in a marine environment in SW France. Construction and Buildings Materials. 17, 165-179.

Carpio, J., Reyes, J., Salazar, J., Parra, A., Martínez, L. (1995) Influence of industrial pollutants on the atmospheric corrosivity of petrochemical installation. Paper LA9635. Proceedings 2nd Latinoamerican Corrosion Congress. Maracaibo, Venezuela.

Cassar, M., Brimblecombe, P., Nixon, T., Price, C., Sabbioni, C., Saiz-Jimenez, C., Van Balen, K. (2004). Sustainable solutions in the conservation and protection of historic monuments and archeological remains: a critical assessment of European research needs. In: Air Pollution and Cultural Heritage. Sáiz-Jiménez Ed. Balkema The Netherlands. ISBN. 905809622.

Cook, D., Van Orden, A., Reyes, j., Oh, S., Balasubramanian R., Carpio, J., Towsend, H. (2000). Atmospheric corrosion in marine environments along the Gulf of Mexico. In: Marine corrosion in tropical environments. Sheldon, W. Dean, Guillermo, Hernandez-Duque Delgadillo, James B, Bushman Eds. American Society for Testing and Materials. 75-97. STP1399.

Corvo, F., Pérez, T., Martin, Y., Reyes, J., Dzib, L., Gonzalez, J., Catañeda, A. (2008). Corrosion Research Frontiers. Atmospheric Corrosion in Tropical Climates". On the concept of time of wetness and its interaction with contaminants deposition. In electroanalytical Chemistry: New research". G. M. Smithe Ed. Nova Science Publishers Inc.62-9.ISBN 978-1-60456-347-4.

Corvo, F., Pérez, T., Reyes, J., Dzib, L., Gonzales-Sanchez, J.,Catañeda, A. (2009). Atmospheric corrosion in tropical humid climate. In: Environmental degradation of infrastructure and cultural heritage in coastal tropical climate.J. Gonzalez-Sanchez, F. Corvo, N. Acuña-González. 1-34. ISBN: 978:81-7895-426-4.

Corvo, F., Reyes, J., Valdes, C., Villaseñor, F., Cuesta, O., Aguilar, D., Quintana, P. (2010). Water, Air and Soil Pollution. 205, 359-375.

Del Monte, M., Sabbioni, C. (1986). Chemical and biological weathering of an historical building: Reggio Emilia Cathedral. Science of the Total Environment. 50, 65-182.

Dzul, B. (2010). Elemental analysis of atmospheric particles, PM_{10} fraction at San Francisco de Campeche City by using PIXE spectroscopy. Undergrade thesis. Autonomous University of Campeche, San Francisco de Campeche, México.

Gongora, H. (2010). Influence of corrosive environmental parameters on degradation of historical and industrial materials in three microclimates of the State of Campeche. Undergrade thesis. Autonomous University of Campeche, San Francisco de Campeche, México

Gobbi, G., Zappia, G., Sabbioni, C. (1998). Sulphite quantification on damaged stones and mortars. Atmospheric Environment. 32(4), 783-789.

Gorbushina, A. (2007). Minireview: Life on the rocks. *Environmental Microbiology* 9(7): 1613–163.

Government of the State of Campeche (2004). Diagnostic of vial and transport system at San Francisco de Campeche City. Campeche, México

Graedel, T., (2000). Mechanisms for the atmospheric corrosion of carbonate stone. Journal of the Electrochemical Society, 147 (3), 106-109.

Guiamet, P., Gómez de Saravia, S., Nuñez G. (2005). *Biodeterioration of buildings stone by cyanobacteria, bacteria and fungi.* In Ortega-Morales, B. O; Gaylarde, C. C.;

Narváez-Zapata y Gaylarde P.M. (Eds). LABS 5 Latín América Biodegradation and Biodeterioration Symposium. Autonomous University of Campeche and National Polytechnic Institute. 51-53. ISBN 968-5722-34-X.

Kuchera, V., Tidblad, J., Kreislova, K., Knotkova, D., Faller, M. Reiss, D. (2007). UN/ECE ICP materials dose-response functions for the multi-pollutant situation. Water Air and Soil Pollution: Focus. 7, 249-258.

Little, B.J., Ray, R. (2005). *The role of fungi microbiologically influenced corrosion.* En: Ortega-Morales, B. O; Gaylarde, C. C.; Narvaez-Zapata y Gaylarde P.M. (Eds). LABS 5 Fifth Latín América Biodegradation and Biodeterioration Symposium. Autonomous University of Campeche and National Polytechnic Institute. 51-53. ISBN 968-5722-34-X.

Lpifert, F. (1989). Atmospheric damage to calcareous stones. Atmospheric Environment. 23, 415-429.

Massey, S. (1999). The effects of ozone and NO_X on the deterioration of calcareous stone. The Science of the Total Environment. 227, 109-121.

Mala, B. (1999). Global transport of anthropogenic contaminants and the consequences for the Arctic marine ecosystem. Marine Pollution Bulletin. 38 (5), 356-379

Miss M. (2008). Characterization of aliphatic hydrocarbons and analysis of spacial and temporal variation of atmospheric particles at San Francisco de Campeche City.

Undergrade thesis. Autonomous University of Campeche, San Francisco de Campeche, México.

Monna, F., Puertas, A., Leveque, F., Losno, R., Fronteau, G., Marin, B. (2008). Geochemical records of limestone facades exposed to urban atmospheric contamination as monitoring tools?. Atmospheric Environment. 227, 109-121.

Moropoulou, A., Konstanti, A. (2004). European strategies for preservation and managements of Historic Cities: knowledge based decision-making approach. In: Air Pollution and Cultural Heritage. Sáiz-Jiménez Ed. Balkema The Netherlands. ISBN. 905809622.

Muñoz, A., Rodriguez, A., Villalobos, P., MUnive, Z., Marttelo, O., Diaz, G., Bravo, C. Gomez, A. (2001). Suspended particles, policyclic aromatic hydrocarbons and mutagenesis at Southwest of Mexico City. Revista Internacional de Contaminación Ambiental. 17, 193-204.

Ortega-Morales, B.O. 2004. Biofilms fouling ancient limestone Maya monuments in Uxmal, México. Current Microbiology 40, 81-85.

Primerano, P., Marino, G., Dipascale, s., Mavilia, L., Corigliano, F. (2000). Possible alteration of monuments caused by particles emitted onto the atmosphere carrying strong primary acidity. Atmospheric Environment. 34, 385-401.

Quirarte, O. (2010). Variation of rain acidity in San Francisco de Campeche City and Calakmul Biosphere Reserve (archeological zone), during 2006 and 2007. Undergrade thesis. Autonomous University of Campeche, San Francisco de Campeche, México.

Rampazzi, l., Andreotti, A., Bonaduce, I., Colombini, M, P.; Colombo, C., Toniolo, L. (2004). Analytical investigation of calcium oxalate films on marble monuments. Talanta. 63, 967-977.

Ravindra, K., Mor, S., Kamgotra, J., Kaushin, C. (2003). Variations in spatial patterns of criteria air pollutants before and during initial rain in Monsoon. *Environmental Monitoring Assessment.* 87, 145-153.

Reyes, J. (1998). Influence of the main atmospheric and air quality factors on metals atmospheric corrosion along the southeast cost of the Gulf of México M. Phil. Thesis. Universidad Veracruzana. Boca del Rio, Veracruz. Mexico.

Reyes, J. (2004). Diagnostic criteria for the identification of particular matter organic compounds from vehicular exhaust and their application to study the deterioration of the Cathedral of Seville. PhD. Thesis. University of Seville. Spain.

Reyes, J. Hermosin, B., Sáiz-Jimenez, C. (2004). Organic analysis of aerosols in Seville Atmosphere (2004). In: Air Pollution and Cultural Heritage. Sáiz-Jiménez Ed. Balkema The Netherlands (ISBN. 905809622).

Reyes, J. (2005a). Research Project CONACYT-46434-Y. National Council for Science and Technology. Autonomous University of Campeche. Mexico.

Reyes, J. (2005b). Research Project FOMIX-CAMP-2005-C01-028. National Council for Science and Technology –Goverment of the State of Campeche. Autonomous University of Campeche. México.

Reyes, J., Hermosín, B., Sáiz-Jimenez, C. (2006). Organic Composition of Seville Aerosols. Organic Geochemistry, 37, 2019-2025,

Reyes, J., Torres, F., Ché, I., Corvol, F., Pérez, T., Bravo, H., Sánchez, P., Aguilar, D., Quintana, P. (2010a), Dissolution of Traditional Mortars under Artificial Rain Conditions: A Laboratory Test. In: José Luis Ruvalcaba Sil, Javier Reyes Trujeque, Jesús A. Arenas Alatorre, Adrián Velázquez Castro Eds. Proceedings of 2nd Latin-American Symposium on Physical and Chemical Methods in Archaeology, Art and Cultural Heritage Conservation & Archaeological and Arts Issues in Materials Science - IMRC 2009, Mexico. 195-200. ISBN: 978-607-02-2017-3.

Reyes, J., Corvo, F., Espinosa-Morales, Y. Valdes, C., Bartolo, P., Aguilar, D., Quintana, P., Hermosín, B., Saiz-Jimenez, C. (2010b). Analysis of Black Crust from a Cuban Historic Building . In: José Luis Ruvalcaba Sil, Javier Reyes Trujeque, Jesús A. Arenas Alatorre, Adrián Velázquez Castro Eds. Proceedings of 2nd Latin-American Symposium on Physical and Chemical Methods in Archaeology, Art and Cultural Heritage Conservation & Archaeological and Arts Issues in Materials Science - IMRC 2009, Mexico. 195-200. ISBN: 978-607-02-2017-3.

Sessa, C. (2004). Cultural Heritage protection with integrated policies. In: Air Pollution and Cultural Heritage, Sáiz-Jiménez Ed. Balkema The Netherlands. ISBN. 905809622.

Schifter, I., Díaz, L., López-Salinas, E. (2006). "Assessment of new vehicles emissions certification standars in the Metropolitan area of México city. *Environmental Monitoring Assessment.* 114: 419-432.

Sosa-Echeverria, R., Bravo-Alvarez, H., Sánchez-Alvarez, P., Soto-Ayala R., Alarcón-Jiménez, A., Khal, J. (2006). Determination of total suspended particles during five research cruises at Gulf of Mexico continental platform. Ingeniería, Investigación y Tecnología. VII (2), 71,-83.

Stefanis, N., Theoaulakis, P., Pilinis, C. Dry deposition effect of marin eaerosol to the building stone of the medieval city of Rhodes, Greece. Building and Environment. 44, 260- 270.

Tercer, L. (1998). Laboratory experiments on the investigation of the effects of sulfuric acid on the deterioration of carbonate stone and surface corrosion. Water, Air and Soil Pollution. 114, 1-12.

Tidblad, J., Mikhailov, A., Kuchera, V. (2000) Application of a model for prediction of atmospheric corrosion in tropical environments. In: Marine corrosion in tropical environments. Sheldon, W. Dean, Guillermo, Hernandez-Duque Delgadillo, James B, Bushman Eds. American Society for Testing and Materials. 18-32. STP1399.

Torres, F. (2009). Effects of rain in stony materials from the Campeche State cultural heritage. M.Phil. thesis. Autonomous University of Campeche, San Francisco de Campeche, México.

Valeroso, I. I., Monteverde, C. A., Estoque, M. A. (1992). Diurnal variations of air pollution over metropolitan Manila. *Atmosfera.* 5: 241-257.

Villaseñor, F. (2008). The use of passive techniques for sampling atmospheric pollutants in the State of Campeche. Undergrade thesis. Autonomous University of Campeche, San Francisco de Campeche, México.

Zappia, G., Sabbioni, C., Riontino, C., Gobbi, G., Favoni, O. (1998). Exposure test of building materials in urban atmosphere. The Science of the Total Environment. 224, 235-244.

Zenddri, E., Biscontin, G., Kosmidis, P. (2001). Effects of condensed water on limestone surfaces in a marine environment. Journal of Cultural Heritage. 4, 283-289.

Part 5

Plasma Technologies for Air Pollution Control

14

Plasma-Based Depollution of Exhausts: Principles, State of the Art and Future Prospects

Ronny Brandenburg[1] et al.,*
[1]*Leibniz Institute for Plasma Science and Technology,*
[2]*Uppsala University,*
[3]*Institute of Nuclear Chemistry and Technology,*
[4]*Szewalski Institute of Fluid Flow Machinery,*
[5]*West Pomeranian University of Technology,*
[6]*University of Tartu,*
[7]*Technical University of Denmark,*
[1]*Germany*
[2]*Sweden*
[3,4,5]*Poland*
[6]*Estonia*
[7]*Denmark*

1. Introduction

Nowadays non-thermal plasma technologies are state of the art for the generation of ozone as an important oxidant for water cleaning or bleaching, the incineration of waste gases or for the removal of dust from flue gases in electrostatic precipitators. Furthermore their possibilities of gas depollution are well known. Plasmas contain reactive species, in particular ions, radicals or other oxidizing compounds, which can decompose pollutant molecules, organic particulate matter or soot. Electron beam flue gas treatment is another plasma-based technology which has been successfully demonstrated on industrial scale coal fired power plants.

This chapter aims a comprehensive description of plasma-based air remediation technologies. The possibilities of exhaust air pollution control by means of non-thermal plasmas generated by gas discharges and electron beams will be summarized. Therefore plasma as the 4th state of matter, its role in technology and the principle of plasma-based depollution of gases the will be described. After an overview on plasma-based depollution technologies the main important techniques, namely electron beam flue gas treatment, gas discharge generated plasmas including plasma-enhanced catalysis and injection methods will be described in separate sections. In these sections selected examples of commercially available or nearly commercialised processes for flue gas treatment or the removal of volatile organic compounds and deodorization will be described, too. Current trends and concepts will be discussed.

---

* Hana Barankova[2], Ladislav Bardos[2], Andrzej G. Chmielewski[3], Miroslaw Dors[4], Helge Grosch[1], Marcin Hołub[5], Indrek Jõgi[6], Matti Laan[6], Jerzy Mizeraczyk[4], Andrzej Pawelec[3], Eugen Stamate[7]

2. Plasmas and plasma-based depollution technologies

In physics and chemistry, plasma is an ionised gas containing free electrons, ions and neutral species (atoms and molecules) characterized by collective behaviour. Plasma is often referred as the "4th state of matter" since it has unique physical properties distinct from solids, liquids and gases. In particular, due to the presence of charge carriers plasmas are electrically conductive and respond strongly to electromagnetic fields. It contains chemically reactive media as well as excited species and emits electromagnetic radiation in various wavelength regions. The majority of matter in the visible universe (stars, interplanetary and interstellar medium) is in the plasma state. Lightnings, sparks, St'Elmos fires and the polar aurorae are examples for natural terrestrial plasmas. Furthermore, since more than 150 years plasmas are generated artificially by supplying energy to gases, liquids or solids. Such plasmas are used and under investigation for various applications, e.g. surface modification, chemical conversion, light generation or controlled nuclear fusion. Natural as well as artificial plasmas cover an extremely wide range of parameters like temperatures, particle densities and pressure. Broadly speaking, plasmas can be distinguished into thermal and non-thermal plasmas. In thermal plasmas all present species (electrons, ions and neutral species) are in the local thermal equilibrium, i.e. all species have the same mean free kinetic energy (temperature). Such plasmas are produced in fusion experiments with temperatures higher than 10^4 K. Contrary, in other situations most of the coupled energy is primarily released to the free electrons which exceed the temperatures of the heavy plasma components (ions, neutrals) by orders of magnitude. Such mixtures of energetic electrons in a relatively cold mass of ions and neutrals are called non-thermal or non-equilibrium plasmas. If the gas temperature stays nearly at or slightly above room temperature the plasma is termed "cold plasma". Even in non-equilibrium plasmas the gas temperature can increase to some 10^3 K. In such cases it is called "hot non-thermal plasma" or "translational plasma" since it marks the transition to the thermal regime. In fact cold as well as translational plasmas are used for gas depollution.

The most common method for plasma generation for technological and technical application is by applying an electric field to a neutral gas. If the applied field exceeds a certain threshold (breakdown field strength) a gas discharge and thus plasma is formed. There are many different designs of plasma sources for depollution and the most important will be described in the next two sub-sections. Alternatively by the interaction of an electron beam with gaseous medium plasma can be generated. Such electron beam generated plasmas are used in the so-called electron beam flue gas treatment, which is further described in section 3 of this chapter.

2.1 Plasma-based depollution by means of "hot" plasmas

Plasma pollution control can be done by an increase of the gas enthalpy by means of hot (i.e. thermal or translational) plasmas. Such plasmas are widely used for the incineration of gaseous but also liquid and solid waste. An overview is given in (Hammer, 1999). Typical examples are high-intensity arc or plasma torches. Electric arcs discharges are driven between two electrodes (see fig. 1 a) by high current (10 to 1000 A). Thus in arc plasmas high energy and current densities are reached (10^7–10^9 J m^{-3}; 10^7–10^9A m^{-2}). High-current arcs at atmospheric pressure can be characterized as thermal plasmas reaching temperatures in the range 5,000–50,000 K (Kogelschatz, 2004), which makes them very useful for material processing (welding, cutting, spraying) and waste treatment.

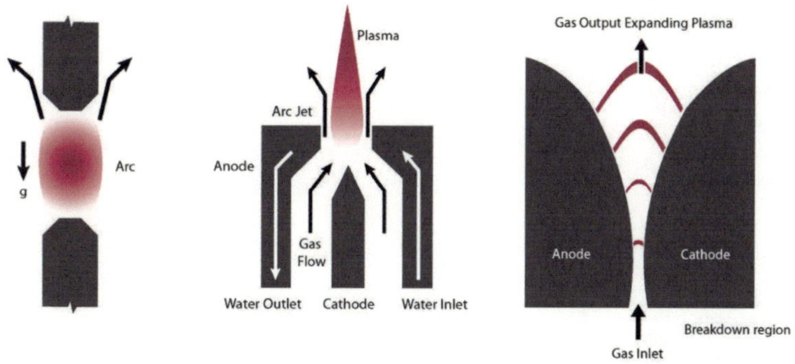

(a) (b) (c)

Fig. 1. General schemes of thermal and translational plasmas (a) free burning arc discharges in vertical and horizontal configurations; (b) plasma torch; (c) gliding arc

In plasma torches (also referred to as plasmatrons or plasma guns) the electrical energy is coupled into the working gas inside a nozzle and a high gas flow leads to the expansion outside the nozzle as a plasma jet (fig. 1 b). A large variety of plasma torches has been developed. The majority of commercial torches uses direct current arc, inductively coupled radio frequency discharges or microwave excited plasmas as the heat source and atmospheric-pressure air as working medium. The power consumption of plasma torches is in the range of several kW up to some MW. As a very rough estimation, the energy costs for conversion of noxious compounds is about 20 eV/molecule. This corresponds to 0.1 to 1 kg/kWh, a value which is comparable to that obtained in non-thermal plasmas (Hammer, 1999). Gliding arcs (fig. 1 c) are another example for translational plasmas studied for gas depollution and other applications. They consist of at least two diverging electrodes which are passed by a gas flow. The discharge starts at nearest distance between the electrodes, is spreading by gliding along the electrodes in the direction of the gas flow which leads to cooling of the plasma.

Microwave driven plasma torches at atmospheric pressure are typical examples for translational plasmas (non-thermal plasmas at elevated gas temperatures up to 4,000 K). However the gas temperature is high enough to decompose stable organic molecules. In particular nozzle-type microwave plasma source (MPS) (see e.g. Jasinski et al., 2002) has been used for the destruction of gaseous pollutants - mainly vapours of organic solvents - of relatively high concentration, up to tens of vol.%. The nozzle-type MPSs first appeared as structures based on microwave coaxial line components (see e.g. Cobine & Wilbur, 1951) where the microwave plasma was induced in the form of a plasma "flame" at the open end of a rigid coaxial line, at the tip of its inner conductor. The power-handling capability of coaxial-line-based microwave discharges is generally limited to much less than 1 kW due to the low thermal strength of the coaxial line components. Parallel with the coaxial-line-based nozzle-type MPSs so-called waveguide-based nozzle-type MPSs have been developed (e.g. Yamamoto & Murayama, 1967; Moisan et al., 1994, 2001). In these applicators the microwave plasma is also induced in the form of a plasma flame at the tip of a field-shaping structure that is similar to that of the coaxial-line based MPSs. However, the microwave power is fed into this structure from a waveguide, usually rectangular at 2.45 GHz. In advanced devices, the microwave power is delivered to the field-shaping structure in form of a conductor with a conical nozzle through a waveguide with a reduced-height section (fig. 2 a).

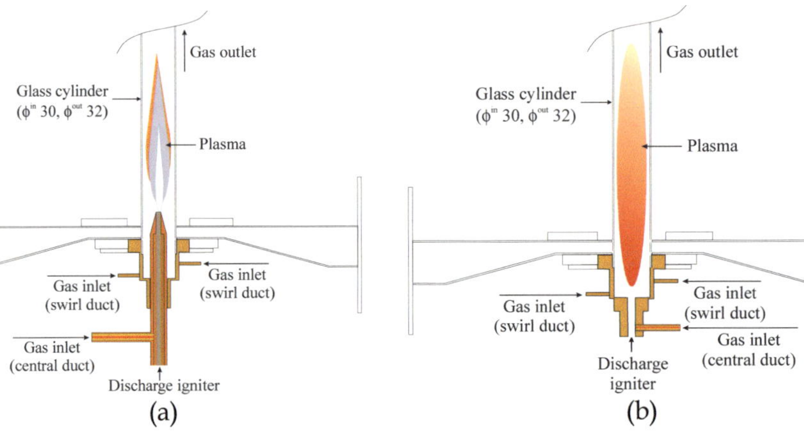

Fig. 2. Sketches of the waveguide-based cylinder-type MPS (a) and waveguide-based nozzle-type MPS (b). Dimensions are given in mm.

Since both microwave discharges, the coaxial-line-based and waveguide-based one, are gas flowing systems, they are particularly suitable for processing various gases or materials carried by gases. Recently, a new MPS was developed (e.g. Uhm et al., 2006) based on the rectangular waveguide with a reduced-height section, where the discharge is generated inside of a dielectric cylinder with a swirl flow of the working gas. There are no nozzles in the system (see fig. 2 b). It was successfully used for destruction of refrigerant HFC 134a (Jasinski et al., 2009) with destruction mass rate and corresponding energetic mass yield of up to 34.5 kg h^{-1} and 34.4 kg per kWh of microwave energy absorbed by the plasma, respectively.

2.2 Plasma-based depollution by means of "cold" non-thermal plasmas

In cold non-thermal plasmas the free energetic electrons are able to produce radicals and other reactive species (e.g. ions) which react with the pollutant molecules or particles. Furthermore, if ions can be extracted from the discharge, fine particles can be charged and thus filtered electrically from the flue gas (Grundmann et al., 2007). Additional a biological decontamination of air due to plasma treatment has been reported (e.g. Müller & Zahn, 2007).

2.2.1 Cold non-thermal plasma sources for the depollution of gases

As already mentioned, non-thermal plasmas in gas streams at atmospheric pressure can be generated in two ways. Either with the injection of a high energetic electron beam (so-called electron beam flue gas treatment, EBFGT) or the generation of a gas discharge by means of a sufficient high voltage applied to two electrodes (gas discharges). In discharge generated plasmas the electrons have lower mean energies than in electron beam produced plasmas. Thus plasma chemical reactions can differ and usually in electron beam generated plasmas the energy efficiency is better. However, discharge generated plasmas give the chance to construct more compact after treatment systems for small and medium size gas streams.

To generate plasmas with electron beams special electron accelerator units are needed. Electrons are produced via thermionic emission from a cathode and accelerated inside the vacuum tube. The electron beam transits from the beam generation environment at vacuum pressure (10^{-5} mbar) into the flue gas stream at atmospheric conditions via a beam window and than through a secondary window (Chmielewski et al., 1995). Due to a beam alignment-steering system the beam will scan across or along the flue gas stream. Beam scanning and window cooling is necessary to avoid destruction of the titanium windows. The beam acceleration ranges from 0.7 to 1.2 MeV, allowing the beam to penetrate the windows without excessive energy loss. The maximum power per accelerator available nowadays is up to 400 kW, total beam power in installations exceed 1 MW (Department of Energy, 2010). Next generation electron beam techniques use radio frequency cavity systems instead of DC transformators (Edinger, 2008). This enables pulsed driven beams with optimized energy control.

To generate plasmas by gaseous discharges several possibilities exists (Becker et al., 2005; Fridman, 2008; Kogelschatz, 2004). The most common discharge types are dielectric barrier discharges (DBDs) and corona discharges. For both types different configurations and geometries, namely cylindrical and planar, exist as shown in fig. 3. DBDs, also referred to as barrier discharges or silent discharges are characterized by the presence of at least one dielectric layer between the electrodes (Kogelschatz, 2004; Wagner et al., 2003). Typical materials for dielectric barriers are glass, quartz and ceramics. Fig. 3 a shows a so-called volume barrier discharge in cylindrical geometry. The discharge gap is usually in the range of 1 mm. Fig. 3 c is a planar surface barrier discharge, i.e. both electrodes (metal meshes) are in direct contact with the dielectric plates. Another type of DBD is the so-called coplanar discharge where both electrodes are embedded in the dielectric material. Due to the capacitive coupling of the insulating material to the gas gap DBDs can only be driven by alternating feeding voltage or pulsed DC voltages. When a sufficient voltage is applied to the electrodes, electrical breakdown occurs most commonly as number of individual discharge filaments or microdischarges (Kogelschatz, 2002). Microdischarges have a small duration (tens of nanoseconds in air), small size (diameter about 100 μm) (Brandenburg et al., 2005) and are distributed over the whole surface area. Due to the local charging of the dielectric surface after microdischarge inception the local electric field is weakened leading to the extinction of the microdischarge after several ten nanoseconds. Thus the barrier prevents the formation of a spark or arc discharge, keeping the plasma in the non-thermal regime. Despite the numerous applications of DBDs the knowledge on microdischarge development and thus plasma parameters and elementary processes within these microplasmas is not sufficient, although the multitude of subsequent microdischarges determines the efficiency and selectivity of the exhaust gas treatment.

Special feature of DBDs are so-called packed bed reactors, where dielectric or ferroelectric pellets (e.g. alumina oxide Al_2O_3, titanium oxide, TiO_2 or barium titanate $BaTiO_3$) are packed between two electrodes (see fig. 4; Holzer et al., 2005; Yamamoto et al., 1992). Due to spontaneous polarization of the ferroelectric a high electric field at the contact points of the pellets is formed resulting in microdischarge inception. The use of pellets is disadvantageous in terms of pressure drop but lead to uniform distribution of gas flow and plasma in the reactor. Furthermore the pellets can be used as catalyst enabling direct interaction between plasma and catalyst.

Corona discharges are characterized by a non-uniform configuration of the electric field, which is achieved by special electrode geometries, e.g. point-to-plane, wire-to-plane (see fig. 3 d) or coaxial wire-in-cylinder configurations (see fig. 3 b). The non-uniformity of the discharge gap enables breakdown at lower voltages allowing low current, non-thermal plasma channels based on the streamer mechanism. Thus coronas often show a filamentary character like DBDs. The electrode gap can be set to several centimetres, which is favourable for large scale applications and minimizes pressure drops. Corona discharges are usually DC-driven discharges, but for environmental applications they are often driven by high voltage pulses with rapid voltage rise (several kV per ns) and short duration (some tens of ns). This concept also referred to as pulsed corona discharges (PCD).

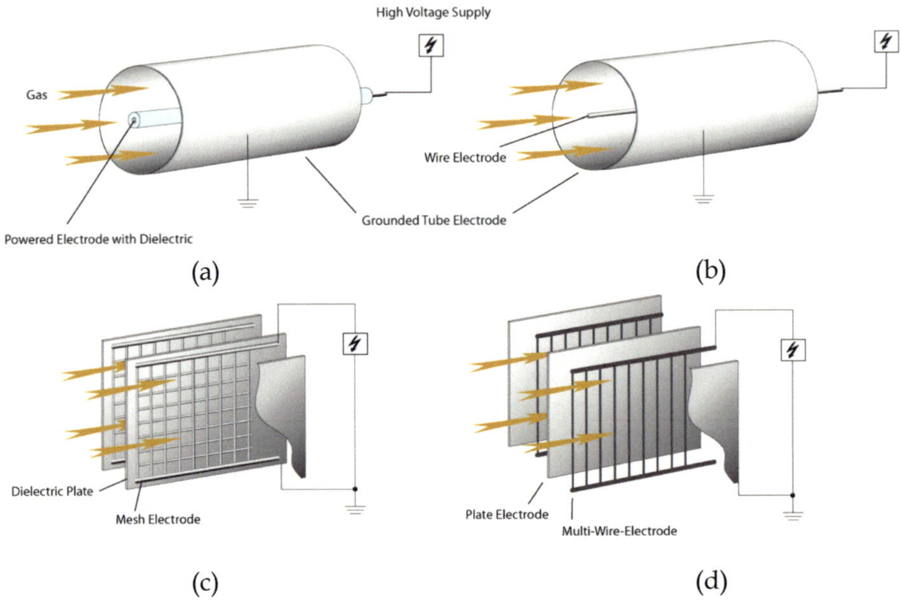

Fig. 3. Typical configurations of barrier (a, c) and corona discharges (b,d) for gas treatment (a) cylindrical asymmetric volume barrier discharge, (b) cylindrical wire-in-tube corona arrangement, (c) plate-like surface barrier discharge, (d) multineedle-plate-corona arrangement

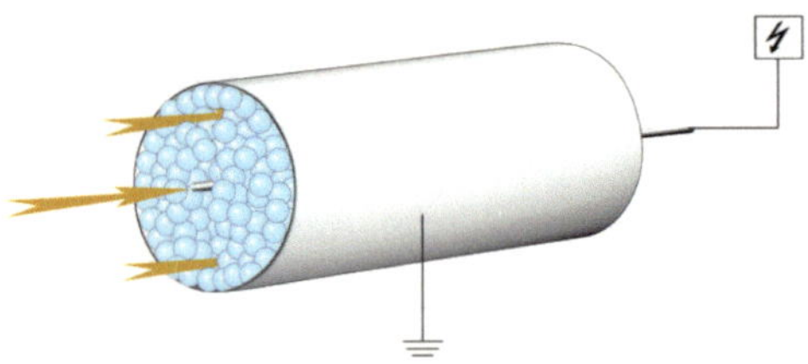

Fig. 4. Example of a packed bed reactor with special pellet filling

DC-driven corona discharges are established in pollution control as electrostatic precipitators (ESP) for dust removal of flue gases. In this application the active plasma is restricted to the region closed around the wire electrode. Between this so-called active zone and the opposite electrode (so-called collecting electrode made as plate or cylinder) a passive zone of low conductivity is formed. Ions generated in the active plasma zone enter the passive zone and drift to the collecting electrode. On their way they charge solid particles or droplets which migrate to the collecting electrode. The charged particles precipitate onto the collecting surfaces, are neutralized, dislodged and removed. Various types of dust, mist, droplet etc. down to submicron size can be removed under dry and wet conditions with high efficiency and low pressure drop (Kogelschatz, 2004). Thus ESP technology uses physical aspects of corona discharge and not the chemical processes, although the promotion of plasma chemistry is possible, too. To overcome the "back corona effect" or to decrease the power consumption pulsed operation was proposed (Mizuno, 2007; H.H. Kim, 2004). The back corona effect is obtained with high resistivity dust (e.g.

cement particles), which leads to the formation of insulating dust layers on the collecting electrode which reduces the emissions of ions. Alternatively sulphur trioxide can be injected into the flue gas stream to lower the resistivity of the particles.

An interesting concept of corona discharge is the (corona) radical shower discharge, which was developed in particular for NOx- and later for combined NOx- and SOx-removal (Ohkubo et al., 1996; J.P. Park et al., 1999). The discharge only treats a portion of the total contaminated exhaust flow. The treated gas with plasma generated active species is then injected in the total exhaust gas flow like a shower.

Typically DBD and PCD reactors require different supply waveforms with efficiencies (i.e. overall consumed plug power vs. power dissipated into the plasma) as high as possible. DBD reactors are most often supplied using alternating, sinusoidal voltage while the corona discharge systems are pulsed supplied. In case of DBD in many cases classical 50 or 60 Hz supplies are used with high-voltage transformers (Sasoh et al., 2007; Kostov et al., 2009). Due to operating conditions higher operation frequency is often necessary in order to increase the discharge power. The average power control is critical for the yield of the chemical processes. Modern supply system designs include power amplifiers with high-voltage transformers (Francke et. al., 2003; Mok et al., 2008) or many solid-state switch based power electronic converter topologies, often resonant ones (Casanueva et al., 2004). Since resonant operation complicates fluent control of the output power, often a time-averaged burst (so-called pulse density modulation - PDM) technique is used (Fujita & Akagi, 1999). Basic configurations of non-thermal plasma supply systems are depicted in fig. 5.

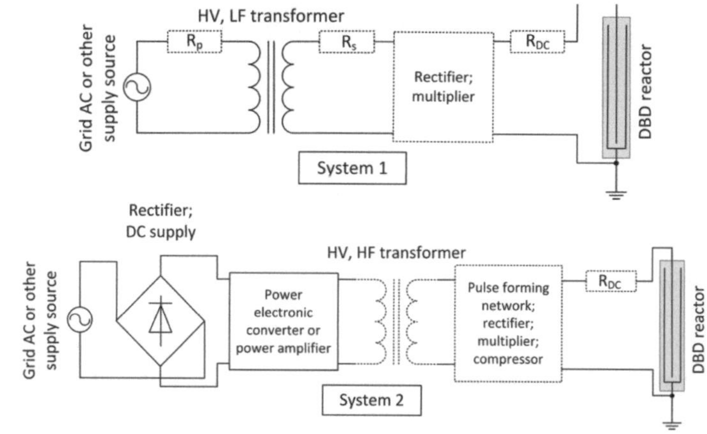

Fig. 5. Basic configurations of power supplies: low frequency systems (left) and high frequency systems (right).

Generally low frequency or high frequency systems are used. In the case of low frequency primary or secondary transformer side current limiting resistors are sometimes used (R_p or R_s), in case of pulsed DC supplies sometimes a reactor current-limiting resistor is implemented (R_{DC}). These types of supplies usually have limited efficiency ratings (about 40% for low power systems) and due to low operating frequency large weight/volume consumption. In case of controllable systems an adjustable transformer is sometimes used. High frequency supplies usually use a rectifier as the first power electronic converter. Then different configurations and topologies are used, in many cases a high frequency – high voltage transformer (HF, HV). Sometimes additional pulse forming networks are implemented in order to shape the output voltage waveform. Considering the supply voltage waveform itself a set of different patterns can be defined. Most common is the use of high voltage, AC, sinusoidal supply. In order to influence the average reactor power pulse density modulation technique is sometimes used. Optimization of effectiveness as well as voltage potential distribution levelling sometimes results in a discontinuous, bipolar waveforms.

Pulsed high voltage power supply systems are constructed in a variety as large as in the case of AC sources. In case of large installations, due to high peak values of voltage (up to several MV) and current (up to 0.5 MA), pulse modulators are constructed implementing pulsed thyristors, gas switches (thyratrons, krytrons) or spark gap switching apparatus. These technologies however, due to the principle of operation allow only a low frequency of operation and a limited lifetime. Classical constructions often implement the so called Marx generator topology (Marx, 1928) and Fitch generator topology (Fitch et al., 1968) in connection with magnetic pulse compression, which reaches efficiency rating of up to 76 %. Solid state technology enables much higher operating frequencies and very long lifetime but have a limitation of maximum allowable blocking voltage and maximal repeatable peak current per single power semiconductor. Typically high voltage MOSFET transistors and HV IGBT transistors are used for power electronic supply systems. In order to overcome single element limitations power switching stacks are produced. Nowadays typical efficiency values of up to 96 % are possible.

New concepts of non-thermal plasma sources for the treatment of gases are fused hollow cathodes (FHC). The FHC cold atmospheric plasma source is based on the simultaneous generation of multiple hollow cathode discharges in an integrated open structure with flowing gas (Barankova & Bardos, 2002; 2003). The hollow cathode discharges are non-thermal because of the population of high energy electrons due to the pendulum motion of accelerated electrons between the repelling space charge sheaths at the opposite walls either in cylindrical or planar configurations. For operation at atmospheric pressure small hollow cathode inner diameters (about 200 to 400 µm) are required. The operational stability of the FHC systems is excellent; the plasma is uniform and does not exhibit streamers. The FHC systems allow generation of cold plasma in both monoatomic and molecular gases and the upstream FHC concept with aerodynamic stabilization was successfully tested for gas conversion. The power consumption of FHC has been reported to be about 1-3 orders lower than for other non-thermal atmospheric plasma sources. The FHC design for conversion experiments is based on experimental results obtained with a tuneable radial cathode slit system and different FHC structures (Barankova & Bardos, 2010). A minimum separation of the cathode walls depends both on the type of the gas (monoatomic or molecular) and on the type of generation (pulsed DC or radio frequency). Beside gas conversion the concept has been successfully used for surface treatment, activation and cleaning of temperature-sensitive materials.

2.2.2 Fundamentals

Chemical processes in non-thermal plasmas are based on non-thermal activation of particles via collisions. The quality and quantity of collisions is determined by the density and the kinetic parameters (e.g. mean velocity, collision frequency). In general three different phases has to be distinguished. The first phase is characterized by the electrical breakdown of the gas (e.g. in form of short-lived microdischarges as described above) where free electrons with high kinetic energies are produced via ionising collisions. These electrons undergo further electron-molecule collisions, namely ionisation (1, 3), dissociation (2, 3), excitation (4) and electron attachment (7). Furthermore Penning-ionisation and dissociation (5, 6); charge transfer (8) and ion reactions are possible. All mechanisms have quite different reaction rates due to its different energy thresholds. For example for dissociation energies between 3 and 10 eV are sufficient, while ionisation requires energies more than 10 eV and electron attachment happens at energies of some eV or lower. Indeed, the exact values are determined by the electronic configuration of the molecule being considered. The reaction rate further depends on the gas temperature which depends on the vibrational excitation level of molecules. The second stage of non-thermal plasma chemistry is the radical formation and removal stage, where a multitude of anorganic reactions takes place. In particular radicals are generated through direct electron impact molecule dissociation and ionization as well as ion-molecule reactions (10), dissociate recombination of ions and electrons (11), attachment and detachment reactions (12) (Chang, 2008).

Ionisation:	$AB + e^- \rightarrow AB^+ + 2e^-$	(1)
Dissociation:	$AB + e^- \rightarrow A + B + e^-$	(2)
Dissociative ionisation:	$AB + e^- \rightarrow A^+ + B + 2e^-$	(3)
Excitation:	$AB + e^- \rightarrow AB^* + e^-$	(4)
Penning-Ionisation:	$M^* + A_2 \rightarrow A_2^+ + M$	(5)
Penning-Dissociation:	$M^* + A_2 \rightarrow 2A + M$	(6)
Attachment:	$AB + e^- \rightarrow AB^-$	(7)
	$AB + e^- \rightarrow A^- + B$	
Charge transfer:	$AB^+ + C \rightarrow AB + C^+$	(8)
Recombination:	$AB^+ + e^- \rightarrow AB$	(9)
	$A^+ + B^- \rightarrow AB$	
Ion-Molecule reaction:	$I^+ + AB \rightarrow \text{products}$	(10)
Dissociate recombination:	$AB^+ + e^- \rightarrow \text{products}$	(11)
Detachment:	$AB^- \rightarrow A + B + e^-$	(12)

In air plasmas reactive oxygen species are generated by direct electron collisions (13-16), via Penning-processes (17-19) and charge exchange (20) with subsequent ion-molecule reaction

(21) from O_2 and H_2O. Furthermore in non-thermal plasmas generated in oxygen containing atmospheres at low gas temperatures ozone, and other a strong oxidizing agents like O, $^{\bullet}OH$ and $HO^{\bullet}_2$ will be formed.

$$e^- + O_2 \rightarrow 2\,O(^3P) + e^- \quad (13)$$

$$e^- + O_2 \rightarrow O(^3P) + O(^1D) + e^- \quad (14)$$

$$e^- + O_2 \rightarrow O_2(^1\Delta) + e^- \quad (15)$$

$$e^- + H_2O \rightarrow O^{\bullet} + {}^{\bullet}OH + e^- \quad (16)$$

$$\begin{array}{c} N(^2D, {}^3P) + O_2 \rightarrow O(^3P) + NO \\ N(^2D) + H_2O \rightarrow {}^{\bullet}OH + NH \end{array} \quad (17)$$

$$O(^1D) + H_2O \rightarrow 2\,{}^{\bullet}OH \quad (18)$$

$$N_2(A) + H_2O \rightarrow {}^{\bullet}OH + H + N_2 \quad (19)$$

$$M^+ + H_2O \rightarrow M + H_2O^+ \quad (20)$$

$$H_2O^+ + H_2O \rightarrow {}^{\bullet}OH + H_3O^+ \quad (21)$$

$$O_3 + {}^{\bullet}OH \rightarrow HO^{\bullet}_2 + O_2 \quad (22)$$

$$H + O_2 + M \rightarrow HO^{\bullet}_2 + M \quad (23)$$

Many molecules are readily attacked by free radicals. Decomposition of hazardous compounds is archived without heating of the flue or off-gas. Due to the presence of oxygen, water vapour and ozone, oxidizing reactions are dominant. The resulting chemistry is quite complex and depends on the gas mixture itself as well as the temperature. A complete description of all processes is outside the scope of this chapter and only the main important aspects will be discussed in the following. For more detailed and comprehensive information the reader is referred to several books and review papers, e.g. (Fridman, 2008; Penetrante & Schultheiss, 1993; H.H. Kim, 2004; Chang, 2008). Regarding the removal of saturated hydrocarbons (denoted as RH, e.g. alkane), the process start with dehydrogenization reactions (24, 25) followed by the oxidation of the remaining organic radical $R^{\bullet}$ (26). The latter reaction result in the formation of peroxy radicals $RO^{\bullet}_2$ (26) which are further oxidized down to CO_2 and H_2O (total oxidation) or trigger a radical chain reaction with alkyl hydroperoxide radicals R-OOH (27). In case of unsaturated hydrocarbons additionally radical addition following oxidation, radical chain reaction or polymerisation of hydrocarbons are taking place.

$$R\text{-}H + O^{\bullet} \rightarrow R^{\bullet} + {}^{\bullet}OH \quad (24)$$

$$R\text{-}H + {}^{\bullet}OH \rightarrow R^{\bullet} + H_2O \tag{25}$$

$$R^{\bullet} + O_2 \rightarrow R\text{-}O\text{-}O^{\bullet} \tag{26}$$

$$R_i\text{-}O\text{-}O^{\bullet} + R_j\text{-}H \rightarrow R_i\text{-}OOH + R_j^{\bullet} \tag{27}$$

In plasma-based flue gas treatment for NO and SO_2 removal desired reductive reaction paths are of minor importance. Oxidative processes (28 - 30) lead to the formation of NO_2. The oxidation up to N_2O_5 is possible (see section 5). If hydrocarbons are present (e.g. ethene, propene, propane) $HO^{\bullet}_2$ and peroxy radicals become the dominant oxidizers (30, 31) and the energy required to oxidize NO molecule can be reduced. However, to remove NOx from the gas a heterogeneous chemical process for NO_2 reduction must follow the plasma treatment. In a similar way SO_2 oxidation to SO_3 by means of plasma treatment is possible, while SO_3 needs to be removed chemically.

$$NO + O(^3P) + M \rightarrow NO_2 + M \tag{28}$$

$$NO + O_3 + M \rightarrow NO_2 + O_2 + M \tag{29}$$

$$NO + HO_2 + M \rightarrow NO_2 + {}^{\bullet}OH + M \tag{30}$$

$$NO + R\text{-}O\text{-}O^{\bullet} \rightarrow NO_2 + R\text{-}O^{\bullet} \tag{31}$$

Following the removal stage aerosol particles are formed through reaction of larger radicals with cluster ions and molecules. Aerosol formation is a quite important process since aerosol surface reaction rate is a few orders of magnitude higher then the electronic, ionic and radical reactions. The removal processes are promoted due to heterogeneous reactions. Regarding SO_2 the stimulation of chain oxidation mechanism by plasmas in liquid droplets or ionic clusters at humid gas conditions is known (see Fridman, 2008).

In order to compare different concepts and technologies different aspects must be considered. The main focus is the efficiency evaluation, but costs for investment and operation (warranty intervals, consumption of additives) need to be taken into account, too. Several examples are described, see (Chang, 2008) and references therein. There is no universal parameter for the energy efficiency and the conditions of operation in research and application vary to a great extend. Most widely used parameters are the Specific Input Energy (SIE, or specific energy density SED) and the G-value. The SIE is the dissipated discharge power divided by the gas flow rate Q (32). In general the gas flow rate Q relates to standard or normal conditions (Temperature T_N= 273.15 K, pressure p_N= 100 kPa) and SIE is given in J/sl or kWh/Nm^3. The SIE is a reliable scaling parameter and together with the energy efficiency of pollutant removal η (also referred too as energy yield, i.e. mass of removed pollutant Δm_{Pol} divided by consumed energy of the plasma E_{PL}) a good economic evaluation can be done by η(SIE) characteristics (Chang, 2008). It should be mentioned again, that a comprehensive evaluation must consider the efficiency of the power supply transformation, too (i.e. $P_{tot} > P_{PL}$).

$$SIE = P_{Plasma} / Q_N \tag{32}$$

$$\eta = \Delta m_{Pol} / E_{PL} \tag{33}$$

$$G(-A) = \beta_A \; p_A \; Q_N \; N_0 / (E\,R\,T) \tag{34}$$

The G-value is adapted from radiolysis and refers to the number of molecules of reactant consumed per 100 eV of energy absorbed (Baird et al., 1990; Penetrante et al., 1996). It is defined as given in (34), where A is removed specie, β_A percentage of destroyed contaminants, p_A partial pressure of A, N_0 Avogadro constant, E used energy and R gas constant. In plasmas G-value gives the number of radicals generated per 100 eV. Another value to be considered is the chemical selectivity S_A of one possible chemical product A. It is given by the ratio of its concentration (or number density of molecules etc.) and the sum of concentrations of all possible products of one reaction.

3. Electron beam flue gas treatment (EBFGT)

Electron beam flue gas treatment technology is one among the most promising advanced air pollution control techniques. EBFGT is a dry-scrubbing process of simultaneous SO_2 and NO_x removal, where no waste (except by-products) is generated. The main components of flue gases are N_2, O_2, H_2O, and CO_2, with SO_x and NO_x in much lower concentrations. Ammonia NH_3 may be present as an additive to support the removal of SOx and NOx. The electron energy is transferred to the gas components present in the mixture in proportion to their mass fraction. The fast electrons slow down by collisions, secondary electrons are formed which plays an important role in overall energy transfer and the plasma is formed in the flue gas. Then, fast electrons interact with gas creating various ions and radicals, the primary species formed include N_2^+, N^+, O_2^+, O^+, H_2O^+, OH^+, H^+, CO_2^+, CO^+, N_2^*, O_2^*, N, O, H, OH, and CO. In case of high water vapor concentration the oxidizing radicals $^\bullet OH$, $HO_2^\bullet$ and $O(^3P)$ as well as excited ions are the most important products. These species take part in a variety of ion-molecule reactions, neutralization reactions, dimerization etc. SO_2, NO, NO_2, and NH_3 cannot compete with the reactions because of very low concentrations, but react with N, O, $^\bullet OH$, and $HO^\bullet_2$ radicals. After humidification and lowering of the temperature, flue gases are guided to reaction chamber, where irradiation by electron beam takes place. NH_3 is injected upstream of the irradiation chamber. There are several pathways of NO oxidation known. In the case of EBFGT the most common are as follows (Tokunaga & Suzuki, 1984):

$$NO + O(^3P) + M \rightarrow NO_2 + M \tag{35}$$

$$O(^3P) + O_2 + M \rightarrow O_3 + M \tag{36}$$

$$NO + O_3 + M \rightarrow NO_2 + O_2 + M \tag{37}$$

$$NO + HO^\bullet_2 + M \rightarrow NO_2 + {}^\bullet OH + M \tag{38}$$

After the oxidation NO_2 is converted to nitric acid in the reaction with $^\bullet OH$ according to the reaction (39) and HNO_3 aerosol reacts with NH_3 giving ammonium nitrate. NO is partly reduced to atmospheric nitrogen.

$$NO_2 + {}^{\bullet}OH + M \rightarrow HNO_3 + M \quad (39)$$

$$HNO_3 + NH_3 \rightarrow NH_4NO_3 \quad (40)$$

There can be also several pathways of SO_2 oxidation depending on the conditions. In the EBFGT process the most important are radio-thermal and thermal reactions. Radio-thermal reactions proceed through radical oxidation of SO_2 in the reaction (41) and HSO_3 creates ammonium sulphate in the following steps (42) and (43).

$$SO_2 + {}^{\bullet}OH + M \rightarrow HSO_3 + M \quad (41)$$

$$HSO_3 + O_2 \rightarrow SO_3 + HO^{\bullet}{}_2 \quad (42)$$

$$SO_3 + H_2O \rightarrow H_2SO_4 \quad (43)$$

$$H_2SO_4 + 2NH_3 \rightarrow (NH_4)_2SO_4 \quad (44)$$

The thermal reaction is based on the following process:

$$SO_2 + 2NH_3 \rightarrow (NH_3)_2SO_2 \quad (45)$$

$$(NH_3)_2SO_2 \xrightarrow{O_2,\, H_2O} (NH_4)_2SO_4 \quad (46)$$

The total yield of SO_2 removal consists of the yield of thermal and radio-thermal reactions that can be written as follows (Chmielewski, 1995).

$$\eta_{SO2} = \eta_1(\phi, T) + \eta_2(D, \alpha_{NH3}, T) \quad (47)$$

Where η, ϕ, T, D and α_{NH3} are process efficiency, gas humidity, gas temperature, dose deposited (amount of energy transferred to gas by means of irradiation) and ammonia stoichiometry (NH_3 concentration in relation to stoichiometric value) respectively. The yield of the thermal reaction depends on the temperature and humidity and decreases with the temperature increase. The yield of the radio-thermal reaction depends on the dose, temperature and ammonia stoichiometry. The main parameter in NOx removal is the dose. The rest of parameters play minor role in the process. Nevertheless in real, industrial process, dose distribution and gas flow conditions are important from the technological point of view.

The technology was originally implemented in coal fired power plants but can be applied for the cleaning of off-gases from various combustion processes. A complete EBFGT installation is schematically shown in fig. 6. After the boiler fly ash is removed from the flue gas by an ESP and cooled down and humidified in spray towers. Cooled and humidified gases are than exposed to the electron beam radiation after the injection of ammonia. The high-energetic electrons are forming the plasma and initiate a series of the above listed reactions which lead to the removal of the SO_x and NO_x by forming ammonium sulphate $(NH_4)_2SO_4$ and ammonium nitrate NH_4NO_3 respectively. The reacted gas then passes through a particulate removal device (e.g. ESP) to remove the ammonium sulphate and ammonium nitrate which are used as fertilizers. Pilot and industrial installations demonstrated the feasibility of this technology for effective flue gas purification. The process was implemented in industrial scale in Pomorzany Power Plant (Poland) for total capacity of 270,000 Nm^3/h of flue gas. SO_2 removal efficiency above 95 % and NOx removal above 75 % were reported for optimal treatment conditions. A dose of up to 10 kGy (1 kGy = 1 kJ/kg flue gas) is required for NOx removal, while SO_2 can be removed in proper conditions at lower energy consumption. Nowadays most technical problems occurred in the prototype installations has been solved (Chmielewski et al., 2004).

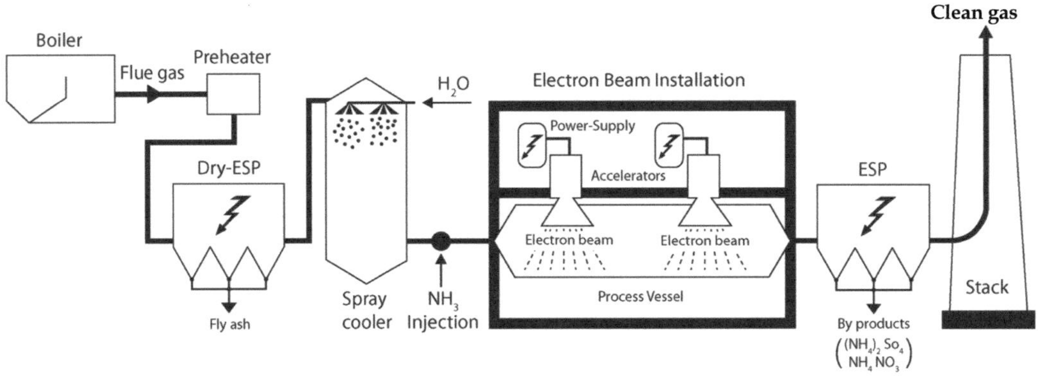

Fig. 6. Scheme of an EBFGT process

In recent investigations the electrical energy consumption could be decreased and the increase of system availability is in progress, too. The new applications concern application of electron beam for flue gases treatment from high sulphur oil fired boiler performed for Saudi Aramco Company (Basfar et al., 2008). In addition the removal of VOCs, dioxins, mercury and other pollutants from flue gases using EBFGT has been investigated. In the case of VOCs, decomposition the process itself is based on the similar principles as primary reactions concerning SO_2 and NOx removal i.e. free radicals attack on organic compounds chains or rings causing VOCs decomposition. For chlorinated aliphatic hydrocarbon' decomposition (e.g. chloroethylene), Cl^- dissociated secondary electron attachment and Cl, OH radicals reaction with VOCs play very important roles (Sun et al., 2006). The most important development concerns application for the reduction of polychlorinated dibenzodioxins (PCDDs, so-called dioxins) and polychlorinated dibenzofurans (PCDFs) emission from municipal solid waste incinerators (Hirota et al., 2003). Electron beam irradiation demonstrated high levels of mercury oxidation at the bench scale, and the technology might help to improve mercury removal in wet scrubbers or wet ESPs when employed as a primary or secondary mercury oxidation technique (J.C. Kim et al., 2010).

4. Air-depollution by means of discharges generated plasmas and plasma-enhanced catalysis

Several examples on the use of gas discharges for depollution of exhaust air will be discussed in the following. This will cover the removal of volatile organic compounds (VOCs) and deodorization, NOx- and SOx removal and removal of particulate matter (PM), e.g. soot.

4.1 VOC-removal and deodorization

Plasmas have been demonstrated to be able to decompose VOCs and thus odour molecules very efficiently for low decontamination levels (< 1 gC_{org}/m^3) as like in deodorization issues. Odour emission is a significant problem in the production of food, for farms, in gastronomy and kitchens as well as in waste management. Typical odour molecules are in fact VOCs, namely aldehydes, fatty acids, alkanes, formic acid, amines or esters. VOC contamination is an issue of increasing importance for the depollution of exhaust air, i.e. off-gases from industrial processes. For such problems plasma can be better suited than classical methods like wet scrubbing, adsorption or thermal processes because of lower energy consumption. In (Rafflenbeul, 2008) a plasma based process with an energy consumption of about 35 kWh for 70,000 m^3/h exhaust air is described, while an odour reduction of up to 99 % is possible. Due to its compactness plasma devices can be easily integrated in existing systems and processes. Furthermore no waste or waste water is generated. However, the application of non-thermal plasma has to be reckoned for every specific exhaust problem and in fact, in industrial practice plasmas are combined with catalysts, absorbing agents and other methods of depollution. E.g. undesirable by-products can be formed since the plasma-chemical conversion is not selective or influenced by the gas composition and properties, such as residual humidity or temperature. Energetic efficiency has been found to be best at low contamination levels and low gas flows. Special attention must be paid to the geometrical properties of the reactor (length, cross section) since it influences and determines the residence time as well as the back pressure. The residence needs be optimised for a successful treatment (1 to 3 sec are given in literature), while the back pressure should be as low as possible in order to ensure a proper integration in an exhaust air system (Rafflenbeul, 2008; Müller & Zahn, 2007).

For deodorization applications NTP is often enhanced with catalyst or absorption methods. The plasmaNorm process (airtec competence GmbH) comprises a three-stage treatment unit (Müller et al., 2006). In the first stage the polluted gas is stripped of solids, aerosols and particulates by means of a pre-filter. Appropriate filter media such as bag filters for damp or oily air are used according to the air impurities to be removed. A surface DBD serves as the second stage, where pre-filtered air is subjected to reactive radicals and ions initiating oxidation reactions and the decomposition of VOCs and other contaminants. Finally compounds not yet oxidised are retained in an activated carbon bed, which is described as a storage reactor that, among other effects, revert residual ozone to atmospheric oxygen. The economical, long serviceable life of the activated carbon, as it regenerates itself during the process is promoted as one of the main special characteristic of this technology. It is successfully used in gastronomy and kitchens (large scale and private households) as well as food processing industry. E.g. the exhaust from 1.5 MW ovens for convenience products made of meat generating an exhaust stream of 8,000 Nm^3/h can be deodorized (Langner, 2009).

In (Rafflenbeul, 1998) a commercial plasma process combined with a biofilter as pre-filter and oxidation catalyst as after-filter is described. Biodegradable compounds in higher concentrations are decomposed in the biofilter, while the subsequent plasma unit partially oxidizes non-biodegradable pollutions which are finally decomposed in the oxidation catalyst section. The same company (Envisolve) describes several commercialised combinations of non-thermal plasmas with catalysts or molecular sieves for waste management facilities, paintshops and other industrial applications generating exhaust streams of up to 300,000 m^3/h.

In case of VOC removal different types of power supplies are used in the terms of voltage type and shape, operation frequency, supply system topology. All of the above mentioned parameters can strongly influence overall system performance and an optimum for most cases can be found. Power supply properties may influence the nature of reactor operation just like the reactor construction itself. First of all the operating frequency influences the breakdown voltage (Valdivia-Barrientos et al. 2006) according to the semi - empiric equation:

$$U_{bd} = 1.4\left(C_d/C_g\right)^2 / \ln(f) \quad (48)$$

where U_{bd} - reactor breakdown voltage, C_d - dielectric barrier capacitance, C_g - gas gap capacitance, f - supply frequency. Second of all for aimed chemical process often an optimal set of supply parameters can be found yielding in maximal destruction and removal efficiency (Magureanu et al., 2007) or productivity (Buntat et al., 2009) when recalculated into SIE. Such dependencies are often hard to follow in industrial cases, where gas composition is complex and varying with time but nevertheless power supply system parameters play an important role in the VOC removal process.

4.2 Flue gas treatment by means of plasma-enhanced catalysis

Non-thermal plasma has been applied for the treatment of exhausts of varying sizes of diesel engines from small cars, heavy trucks and marines (Miessner et al., 2002; Bröer & Hammer, 2000; Mok & Huh, 2005; Mizuno, 2007; Cha et al., 2007; McAdams et al., 2008). Same technology has also tested for oil fired boilers (Park et al., 2008). Typically the flue gases of these sources contains 200-1,000 ppm NO_x, 10-200 ppm hydrocarbons, 200-700 ppm CO, 2-8 % CO_2, 1-5 % H_2O and 10-18 % O_2. The exhaust gas contains also particles with varying sizes.

A great deal of effort was devoted to the treatment of particulate matter in flue gas from diesel engines. In (Müller & Zahn, 2007) a reactor combining a DBD with a wall flow filter for soot reduction is described. In this system one electrode is porous and gas-permeable. The flue gas is let out through the porous electrode, which filters and holds back the soot particles. Thus soot-particles are stored on this electrode which faces to an electrode surrounded by a dielectric material. Toxic and soot-containing harmful substances are decomposed in the plasma. The accumulated soot is decomposed due to cold oxidation process initiated by active plasma species leading to constant regeneration of the filter at low temperatures during all engine operation conditions. In (Yamamoto et al., 2003) the diesel particulate filter (DPF) regeneration for real diesel engine emissions at low temperatures by means of indirect or direct non-thermal plasma treatment was demonstrated. In other studies (Chae et al., 2001, Mok & Huh, 2005) corona and DBD reactors were successfully used for the removal of smoke and particulate maters from diesel engines.

For the reduction of NOx from diesel engine exhausts selective catalytic reduction is used but the catalysts do not work properly at low temperatures below (200-300°C) (Penetrante et al., 1998; Bröer & Hammer, 2000; Tonkyn et al., 2003). For improvement of the reduction efficiency at lower temperatures, plasma enhanced selective catalytic reduction (PE-SCR) has been investigated (Penetrante et al., 1998; Bröer & Hammer, 2000; Tonkyn et al., 2003; Miessner et al., 2002; Mizuno 2007; Mok & Huh, 2005; Cha et al. 2007; McAdams et al., 2008; Hammer et al., 1999). In PE-SCR the plasma serves for the oxidation (NO to higher nitride oxides and hydrocarbons to partially oxidized ones). Oxidation is needed because many NOx-reduction catalysts have a higher activity toward NO_2 and thus the removal efficiency at low temperatures is significantly enhanced (Penetrante et al., 1998; Bröer & Hammer,

2000). Further enhancement of NOx reduction on catalyst is achieved by hydrocarbon radicals generated in the plasma (Penetrante et al., 1998; Miessner et al., 2002; Tonkyn et al., 2003). In the oxidizing environment of diesel exhaust, an effective reduction of NOx to N_2 on catalyst takes place only when there are enough reducing agents (NH_3, hydrocarbons like propene). In the case of optimized burning process, there are usually not enough hydrocarbons in the engine exhaust for efficient reduction of NOx (Tonkyn et al., 2003; Miessner et al., 2002; Cha et al., 2007). Thus it is necessary to inject additional reducing agent to the exhaust gas. When NH_3 or urea is used as reducing agent, it has to be carried in a separate tank while hydrocarbons can also be obtained directly from the fuel. Presence of hydrocarbons in the plasma stage enhances the oxidation of NO to NO_2 and additionally inhibits the oxidation of SO_2 and the formation of HNO_3 (Penetrante et al., 1998). The production of HCN can be problematic when hydrocarbons are used (Tonkyn et al., 2003) while the ammonia slip and catalyst poisoning by NH_4NO_3 have to be considered when ammonia is used (Dors & Mizeraczyk, 2004).

For the removal of diesel exhaust, the additional fuel consumption due to plasma treatment should not exceed 5 % which corresponds to SIE of 15-60 J/sl for different diesel engines (Tonkyn et al., 2003; Mizuno 2007). In experiments with synthetic exhaust gas without particles, 80 % of NOx reduction has been achieved with energy input of 27-32 J/sl when hydrocarbons and NH_3 were used as reducing agent and temperature was 150 °C (Bröer & Hammer, 2000; Mizuno 2007; Lee et al., 2007). V_2O_5/TiO_2 based catalysts or Co-ZSM5 were used and space velocities were 15,000 and 2,000 h^{-1}. When only hydrocarbons have been used as catalysts, the NOx reduction above 70 % has been achieved in temperature range of 170°C to 260°C with BaY zeolite (space velocity 12,000 h^{-1}) and by using both BaY and γ-Al_2O_3 the temperature range has been extended to 500 °C (Kwak et al., 2004). For real diesel exhaust gases, the reduction efficiencies are usually smaller because of the presence of particles which reduce NO_2 back to NO (Dorai et al., 2000). Diesel exhaust gas of an Multicar M25-10 engine (1,997 cm^3, without catalyst) having gas flow of 10 sl/min (space velocity 20,000 h^{-1}) and typically 434 ppm NOx was treated with plasma-enhanced catalysis where catalyst was placed downstream from the plasma reactor (Miessner et al., 2002). Catalyst alone (γ-Al_2O_3 + oxidation catalyst at 250°C) removed only CO and hydrocarbons (10-50 ppm in the exhaust). The NOx removal by plasma (54 J/sl) and catalyst was only 7% while the injection of additional propene (1.2 sl/h or 2,000 ppm) increased the NOx reduction to 56 %. Further increase of energy density did not improve NOx removal. For another diesel engine exhaust the NOx reduction of 73 % was achieved at energy densities 43 J/sl and at low temperature of 150 °C (Mizuno 2007). The gas flow rate was of 6 sl/min (space velocity of 36,000 h^{-1}) while inlet gas had about 313 ppm of NOx and 881 ppm acetylene. In this experiment, Pt-Al_2O_3 pellets were used as catalyst inside the backed bed plasma reactor. For both systems, the estimated additional fuel consumption due to plasma generation was 4-5 % (Miessner et al., 2002; Mizuno 2007). In the second experiment, the removal of particulate matter was determined to be 95 % (5-7 mg/m^3).

NH_3 as reducing agent was used for a commercial diesel engine from a used truck (Mok & Huh, 2005). Part of the diesel exhaust (10 sl/min) at no load condition with 180 ppm NOx and around 0.6 mg/m^3 particulate matters was introduced to the reactor system where monolithic V_2O_5/TiO_2 catalyst was placed downstream of the reactor. The effect of plasma SIE was tested in the temperature range of 100 to 200°C with the ratio of NH_3 and inlet NOx concentration set to 0.9. Plasma had strongest effect at 150 °C where the NOx reduction increased from 45 % to 80 % at input energy density of 25 J/sl whereas further increase in input energy did not improve the reduction. At 200 °C, the reduction was above 65 % already without plasma. Same system was also used for the removal of PM and it was possible to remove 50 % and 80 % of PM at SIE of 20 and 40 J/sl respectively (Mok and Huh, 2005). Up to 85 % of NOx reduction with 2 % of fuel penalty has also been achieved with similar dielectric-barrier discharge/urea-SCR hybrid system applied to VW Passat TDI engine exhaust (cold start and urban driving condition) (Hammer, 2002). An earlier experiment with Hatz 1D30 engine resulted in more than 75 % of NOx reduction with 17 J/sl and at catalyst temperature 170 °C (Hammer et al., 1999).

Plasma-enhanced catalysis has also used to improve the cleaning of marine diesel exhaust at low temperatures below 200 °C where commercial NH_3 SCR catalysts do not work properly (Cha et al., 2007). In this study, 1/10th of the exhaust (100 Nm^3/h) from 300 hp Yanmar engine was directed to hybrid plasma-catalyst system and the engine load was 25 % (550 ppm NOx, 116 ppm C_3H_6). The plasma reactor operated properly even after more than 1000 hr of work in highly humid and sooting conditions. The NOx reduction efficiency on catalyst (space velocity of 450,000 h^{-1}) increased from 20 to 80 % at 100 °C and from 55 to 90% at 200 °C at energy density of 40 J/sl with additional C_3H_6 above 1.5 times the NOx concentration injected to the exhaust and 550 ppm NH_3 injected after the plasma reactor. The estimated power consumption of plasma device for the warming period of the engine (500 Nm^3/h) was 5-6 kW and this corresponds for about 2 % of the engine power. The plasma reactor reduced also 45 % of the particulate maters (Cha et al., 2007). For larger NOx concentrations of 1,200 ppm, simulated marine diesel exhaust experiments have been carried out (McAdams et al., 2008). At 250°C and with Ag/Al_2O_3 catalyst, it was possible to obtain 50% of NOx reduction at energy densities of 60 J/sl (with the C_3H_6:NOx ratio of 2). At 350°C, above 90% of NOx reduction was measured at same energy density values. The catalysts were sulphur tolerant up to the concentrations of 1 %. The fuel penalty of 10 % was estimated for the type of engines simulated in the experiment.

When hydrocarbons are used as reducing agents, some of the NO_2 is reduced back to NO on the catalysts and this limits the maximum achievable reduction efficiency (Tonkyn et al., 2003). Use of multiple stages of plasma reactors and catalysts can overcome this limitation and increase of the reduction up to 90 % has been demonstrated in simulated exhaust gases (Tonkyn et al., 2003). Hybrid plasma-catalysis reactor with modular design was recently also tested for removal of NOx in oil-fired boiler (Park et al., 2008). The reactor consisted of four consequent plasma/catalyst modules where catalysts could be either TiO_2 or Pd/ZrO_2. The hybrid system with two first catalyst modules from Pd/ZrO_2 and last two from TiO_2 allowed to obtain the best results giving 74 % of NOx reduction with stoichiometric amount of C_3H_6 at 150 °C and space velocity of 3,300 h^{-1}. Initial NOx concentration was 500 ppm.

In addition to NOx reductions, several examples for large- and full-scale demonstration installations for flue-gas cleaning of SOx, dioxin and some VOCs are given in (H.H. Kim 2004; Mizuno 2007). Most recent review of research on catalytic processes enhanced by non-thermal plasma are presented in (Van Durme et al., 2008).

Pulsed DC-driven FHC discharges with aerodynamic stabilization were used for conversion of NOx in air mixtures. The discharge works as 100% oxidation catalyst, converting NO to NO_2, without any additives (Barankova & Bardos 2010). The electrode material plays an important role in the plasma chemical kinetics as it brings about its own material constants, e.g. work function, secondary electron emission coefficient and catalytic activity. Due to an optimized geometry and efficient transfer of power to the electrons in the system, the power consumption for gas conversion is extremely low. Typical specific energy densities within the processing window are around 5 J/sl, i.e. 0.14 kWh/100 Nm^3.

5. Injection methods and scrubbing-combined plasma processes

Alternatively to the direct treatment (gas is passing the plasma reactor completely) indirect, remote or so-called injection methods are possible. In this case clean gas will be treated by the plasma and then admixed to the flue gas. The most well known example is the low temperature oxidation (LTO) of NOx by ozone injection. The idea of LTO is to oxidize relatively insoluble NOx to higher oxides such us N_2O_5 that are highly soluble and can easily be removed in wet scrubbers (Jarvis et al., 2003; Ferrell, 2000).

In non-thermal plasmas in oxygen, ozone formation starts by dissociation of O_2 via electron impact (49). Resulting oxygen atoms form ozone in three-body collisions (50, M is a third partner), while ozone production is balanced by the decomposition reaction (51) and thermal dissociation (52) at steady state conditions.

$$e^- + O_2 \rightarrow 2\,O(^3P, ^1D) \tag{49}$$

$$O(^3P, ^1D) + O_2 + M \rightarrow O_3 + M \tag{50}$$

$$O(^3P) + O_3 \rightarrow 2\,O_2 \tag{51}$$

$$O(^3P) + O_3 \rightarrow O_2 + O \tag{52}$$

For example, exhaust NO can be oxidized by O_3 to form NO_2, NO_3 and, subsequently, N_2O_5 (53-55). Reaction (54) is the slowest reaction in this chain. Nitric pentoxide (N_2O_5, anhydride of nitric acid) can be efficiently removed from the exhaust by a washing bottle or scrubbing forming nitric acid (HNO_3, 56). In humid exhaust gases, HNO_3 may be formed in the exhaust gas itself. It can be used as chemical feedstock or it can be neutralized and used e.g. as fertilizer, similar as in the EBFGT process.

$$O_3 + NO \rightarrow NO_2 + O_2 \tag{53}$$

$$O_3 + NO_2 \rightarrow NO_3 + O_2 \tag{54}$$

$$\begin{aligned} NO_3 + NO_2 &\rightarrow N_2O_5 \\ O_3 + 2\,NO_2 &\rightarrow N_2O_5 + O_2 \end{aligned} \tag{55}$$

$$N_2O_5(g) + H_2O \rightarrow 2\,HNO_3(aq) \tag{56}$$

The deNOx efficiency was found to be maximum at 100 °C and the addition of small water droplets improves the NOx oxidation rate (Stamate et al., 2010). Advantages of LTO NOx is that the plasma discharge is kept clean and the removal rate of NO is higher than direct oxidation methods where the reverse reactions occur to reform NO and NO_2 by the O radical (Yoshioka et al., 2003; Eliasson & Kogelschatz, 1991 as cited in Stamate et al., 2010).

A commercial system applying ozone injection is available under the trademark LoTox (Low Temperature Oxidation). The process works within the Electro-dynamic Venturi (EDV) wet scrubbing system in order to achieve a combined reduction of PM, SOx and NOx of stationary emission sources, especially refinery applications (Confuorto & Sexton, 2005).

Ozone is generated on site and on demand and injected after the dry ESP directly into the wet scrubber. N_2O_5 is converted to HNO_3 and finally neutralized by the scrubbers alkali reagent to $NaNO_3$. Other pollutants such as SO_2 and HCl are removed in the wet scrubbing process simultaneously. There exist a number of commercial installations in the USA and in Asia on different emission sources. NOx removal higher than 90 % has been reported. The removal of mercury was demonstrated, too. Several refinery installations have demonstrated LoTox performance and reliability on an applicable scale, the process is available from DuPont BELCO Clean Air Technologies.

There are several advantages combining plasma treatment of gases with scrubbing processes. Gutsol et al. reports on a wet or spray pulsed corona discharges studied for the VOC-removal from paper mill exhaust gases (Fridman, 2008). In case of spay corona water is injected to the corona discharge like a shower, while in wet corona a thin water film rinse on the outer wall electrode. In such arrangement soluble VOCs adsorb on the water droplets or film while non-soluble VOCs can be converted to soluble compounds (e.g. peroxides and peroxide radicals) by means of plasma treatment and subsequently scrubbed within the same arrangement. This results in much lower energy requirements. Furthermore, plasma-stimulated oxidation continuous after adsorption resulting in a larger adsorbing capacity of the water and thus water consumption. However such process is only applicable where already large amounts of polluted water are generated and which requires effective water cleaning.

The ECO (Electro-Catalytic Oxidation) process is another example for a commercialized plasma-assisted depollution process combined with scrubbing (Boyle, 2005). The process is designed for installation downstream of a dry ESP or fabric filter (ash removal). The flue gas is directly exposed to DBD and oxidizes pollutants to soluble or capturable compounds (e.g. NO to NO_2; SO_2 to SO_3; Hg to mercury oxide HgO) and form particulate matter and aerosol mist. SO_2, NO_2 and HgO are removed in a subsequent absorber vessel (two-loop scrubber). Ammonia is added to the scrubber to maintain the pH of the solution for keeping high SO_2 scrubbing rate. NO_2 formed in the ECO reactor is scrubbed by sulphite ions, which are formed by SO_2. Finally $(NH_4)_2SO_4$ and NH_4NO_3 are formed as well. Several preliminary designs for coal-fired electric utility applications ranging from 175 – 1,000 MW has been developed and long time performance and reliability test were successfully completed. The process is available by the company Powerspan Corporation and has recently combined with post-combustion CO_2 capture technology.

6. Conclusion

To a great extend non-thermal plasma processes were demonstrated for commercial pollution control applications having following peculiarities: The decomposition of contaminants without heating of the gas can be achieved, while a wide range of pollutants (gases and particulate matter/aerosols) can be treated. Organic particles can be decomposed due to oxidation at low temperatures. The best efficiency is reached in low contaminated gases making them well suited for deodorization issues, too. An advantage of plasmas for gas depollution is that the energy consumption of the plasma stage can be regulated easily with the pollutant mass flow by the electrical parameters.

However, in all examples the plasma is one part of a complete depollution system, since plasma-chemical conversion is not selective and mainly oxidative Furthermore, energetic efficient treatment is achieved in case of low contaminated exhaust air and the formation of undesirable by-products has to be taken into account. Plasma processes combined with other treatment processes give synergetic effects. The addition of ammonia-based substances as reducing agents for plasma generated higher nitrogen oxides can be considered as state-of-the-art since it is already used in several processes (EBFGT, LoTox, ECO). Theses process has been successfully demonstrated on an industrial scale, e.g. for the flue gas treatment of coal fired power plants. New developments of EBFGT technology concern the treatment of flue gases from high sulphur oil fired boilers and the removal of (poly)chlorinated VOCs like dioxins from municipal solid waste incinerators. In this context the range of removable contaminants will be extended. Several VOCs, dioxins and mercury are under investigation with promising results at bench scale. The combination of non-thermal plasma with catalysts, absorbing agents or scrubbing techniques are promising approaches.

Hybrid systems and especially plasma-driven catalysis will be one of the major prospects for future developments. Therefore the interaction of plasmas with catalysts has to be investigated more detailed and a profound understanding of the development, physics and chemistry in polluted gases is desired. In this context more efforts on the understanding of the physics of filamentary plasmas consisting of microdischarges is necessary. Furthermore the power supply system parameters play an important role in the removal process and novel topologies with high potential for further improvement are under development.

7. Acknowledgment

This chapter is dedicated to Prof. Jen-Shih Chang (McMaster University, Ontario/Canada), who passed away 2011. His work is an outstanding contribution to the present knowledge on plasma science and its application to environmental problems.

The contribution was prepared within the transnational project "PlasTEP - Dissemination and fostering of plasma based technological innovation for environment protection in the Baltic Sea region "part-financed by the European Union (European Regional Development Fund)." The authors like to express their gratitude to all involved partners and colleagues within this project, in particular Alexander Schwock, Justyna Jaskowiak and Jane Schmidt from the Technology Centre of Western Pomerania in Greifswald for support.

8. References

Baird, J.K.; Miller, G.P. & Li N. (1999). The G value in plasma and radiation chemistry. Journal of Applied Physics, Vol. 68, 7, pp. 3661 - 3668, ISSN: 0021-8979

Baránková, H. & Bárdos, L. (2002). Fused hollow cathode cold atmospheric plasma source for gas treatment, *Catalysis Today*, Vol. 72, pp. 237–241, ISSN 0920-5861

Baránková, H. & Bárdos, L. (2003). Hollow cathode cold atmospheric plasma sources with monoatomic and molecular gases, *Surface Coating Technology*, Vol. 649, pp. 163 –164

Baránková, H. & Bárdos, L. (2010). Effect of the electrode material on the atmospheric plasma conversion of NO in air mixtures, *Vacuum*, Vol. 84, pp. 1385-1388

Basfar, A.A.; Fageeha, O.I.; Kunnummal, N.; Chmielewski, A.G; Pawelec, A.; Licki, J. & Zimek, Z. (2008). Electron beam flue gas treatment (EBFGT) technology for simultaneous removal of SO_2 and NO_x from combustion of liquid fuels, *FUEL*, Vol. 87, 8-9, pp. 1446-1452, ISSN 0016-2361

Becker, K.H.; Kogelschatz, U.; Schoenbach, K.H. & Barker, R.J. (2005). *Series in Plasma Physics: Non-Equilibrium Air Plasmas at Atmospheric Pressure,* Institute of Physics Publishing Ltd, Bristol and Philadelphia, USA, ISBN 0-7503-0962-8

Boyle, P. D. (2005). Multi-Pollutant Control Technology for Coal-Fired Power Plants, *Proceeding of the Clean Coal and Power Conference,* Washington, DC, November 21-22, 2005; available at http://www.powerspan.com/technology/eco_overview.shtml

Brandenburg, R.; Wagner, H.-E.; Morozov, A.M. & Kozlov, K.V. (2005). Axial and radial development of the microdischarges of barrier discharge in N_2/O_2 mixtures at atmospheric pressure. *Journal Physics D: Applied Physics,* Vol. 38, 11, pp. 1649-1657, ISSN

Bröer, S. & Hammer, T. (2000). Selective Catalytic Reduction of Nitrogen Oxides by Combining a Non Thermal Plasma and a V2O5-WO3/TiO2 Catalyst. *Applied Catalysis B: Environmental,* Vol. 28, pp. 101-111, ISSN 0926-3373

Buntat, Z.; Smith, I. R. & Razali, N. A. M. (2009). Ozone generation using atmospheric pressure glow discharge in air. *Journal Physics D: Applied Physics,* Vol. 42, 235202 (5pp)

Casanueva, R.; Azcondo, F. J. & Bracho S. (2004). Series–parallel resonant converter for an EDM power supply. *Journal of Materials Processing Technology* Vol. 149, pp. 172 - 177, ISSN 0924-0136

Cha, M.S.; Song, Y.-H.; Lee, J.-O. & Kim, S.J. (2007). NOx and Soot Reduction Using Dielectric Barrier Discharge and NH_3 Selective Catalytic Reduction in Diesel Exhaust. *InternationalJournal on Plasma Environmental Science and Technology,* Vol. 1, pp. 28-33, ISSN 1881-8692

Chae, J.O.; Hwang, J.W.; Jung, J.Y.; Han, J.H.; Hwang, H.J.; Kim, S. & Demidiouk, V.I. (2001). Reduction of the particulate and nitric oxide from the diesel engine using a plasma chemical hybrid system. *Physics Plasmas,* Vol. 8, pp. ISSN 1403-1410, ISSN 1070-664X

Chmielewski, A. G.; Zimek, Z.; Panta, P. & Drabik, W. (1995). The double window for electron beam injection into the flue gas process vessel. *Radiation Physics and Chemistry,* Vol. 45, No. 6, pp. 1029-1033, □ISSN 0969-806X

Chmielewski, A.G., (1995a) Technological development of eb flue gas treatment based on physics and chemistry of the process. *Radiation Physics and Chemistry,* Vol. 46, No. 46, pp. 1057-1062, ISSN 0969-806X

Chmielewski, A.G.; Licki, J.; Pawelec, A.; Tyminski, B. & Zimek, Z. (2004). Operational experience of the industrial plant for electron beam flue gas treatment. *Radiation Physics and Chemistry,* Vol. 71, pp. 439–442, ISSN 0969-806X

Chang, J. S. (2008). Physics and chemistry of plasma pollution control technology. *Plasma Sources Science and Technology,* Vol. 17, 045004 (6 pp), ISSN 0963-0252

Cobine, J.D. & Wilbur, D.A. (1951). The electronic torch and related high frequency phenomena. *Journal Applied Physics,* Vol. 22, No. 6, pp. 835-841, ISSN 0021-8979

Confuorto, N. & Sexton, J. (2007). Wet Scrubbing Based NOx Control Using LoTOx™ Technology - First Commercial FCC Start-up Experience. *Proceedings of NPRA 2007 Environmental Conference,* Austin/Texas, September 24-25, 2007

Department of Energie [DOE] (2010) Accelerators for America's Future, Washington, DC http://www.acceleratorsamerica.org/report/index.html

Dorai, R.; Hassouni, K. & Kushner, M.J. (2000) Interaction between soot particles and NOx during dielectric barrier discharge plasma remediation of simulated diesel exhaust. *Journal Applied Physics,* Vol. 88, pp. 6060-6071, ISSN 0021-8979

Dors, M. & Mizeraczyk, J., (2004). NOx removal from a flue gas in a corona discharge-catalyst hybrid system. *Catalysis Today,* Vol. 89, pp. 127-133, ISSN 0920-5861

Edinger, R. (2008). Reduction of SOx and NOx emissions by electron beam flue gas treatment. available at http://www.ebfgt.com/file_library/userfiles/file/COM2008_WINNIPEG_20080529.pdf

Ferrell, R. (2000). Controlling NOx emissions: A cooler alternative - An LTO system uses ozone injection to reduce NOx emissions. *Pollution Engineering,* Vol. 32, 4, pp. 50-52, ISSN 0032-3640

Fitch R. A. (1968). Electrical Pulse Generators. US Patent US 3,366,799. 30 Jan. 1968

Francke, K. P.; Rudolph, R. & Miessner, H. (2003). Design and operating characteristics of a simple and reliable DBD reactor for use with atmospheric air. *Plasma Chemistry and Plasma Processing*. Vol. 23, No. 1, ISSN 0272-4324

Fridman, A. (2008). *Plasma Chemistry,* Cambridge University Press, Cambridge, USA, ISBN-13 978-0-521-84735-3

Fujita, H. & Akagi, H. (1999). Control and performance of a pulse-density-modulated series-resonant inverter for corona discharge processes. *IEEE Transactions on Industry Applications*. Vol. 35, 3, pp. 621 - 627, ISSN 0093-9994

Grundmann, J.; Müller, S. & Zahn, R.-J. (2007). Extraction of Ions from Dielectric Barrier Discharge Configurations. *Plasma Processesses and Polymers,* Vol. 4, S1004–S1008, ISSN 1612-8850

Hammer, T. (1999). Applications of plasma technology in environemtal techniques. *Contributions to Plasma Physics,* Vol. 39, No. 5, pp. 441-462, ISSN 0863-1042

Hammer, T.; Kishimoto, T.; Miessner, H. & Rudolph, R. (1999). Plasma Enhanced Selective Catalytic Reduction: Kinetics of Nox-Removal and Byproduct Formation. *SAE Technical Paper 1999-01-3632*, ISSN 0148-7191

Hammer, T. (2002). Non-thermal plasma application to the abatement of noxious emissions in automotive exhaust gases. *Plasma Sources Science and Technology,* Vol. 11, A196–A201, ISSN 0963-0252

Hirota, K.; Hakoda, T.; Taguchi, M.; Takigami, M.; Kim H. & Kojima, T. (2003) Application of electron beam for the reduction of PCDD/F emission from municipal solid waste incinerators. *Environmental Science and Technology,* Vol. 37, No. 14, pp. 3164-3170, ISSN 0013-936X

Holzer, F.; Kopinke, F.D. & Roland, U. (2005). Influence of ferroelectric materials and catalysts on the performance of non-thermal plasma (NTP) for the removal of air pollutants. Plasma Chemistry and Plasma Processing, Vol. 25, 6, pp. 595-611, ISSN 0272-4324

Jarvis, J.B.; Naresh, A.T.D. & Suchak, J. (2003). $LoTO_x$ process flexibility and multi-pollutant control capability, *Presented at the Combined Power Plant Air Pollutant Control Mega Symposium,* Washington, DC, May 2003, ISBN 0923204547, 9780923204549

Jasinski, M.; Mizeraczyk, J.; Zakrzewski, Z.; Ohkubo, T., & Chang, J.S. (2002). CFC-11 Destruction by Microwave Torch Generated Atmospheric-Pressure Nitrogen Discharges. *Journal Physics D: Applied Physics,* Vol. 35, No. 18., pp. 2274-2280, ISSN 0022-3727

Jasinski, M.; Dors, M. & Mizeraczyk, J. (2009). Destruction of Freon HFC-134a using a nozzleless microwave plasma source. *Plasma Chemistry and Plasma Processessing,*, Vol. 29, No. 5 , pp. 363-372, ISSN 0272-4324,

Kim, H.-H. (2004). Nonthermal plasma processing for air-pollution control: A historical review, current issues, and future prospects. *Plasma Processes and Polymers,* Vol. 1, pp. 91-110, ISSN 1612-8850

Kim, J.C.; Kim, K.H.; Al Armendariz & Al-Sheikhly, M. (2010). Electron Beam Irradiation for Mercury Oxidation and Mercury Emissions Control. *Journal of Environmental Engineering,* pp. 554 – 559, ISSN 0733-9372

Kogelschatz, U. (2002). Filamentary, patterned, and diffuse barrier discharges. *IEEE Transactions on Plasma Science,* Vol. 30, 4, pp. 1400-1408, ISSN 0093-3813

Kogelschatz, U. (2004). Atmospheric-pressure plasma technology. *Plasma Physics and Controlled Fusion,* Vol. 46, 12B, pp. B63-B75, ISSN 0741-3335

Kostov, K.G.; Honda, R. Y.; Alves, L.M.S. & Kayama, M.E. (2009). Characteristics of dielectric barrier discharge reactor for material treatment. *Brazilian Journal of Physics,* Vol. 39, No.2, ISSN 0374-4922, 0103-9733

Kwak, J.H.; Szanyi, J. & Peden C.H.F. (2004). Non-thermal plasma-assisted NOx reduction over alkali and alkaline earth ion exhanged Y, FAU zeolites. *Catalysis Today,* Vol. 89, pp. 135-141, ISSN 0899-8388

Langner, M.H. (n.d.). Image brochure of airtec competence GmbH, Recke, Germany; available at http://www.plasmanorm.de

Lee, J.O.; Song, Y.-H.; Cha, M.S. & Kim, S.J. (2007). Effects of Hydrocarbons and Water Vapor on NOx Using V_2O_5-WO_3/TiO_2 Catalyst Reduction in Combination with Nonthermal Plasma, *Industrial & Engineering Chemistry Research,* Vol. 46, No. 17, pp. 5570-5575, ISSN 0888-5885

Licki, J.; Chmielewski, A. G.; Iller, E.; Zimek, Z.; Mazurek, J. & Sobolewski, L. (2003). Electron beam flue-gas treatment for multicomponent air-pollution control. *Applied Energy,* Vol. 75, pp. 145-154, ISSN 0306-2619

Magureanu, M.; Mandache, N.B.; Parvulescu, V.I.; Subrahmanyam, Ch.; Renken, A. & Kiwi-Minsker, L. (2007). Improved performance of non-thermal plasma reactor during decomposition of trichloroethylene: Optimization of the reactor geometry and introduction of catalytic electrode. *Applied Catalysis B: Environmental,* Vol. 74, pp. 270–277, ISSN 0926-3373

Marx, E. (1928). Verfahren zur Schlagpruefung von Isolatoren und anderen elektrischen Vorrichtungen. Patentschrift Nr. 455933. 13 Feb. 1928

McAdams, R.; Beech, P. & Shawcross, J.T. (2008). Low temperature plasma assisted catalytic reduction of NOx in simulated marine diesel exhaust. *Plasma Chemistry and Plasma Processing,* Vol. 28, pp. 159-171, ISSN 0272-4324

Miessner, H.; Francke, K.-P. & Rudolph, R. (2002) Plasma-enhanced HC-SCR of NOx in the presence of excess oxygen. *Applied Catalysis B: Environmental,* Vol. 36, pp. 53-62, ISSN 0926-3373

Mizuno, A. (2007). Industrial applications of atmospheric non-thermal plasma in environmental remediation. *Plasma Physics and Controlled Fusion,* 49, A1-A15, ISSN 0741-3335

Moisan, M.; Sauve, G.; Zakrzewski, Z. & Hubert, J. (1994). An atmospheric pressure waveguide-fed microwave plasma torch: the TIA design. *Plasma Sources Science and Technology*, Vol. 3, No. 4, (November 1994), pp. 584-592, ISSN 0963-0252

Moisan, M.; Zakrzewski, Z. & Rostaing, J.C. (2001). Waveguide-based single and multiple nozzle plasma torches: the TIAGO concept. *Plasma Sources Science and Technology*, Vol. 10, No. 3, (August 2001), pp. 387-394, ISSN 0963-0252

Mok, Y.S. & Huh, Y.J. (2005). Simultaneous Removal of Nitrogen Oxides and Particulate Matters from Diesel Engine Exhaust Using Dielectric Barrier Discharge and Catalysis Hybrid System. *Plasma Chemistry and Plasma Processing*, Vol. 25, pp. 625-639, ISSN 0272-4324

Mok, Y. S.; Lee, S.-B.; Oh, J.-H.; Ra K.-S. & Sung B.-H. (2008). Abatement of Trichloromethane by Using Nonthermal Plasma Reactors. *Journal of Plasma Chemistry and Plasma Processing*. Vol. 28, No. 6, ISSN 0272-4324

Müller, S.; Zahn, R.-J.; Grundmann, J. & Langner, M. (2006). Plasma Treatment of Aerosols and Odours. *Proceedings 3rd International Workshop on Microplasmas IWM3*, Greifswald, Germany, May 9 - 11, 2006, p. 121

Müller, S. & Zahn, R.-J. (2007). Air Pollution Control by Non-Thermal Plasma. *Contributions to Plasma Physics*, Vol. 47, No. 7, pp. 520 – 529, ISSN 0863-1042

Ohkubo, T.; Kanazawa, S.; Nomoto, Y.; Chang, J. S. & Adachi, T. (1996). Time dependence of NOx removal rate by a corona radical shower system. *IEEE Transaction on Industry Applications*, Vol. 32, pp. 1058–1062, ISSN 0093-9994

Park, B. R. & Deshwal, S.H., Moon (2008) NOx removal from the flue gas of oil-fired boiler using a multistage plasma-catalyst hybrid system, *Fuel Processing and Technology*, Vol. 89, pp. 540-548, ISSN 0016-2361, 0378-3820

Park, J.Y.; Tomicic, I.; Round, G.F. & Chang J.S. (1999). Simultaneous removal of NOx and SO2 from NO-SO2-CO2-N-2-O-2 gas mixtures by corona radical shower systems. *Journal Physics D: Applied Physics*, Volume: 32, 9, pp. 1006-1011, ISSN: 0022-3727

Penetrante, B. M. & Schultheis, S. E. (1993). *Non-Thermal Plasma Techniques for Pollution Control*, NATO ASI Series, Vol. G34, Springer, Part A and B, Berlin, Germany, ISBN 0258-2023

Penetrante B.M.; Hsiao, M.C,; Merritt, B.T.; Vogtlin, G.E.; Wallman, P.H.; Neiger, M.; Wolf O.; Hammer, T. & Broer, S. (1996). Pulsed corona and dielectric-barrier discharge processing of NO in N2. Applied Physics Letters, Vol. 68, 26, pp. 3719-3721, ISSN: 0003-6951

Penetrante, B.M.; Brusasco, R.M.; Merritt, B.T.; Pitz, W.J.; Vogtlin, G.E.; Kung, M.C.; Kung, H.H.; Wan, C.Z. & Voss, K.E. (1998). Plasma-Assisted Catalytic Reduction of NOx. *SAE Technical Paper 982508* (October 1998)

Rafflenbeul, R. (1998). Nicht-thermische Plasmaanlagen (NTP) zur Luftreinhaltung in der Abfallwirtschaft. *Müll und Abfall* 1/1998, pp. 38-44 (in german), ISSN 0027-2957

Rafflenbeul, R. (2008). Geringe Kosten für gering konzentrierte Abluft. *Wasser, Luft und Boden (wlb)* 6/2008, pp. 36-41, (in german); available at http://www.envisolve.com, ISSN 1421-8615

Sasoh, A.; Kikuchi K. & Sakai, T. (2007). Spatio-temporal filament behaviour in a dielectric barrier discharge plasma actuator. *Journal Physics D: Applied Physics*, Vol. 40, pp. 4181–4184, ISSN 1361-6463

Stamate, E.; Chen, W.; Jørgensen, L.; Jensen, T.K., Fateev, A. & Michelsen, P.K. (2010). IR and UV gas adsorption measurements during NO_x reduction on an industrial gas fired power plant. Fuel, Vol. 89, 5, pp. 978-985, ISSN 0016-2361

Sun, Y.X.; Chmielewski, A.G. & Bulka, S. (2006). Influence of base gas mixture on decomposition of 1,4-dichlorobenzene in an electron beam generated plasma reactor. *Plasma Chemistry and Plasma Processing,* Vol. 26, 4, pp. 347-359, ISSN 0272-4324

Tonkyn, R.G.; Barlow, S.E. & Hoard, J.W. (2003). Reduction of NOx in synthetic diesel exhaust via two-step plasma-catalysis treatment. *Applied Catalysis B: Environmental,* Vol. 40, p. 207, ISSN 0926-3373

Tokunaga, O. & Suzuki, N. (1984). Radiation chemical reactions in NO_x and SO_2 removals from flue gas. *Radiation Physics and Chemistry,* Vol. 24, No. 1, pp. 145-165, ISSN 0020-7055, 0146-5724, 0969-806X

Uhm, H.S.; Hong, Y.C. & Shin, D.H. (2006). A microwave plasma torch and its applications. *Plasma Sources Science and Technology,* Vol. 15, No. 2, (May 2006), pp. S26-S34, ISSN 0963-0252

Van Durme, J.; Dewulf, J.; Leys, Ch. & Van Langenhove, H. (2008). Combining non-thermal plasma with heterogeneous catalysis in waste gas treatment: A Review. *Applied Catalysis B: Environmental,* Vol. 78, No. 3-4, pp. 324-333, ISSN 0926-3373

Wagner, H.-E.; Brandenburg , R.; Kozlov, K.V.; Sonnenfeld, A.; Michel, P. & Behnke, J.F. (2003). The barrier discharge: basic properties and applications to surface treatment *Vacuum,* Vol. 71, pp. 417–436, ISSN: 0042-207X

Validivia-Barrientos, R.; Pacheco-Sotelo, J.; Pacheco-Pacheco, M.; Benitez-Read, J. S. & Lopez-Callejas R. (2006). Analysis and electrical modelling of a cylindrical DBD configuration at different operating frequencies. *Plasma Sources Science and Technology,* Vol. 15, pp. 237-245, ISSN 0963-0252

Yamamoto, T.; Ramanathan, K.; Lawless, P.A.; Ensor, D.S.; Newsome, J.R.; Plaks, N. & Ramsey, G.H. (1992). Control of volatile organic-compounds by an ac energized ferroelectric pellet reactor and a pulse corona reactor. IEEE Trans. on Industry Applications, Vol. 28, 3, pp. 528-534, ISSN: 0093-9994

Yamamoto, M. & Murayama, S. (1967). UHF torch discharge as an excitation source. *Spectrochimia Acta A,* Vol. 23, No. 4, pp. 773-776, ISSN 0370-8322, 0584-8539, 1386-1425

Yamamoto, T.; Okubo, M.; Kuroki, T. & Miyairi, Y. (2003). Nonthermal plasma regeneration of diesel particulate filter. *SAE Technical Paper 2003-01-1182, SAE International,* ISSN 0148-7191